# Advanced Crystalline Materials, Mechanical Properties and Innovative Production Systems

# Advanced Crystalline Materials, Mechanical Properties and Innovative Production Systems

Editors

**Daniel Medyński**
**Grzegorz Lesiuk**
**Anna Burduk**

Basel • Beijing • Wuhan • Barcelona • Belgrade • Novi Sad • Cluj • Manchester

*Editors*

Daniel Medyński
Faculty of Technical and
Economic Sciences, Witelon
Collegium State University
Legnica
Poland

Grzegorz Lesiuk
Faculty of Mechanical
Engineering, Wrocław
University of Science and
Technology,
Wrocław
Poland

Anna Burduk
Faculty of Mechanical
Engineering, Wroclaw
University of Science and
Technology
Wroclaw
Poland

*Editorial Office*
MDPI
St. Alban-Anlage 66
4052 Basel, Switzerland

This is a reprint of articles from the Special Issue published online in the open access journal *Crystals* (ISSN 2073-4352) (available at: https://www.mdpi.com/journal/crystals/special_issues/PU6O8G1NQ7).

For citation purposes, cite each article independently as indicated on the article page online and as indicated below:

Lastname, A.A.; Lastname, B.B. Article Title. *Journal Name* **Year**, *Volume Number*, Page Range.

**ISBN 978-3-7258-1155-7 (Hbk)**
**ISBN 978-3-7258-1156-4 (PDF)**
doi.org/10.3390/books978-3-7258-1156-4

© 2024 by the authors. Articles in this book are Open Access and distributed under the Creative Commons Attribution (CC BY) license. The book as a whole is distributed by MDPI under the terms and conditions of the Creative Commons Attribution-NonCommercial-NoDerivs (CC BY-NC-ND) license.

# Contents

**Zhenxue Shi and Shizhong Liu**
A Study on the Co-Content Optimization of the DD15 Single-Crystal Superalloy
Reprinted from: *Crystals* **2023**, *13*, 389, doi:10.3390/cryst13030389 . . . . . . . . . . . . . . . . 1

**Xing Wang, Zhibin Yang and Lingzhi Du**
Research on the Microstructure and Properties of a Flux-Cored Wire Gas-Shielded Welded Joint of A710 Low-Alloy High-Strength Steel
Reprinted from: *Crystals* **2023**, *13*, 484, doi:10.3390/cryst13030484 . . . . . . . . . . . . . . . . 13

**John Harrison and Paul A. Withey**
Precipitation of Topologically Closed Packed Phases during the Heat-Treatment of Rhenium Containing Single Crystal Ni-Based Superalloys
Reprinted from: *Crystals* **2023**, *13*, 519, doi:10.3390/cryst13030519 . . . . . . . . . . . . . . . . 24

**Wanshuo Sun and Shunzhong Chen**
Phase Transition of $Nb_3Sn$ during the Heat Treatment of Precursors after Mechanical Alloying
Reprinted from: *Crystals* **2023**, *13*, 660, doi:10.3390/cryst13040660 . . . . . . . . . . . . . . . . 36

**Hongchang Zhang, Wenhu He, Huaibei Zheng, Jiang Yu, Hongtao Zhang, Yinan Li, Jianguo Gao, et al.**
Plasma-Pulsed GMAW Hybrid Welding Process of 6061 Aluminum and Zinc-Coated Steel
Reprinted from: *Crystals* **2023**, *13*, 723, doi:10.3390/cryst13050723 . . . . . . . . . . . . . . . . 48

**Guobo Wang, Hao Zhao, Yu Zhang, Jie Wang, Guanghui Zhao and Lifeng Ma**
Friction and Wear Behavior of NM500 Wear-Resistant Steel in Different Environmental Media
Reprinted from: *Crystals* **2023**, *13*, 770, doi:10.3390/cryst13050770 . . . . . . . . . . . . . . . . 63

**Juan Cao, Fulei Jing and Junjie Yang**
Thermal-Mechanical Fatigue Behavior and Life Assessment of Single Crystal Nickel-Based Superalloy
Reprinted from: *Crystals* **2023**, *13*, 780, doi:10.3390/cryst13050780 . . . . . . . . . . . . . . . . 77

**Yan Gao, Wenjiang Feng, Chuang Wu, Lu Feng and Xiuyan Chen**
Investigation on Structural, Tensile Properties and Electronic of Mg–X (X = Zn, Ag) Alloys by the First-Principles Method
Reprinted from: *Crystals* **2023**, *13*, 820, doi:10.3390/cryst13050820 . . . . . . . . . . . . . . . . 88

**Anqi Liu, Fei Zhao, Wensen Huang, Yuanbiao Tan, Yonghai Ren, Longxiang Wang and Fahong Xu**
Effect of Aging Temperature on Precipitates Evolution and Mechanical Properties of GH4169 Superalloy
Reprinted from: *Crystals* **2023**, *13*, 964, doi:10.3390/cryst13060964 . . . . . . . . . . . . . . . . 100

**Mihály Réger, József Gáti, Ferenc Oláh, Richárd Horváth, Enikő Réka Fábián and Tamás Bubonyi**
Detection of Porosity in Impregnated Die-Cast Aluminum Alloy Piece by Metallography and Computer Tomography
Reprinted from: *Crystals* **2023**, *13*, 1014, doi:10.3390/cryst13071014 . . . . . . . . . . . . . . . . 113

**Urška Klančnik, Peter Fajfar, Jan Foder, Heinz Palkowski, Jaka Burja and Grega Klančnik**
Grain Size Distribution of DP 600 Steel Using Single-Pass Asymmetrical Wedge Test
Reprinted from: *Crystals* **2023**, *13*, 1055, doi:10.3390/cryst13071055 . . . . . . . . . . . . . . . . 125

**Włodzimierz Dudziński, Daniel Medyński and Paweł Sacher**
Effect of Machine Pin-Manufacturing Process Parameters by Plasma Nitriding on Microstructure and Hardness of Working Surfaces
Reprinted from: *Crystals* **2023**, *13*, 1091, doi:10.3390/cryst13071091 . . . . . . . . . . . . . . . . . . **143**

**Zhong-Liang Wang, Tian-Le Song, Li-Hua Zhao and Yan-Ping Bao**
Study on Efficient Dephosphorization in Converter Based on Thermodynamic Calculation
Reprinted from: *Crystals* **2023**, *13*, 1132, doi:10.3390/cryst13071132 . . . . . . . . . . . . . . . . . . **162**

**Xiaoming Sun, Jingyi Cui, Shaofu Li, Zhiyuan Ma, Klaus-Dieter Liss, Runguang Li and Zhen Chen**
In-Situ Study of Temperature- and Magnetic-Field-Induced Incomplete Martensitic Transformation in Fe-Mn-Ga
Reprinted from: *Crystals* **2023**, *13*, 1242, doi:10.3390/cryst13081242 . . . . . . . . . . . . . . . . . . **176**

**Zhikang Ji, Xiaoguang Qiao, Shoufu Guan, Junbin Hou, Changyu Hu, Fuguan Cong, Guojun Wang, et al.**
The Tensile Properties and Fracture Toughness of a Cast Mg-9Gd-4Y-0.5Zr Alloy
Reprinted from: *Crystals* **2023**, *13*, 1277, doi:10.3390/cryst13081277 . . . . . . . . . . . . . . . . . . **189**

**Guanghui Zhao, Yu Zhang, Juan Li, Huaying Li, Lifeng Ma and Yugui Li**
Numerical Simulation of Temperature Field during Electron Beam Cladding for NiCrBSi on the Surface of Inconel 718
Reprinted from: *Crystals* **2023**, *13*, 1372, doi:10.3390/cryst13091372 . . . . . . . . . . . . . . . . . . **202**

**Jozef Petrík, Peter Blaško, Dagmar Draganovská, Sylvia Kusmierczak, Marek Šolc, Miroslava Ťavodová and Mária Mihaliková**
The Relationship between Polishing Method and ISE Effect
Reprinted from: *Crystals* **2023**, *13*, 1633, doi:10.3390/cryst13121633 . . . . . . . . . . . . . . . . . . **219**

Article

# A Study on the Co-Content Optimization of the DD15 Single-Crystal Superalloy

Zhenxue Shi * and Shizhong Liu

Science and Technology on Advanced High Temperature Structural Materials Laboratory, Beijing Institute of Aeronautical Materials, Beijing 100095, China
* Correspondence: shizhenxue@126.com

**Abstract:** The fourth-generation single-crystal superalloy DD15 with 6% Co, 9% Co and 12% Co was cast using the vacuum directionally solidified furnace, while other alloying element's content remained unchanged. The long-term aging experiment was conducted at 1100 °C for 1000 h after standard heat treatment. The stress rupture tests of the alloy were conducted at 1100 °C/137 MPa and 1140 °C/137 Mpa. The influence of Co content on the microstructure and stress rupture properties of DD15 alloy had been investigated to optimize the Co content to obtain excellent comprehensive performance. The results showed that the primary dendrite arm spacing of the alloy decreases at first and increases afterwards, and the volume fraction of $\gamma$-$\gamma'$ eutectic decreases with the growth of Co content in the as-cast microstructures. The size, cubic degree and volume fraction of the $\gamma'$ phase of the alloy after standard heat treatment all decrease with the increase in Co content. The microstructure stability of the alloy is enhanced with the increase in Co content. No TCP phase was present in the alloy with 12% Co precipitate even after aging 1000 h. The stress rupture lives at two conditions, both reduced in different degrees with the increase in Co content. The effect of Co on the stress rupture life of the alloy improves with the increase in Co content or test temperature. The acicular TCP phase appeared in the 6% Co alloy and 9% Co alloy in the microstructure of the ruptured specimens with different Co contents. Moreover, the TCP phase content in the 6% Co alloy is much more than that in the 9% Co alloy. There is no TCP phase precipitation in the 12% Co alloy. At last, the relationship between microstructure stability, stress rupture properties and Co content of the alloy is discussed. The alloy containing 9% Co is the best choice considering the microstructure stability and stress rupture properties.

**Keywords:** single-crystal superalloy; the fourth generation; Co content; microstructure stability; stress rupture properties

## 1. Introduction

The Ni-based single-crystal superalloy has been extensively used as one of the key materials for the blade part in advanced aero-engines because of their excellent comprehensive performance [1]. The temperature capability of the turbine blade has increased significantly in the past several decades. Some advancements have been achieved by improving the content of the alloying elements.

Co is an essential alloying element in high-generation single-crystal superalloys. It is mainly soluble in the $\gamma$ phase and only a small amount of Co enters the $\gamma'$ phase. Co can lower the stacking fault energy of the $\gamma$ phase and form secondary carbides [2]. Co can increase the solid solubility of Cr, Mo and W elements in the $\gamma$ matrix [3]. Excessive Co may lower the $\gamma'$ phase content, mechanical strength and initial melting temperature of the alloy [4]. The oxidation and corrosion resistance, stress rupture life and dendrite segregation of the alloy may be improved with the appropriate amount of Co [5–7]. With the increase in the refractory alloying elements, the alloy tends to precipitate TCP phases, which leads to a reduction in the mechanical properties [8–10]. It has become increasingly

difficult to develop new single-crystal superalloys with a balanced combination of strength, environmental resistance, castability and microstructural stability.

There are different opinions about the Co content in the alloy to achieve good microstructure stability. The Co content is 3.3% in the third-generation single-crystal superalloy CMSX-10 [11]. However, the increase in Co to 12.5% can also make another same-generation alloy, René N6, which has good microstructural stability [12]. The Co contents of the fourth-generation single-crystal superalloys EPM-102, NG-MC and TMS-138 developed, respectively, by America, France and Japan are 16.5%, <0.2% and 5.9% [13–15]. Moreover, there are few studies on the effect of Co element on the fourth-generation single-crystal superalloy in the public literature.

The DD15 alloy studied in this paper was developed for aeroengine turbine blade application by the Beijing Institute of Aeronautical Materials. The properties of the alloy are equivalent to those of other fourth-generation single-crystal superalloys. The influence of Co content on the microstructure and mechanical properties at high temperatures of the fourth-generation single-crystal superalloy DD15 was investigated in this paper with the aim to optimize its chemical composition, microstructure and properties.

## 2. Materials and Methods

Commercially pure raw materials were used to prepare master alloy heat in a vacuum-induction furnace. Three single-crystal samples with the size of φ 15 mm × 180 mm were cast using a crystal selection method in the vacuum directionally solidified furnace with a high-temperature gradient by adding different Co contents in the alloy. The Co contents of three alloys were 6%, 9% and 12%, respectively, and the other alloying element's content was the same. The nominal chemical components of the three alloys are listed in Table 1. The crystal orientations of the samples were analyzed with the Laue X-ray back-reflection method, and the growing direction was within 15 degrees deviating from the [001] orientation.

**Table 1.** Nominal chemical components of the three alloys (mass fraction, %).

| Alloy | Cr | Co | Mo | W | Ta | Re | Ru | Nb | Al | Hf | Ni |
|---|---|---|---|---|---|---|---|---|---|---|---|
| 6% Co | 3.0 | 6.0 | 1.0 | 7.0 | 7.5 | 5.0 | 3.0 | 0.5 | 5.6 | 0.1 | Bal. |
| 9% Co | 3.0 | 9.0 | 1.0 | 7.0 | 7.5 | 5.0 | 3.0 | 0.5 | 5.6 | 0.1 | Bal. |
| 12% Co | 3.0 | 12.0 | 1.0 | 7.0 | 7.5 | 5.0 | 3.0 | 0.5 | 5.6 | 0.1 | Bal. |

The heat treatment of the samples proceeded according to their different heat treatment regimes. The single-crystal specimens received a standard heat treatment comprising a solution treatment and a two-step aging treatment. The solution treatments of the alloy with 6% Co, 9% Co and 12% Co were 1340 °C/5 h/AC, 1335 °C/5 h/AC and 1330 °C/5 h/AC, respectively. The two-step aging treatment of the alloy was 1140 °C/4 h/AC + 870 °C/32 h/AC. All the samples were kept for 200 h, 400 h, 600 h, 800 h and 1000 h at 1100 °C for long-term aging to check the effect of Co content on the microstructural stability of the alloy.

The standard cylinder samples for stress rupture tests were prepared after standard heat treatment. The shape and size of the stress rupture specimen are shown in Figure 1. The stress rupture test experiment was carried out at 1100 °C/137 MPa and 1140 °C/137 MPa in air. Every datum is the mean value of three specimens. The longitudinal profile microstructure of the stress rupture samples was observed.

The microstructures of the samples with different states were analyzed with a Leica DM4000M optical microscope (OM) made in Germany, a Zeiss supra 55 field-emission scanning electron microscope (SEM) made in Germany and a JEOL JEM-2100F transmission electron microscope (TEM) made in Japan. The image analysis software was used to analyze the primary dendrite arm spacing and the volume fraction of eutectic. The samples for SEM were etched for 6~9 s with 5 g $CuSO_4$ + 25 mL HCl + 20 mL $H_2O$ + 5 mL $H_2SO_4$, which dissolves the $\gamma'$ phase. Foils for TEM of the creep-ruptured samples analysis were obtained by cutting 0.2 mm thick discs perpendicular to the tensile axis of the specimens using an electric discharge machine. Thin foils were prepared by using twin-jet thinning electrolytically in a solution of 10 vol% perchloric acid and 90 vol% ethanol at −10 °C using liquid nitrogen.

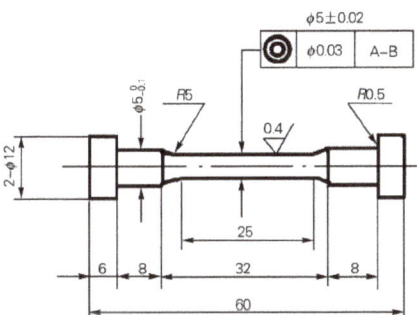

**Figure 1.** Shape and size of the stress rupture property specimen (mm).

## 3. Results

### 3.1. Heat Treatment Microstructure

Figure 2 shows the as-cast microstructure of the alloy with different Co contents. These samples were sectioned normally to the solidification direction and show a typical dendritic microstructure. The primary dendrite arm spacing of the alloy with 6% Co, 9% Co and 12% Co are 365 µm, 342 µm and 351 µm, respectively. The primary dendrite arm spacing decreases at first and increases afterwards with the increase in Co content. In addition, the structure contains pools of $\gamma$-$\gamma'$ eutectic in the interdendritic region. The eutectic fraction represents the remaining liquid at the last stage of solidification. After quantitative calculation, the eutectic fractions in the alloy with 6% Co, 9% Co and 12% Co are 11.8%, 11.2 and 10.5%, respectively. The volume fraction of the $\gamma$-$\gamma'$ eutectic decreases with the increase in Co content.

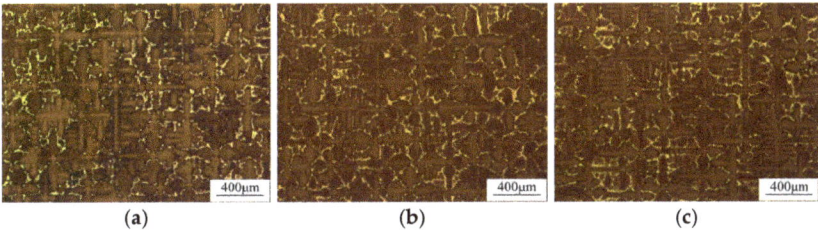

**Figure 2.** As-cast dendritic microstructure of the alloy with different Co contents: (**a**) 6% Co; (**b**) 9% Co and (**c**) 12% Co.

The eutectic fraction represents the remaining liquid at the last stage of solidification. The $\gamma$ forming elements, such as Co, Cr, Re, Mo and W all tend to segregate towards the dendrite core during solidification. This indicates that these elements are the first to solidify within the dendrite cores during the single-crystal superalloy withdrawal process. The $\gamma'$ forming elements, such as Al, Hf, Nb and Ta all tend to segregate into the interdendritic region. These elements are present in greater quantity in the last liquid phase to solidify and result in large eutectic $\gamma$-$\gamma'$ phases forming within the interdendritic regions. Much less residual liquid may be forced to reach the eutectic point with the increase in the concentration of Co. This may result in the formation of much less eutectic. So, the volume fraction of $\gamma$-$\gamma'$ eutectic of the alloy decreases with the increase in Co content in the as-cast microstructures.

## 3.2. Heat Treatment Microstructure

Figure 3 shows the heat treatment microstructures of the alloy with different Co contents. It illustrates that they consist of the cubic $\gamma'$ phase precipitated coherently in the $\gamma$ phase. The $\gamma$-$\gamma'$ eutectic and coarse $\gamma'$ phase completely disappeared after heat treatment at high temperatures. The new fine $\gamma'$ phase precipitated from the supersaturated $\gamma$ solid solution during the following cooling process. The homogeneous distribution of strengthening the cubic $\gamma'$ phase was obtained after two stages of aging treatment. Comparing the three pictures, it can be seen that the cubic degree of $\gamma'$ phase slightly declines as the Co content rises. Figure 4 shows the effect of Co content on the size and volume fraction of the $\gamma'$ phase of the alloy calculated using data statistics. The $\gamma'$ phase size is about 0.3–0.4 μm and the volume fraction of $\gamma'$ phase is about 55–65% for the different Co content alloys. It is shown that the size and volume fraction of the $\gamma'$ phase of the alloy all decrease with the increase in Co content.

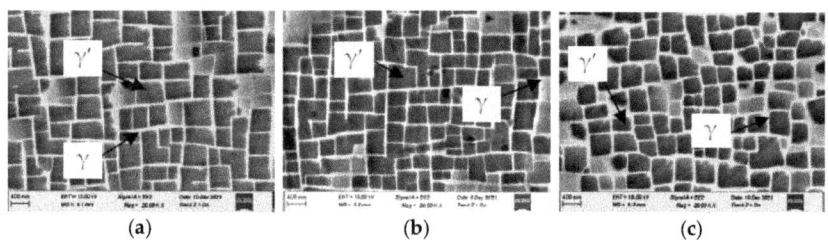

**Figure 3.** The microstructures of the alloy with different Co content after full heat treatment: (**a**) 6% Co; (**b**) 9% Co and (**c**) 12% Co.

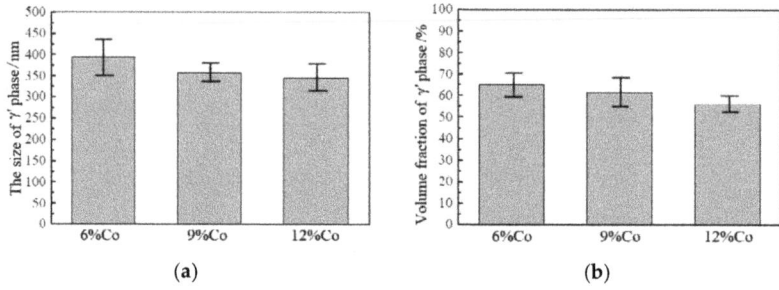

**Figure 4.** The effect of Co content on the size and volume fraction of $\gamma'$ phase: (**a**) size of $\gamma'$ phase and (**b**) volume fraction of $\gamma'$ phase.

## 3.3. Long-Term Aging Microstructure

Figure 5 illustrates the long-term aging microstructures of the alloy with different Co contents at 1100 °C at various times. It is shown in Figure 5 that the $\gamma'$ phases are no longer in cubic shape in the 6% Co alloy after long-term aging 200 h. They merge and grow together to form rafts along the [100] or [010] direction. There is no TCP phase precipitated. However, a lot of acicular TCP phase is observed with a 45° angle relative to the rafted orientation after aging 400 h. The amount of TCP phase greatly increases and the rafting degree of the $\gamma'$ phase does not change greatly after aging 1000 h. The rafted structure has formed and there is no TCP phase observed in the 9% Co alloy after aging 800 h. A small amount of acicular TCP phase appears with a 45° angle relative to the rafted orientation after aging 1000 h. The rafted structure has formed, and no TCP phase precipitates in the alloy with 12% Co even after long-term aging 1000 h. It may be concluded that the microstructural stability of the alloy is enhanced as Co content increases.

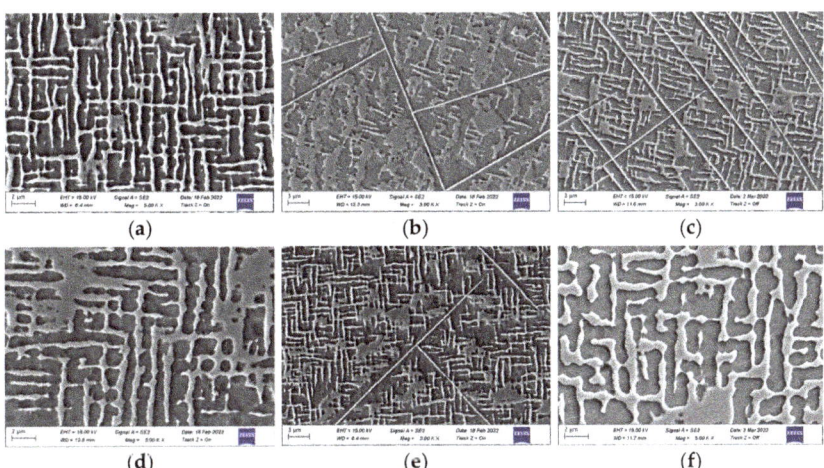

**Figure 5.** Microstructures of the alloy with different Co content after long-term aging at 1100 °C: (**a**) 200 h, 6% Co; (**b**) 400 h, 6% Co; (**c**) 1000 h, 6% Co; (**d**) 800 h, 9% Co; (**e**) 1000 h, 9% Co and (**f**) 1000 h, 12% Co.

The chemical components of the TCP phase in the 6% Co alloy and 9% Co alloy aging for 1000 h are shown in Table 2. In both samples, the TCP phase is rich in Re and W elements. In contrast, the TCP phase in 9% Co alloy contains fewer Re and W elements and much more Co and Ta elements compared with that in the 6% Co alloy.

**Table 2.** Chemical components of the TCP phase in the 6% Co alloy and 9% Co alloy after aging 1000 h at 1100 °C (wt%).

| Alloy  | Al  | Cr  | Co  | Ta  | Ru  | W    | Re   | Mo  | Ni   |
|--------|-----|-----|-----|-----|-----|------|------|-----|------|
| 6% Co  | 2.3 | 2.6 | 3.6 | 3.4 | 2.5 | 16.5 | 31.1 | 1.9 | Bal. |
| 9% Co  | 4.5 | 2.2 | 7.2 | 6.8 | 2.3 | 13.4 | 16.6 | 1.4 | Bal. |

### 3.4. Stress Rupture Properties

The influence of Co content on the stress rupture properties of the alloy at different conditions is shown in Figures 6 and 7.

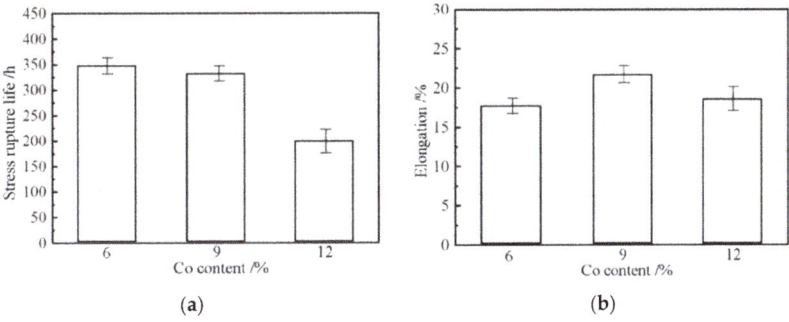

**Figure 6.** Influence of Co content on the stress rupture properties of the alloy at 1100 °C/137 MPa: (**a**) stress rupture life and (**b**) elongation.

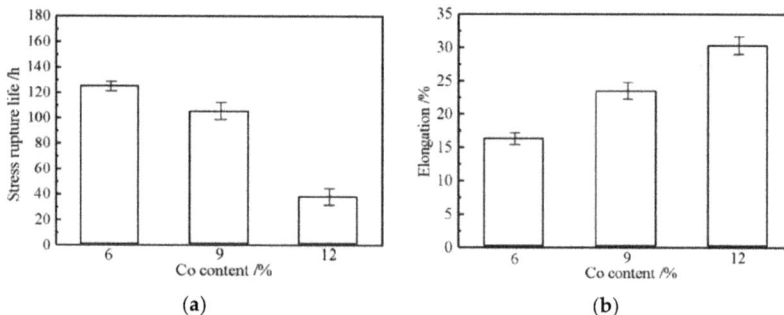

**Figure 7.** Influence of Co content on the stress rupture properties of the alloy at 1140 °C/137 Mpa: (**a**) stress rupture life and (**b**) elongation.

Figures 6 and 7 illustrate that with increasing Co content, the stress rupture life of the alloy at 1100 °C/137 MPa and 1140 °C/137 MPa all reduce in different degrees; the elongation at 1100 °C/137 MPa increases at first and decreases afterwards, but then at 1140 °C/137 MPa, monotonously increases. When Co content increases from 6% to 9% and 12%, the stress rupture life at 1100 °C/137 MPa decreases by 4.5% and 42.2%, then at 1140 °C/137 MPa and declines by 16.7% and 76.9%, respectively. This indicates that the effect of the Co element on the stress rupture life of the alloy improves with the increase in Co content or test temperature.

*3.5. Microstructure of Stress Ruptured Samples*

The longitudinal section microstructure of the stress ruptured samples with different Co content at 1100 °C/137 MPa and 1140 °C/137 MPa was observed. Figures 8 and 9 show the microstructure at 1.5 cm from the fracture surface of the ruptured samples at different conditions, while Figures 10 and 11 show the microstructure adjacent to the fracture surface of the ruptured samples at different conditions.

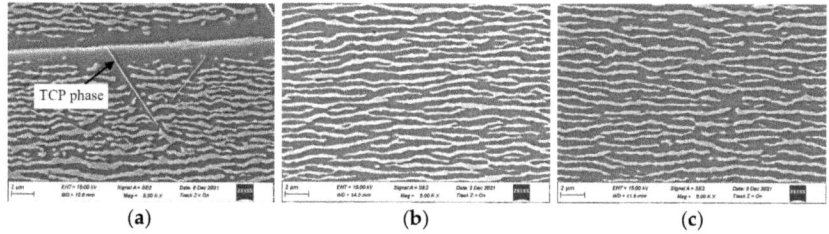

**Figure 8.** Microstructure 1.5 cm from the fracture surface of the ruptured samples at 1100 °C/137 MPa (353.6 h): (**a**) 6% Co; (**b**) 9% Co and (**c**) 12% Co.

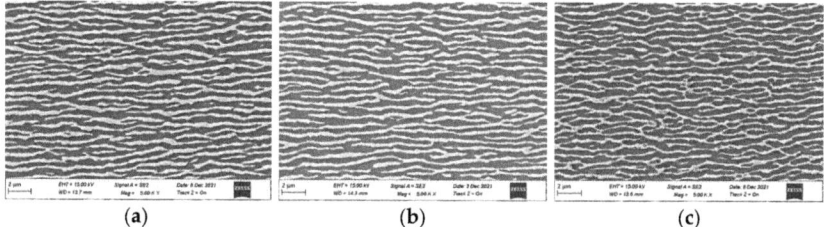

**Figure 9.** Microstructure 1.5 cm from the fracture surface of the ruptured samples at 1140 °C/137 MPa (126.2 h): (**a**) 6% Co; (**b**) 9% Co and (**c**) 12% Co.

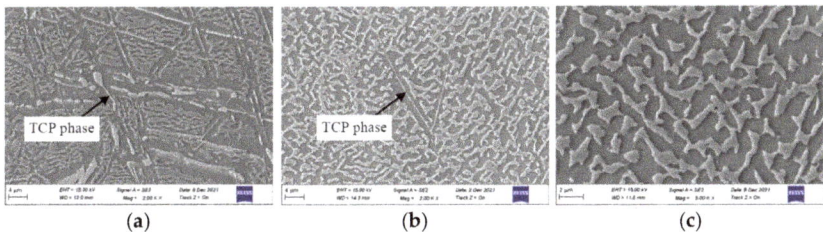

**Figure 10.** Microstructure adjacent to the fracture surface of the ruptured samples at 1100 °C/137 MPa (353.6 h): (**a**) 6% Co; (**b**) 9% Co and (**c**) 12% Co.

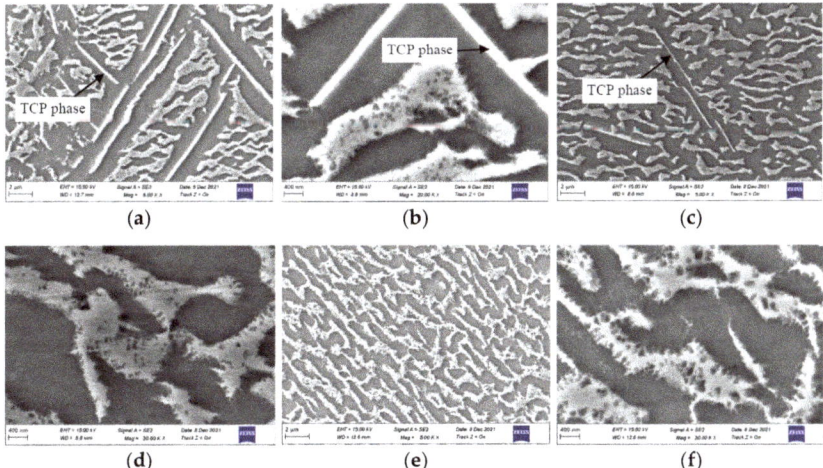

**Figure 11.** Microstructure adjacent to the fracture surface of the ruptured samples at 1140 °C/137 MPa (126.2 h): (**a,b**) 6% Co; (**c,d**) 9% Co and (**e,f**) 12% Co.

It is shown in Figure 8 that the $\gamma'$ particles degenerate into the rafts in the orientation vertical to the tensile stress for three alloys at the condition of 1100 °C/137 MPa. The difference is that the thickness of the raft decreases slightly with the increase in Co content in addition to the presence of the TCP phase in the 6% Co alloy. The continuity of the rafted structure is destroyed by the acicular TCP phase. It can be seen in Figure 10 that the $\gamma$ matrix is not continuous and has become islands that are completely surrounded by the $\gamma'$ precipitates, which is called "topological inversion" [16]. The raft displays a slightly twisted configuration because the stress state near the fracture surface is different to that apart from the fracture surface. The acicular TCP phase has presented in the 6% Co alloy and 9% Co alloy. Moreover, the amount of TCP phase in 6% Co alloy is much more than that in 9% Co alloy.

It is shown in Figure 9 that the $\gamma'$ particles also degenerate into the rafts in the orientation vertical to the tensile stress for three alloys at 1140 °C/137 MPa. The difference is that the thickness of the raft decrease slightly with the increase in Co content, but there is no TCP phase formed in the 6% Co alloy. It is shown in Figure 11 that the $\gamma$ phase presented similar characteristics and the so-called topological inversion can also be seen. The acicular TCP phase forms in the 6% Co alloy and 9% Co alloy. The amount of TCP phase in the 6% Co alloy is much more than that in the 9% Co alloy. Compared with 1100 °C/137 MPa, a new discovery is that the secondary fine $\gamma'$ particles precipitated after the stress rupture process. The integrity of the dislocation network is damaged by the secondary $\gamma'$ particles, so this is an adverse impact on the mechanical properties of the alloy at high temperatures [17].

A conclusion can be drawn from the above analysis, which is as the Co content rises, the microstructure stability of the alloy is enhanced. This is consistent with the results of the long-term aging experiment.

## 4. Discussion

### 4.1. Microstructure Evolution of the $\gamma'$ Phase

The directional growth of $\gamma'$ particles in long-term aging is the result of the combined actions of thermodynamics and kinetics. The driving force for the growth and coarsening is the decrease in the interfacial energy between the $\gamma'$ particles and $\gamma$ matrix [18]. The larger $\gamma'$ particles grow and smaller $\gamma'$ particles dissolve with the increase in aging time. In LSW theory [19], if the coarsening of the $\gamma'$ phase is diffusion controlled, the following formula will be valid:

$$(r_t^3 - r_0^3)^{1/3} = Kt^{1/3} \qquad (1)$$

where $r_0$ is the initial particle radius, $r_t$ is the instantaneous particle radius, $K$ is the rate constant and $t$ is the aging time.

In the stress rupture process, the cubic $\gamma'$ phase gradually changed into a raft structure because of the directional diffusion of the elements. The diffusion and redistribution of the alloying elements in the $\gamma'$ and $\gamma$ phases have occurred [20]. With the action of the applied stress and the misfit stress at high temperature, the $\gamma'$ phase forming elements Al, Ta, Nb and Hf diffuse to the vertical channels to promote the $\gamma'$ phase growth perpendicular to [001] direction. At the same time, the $\gamma$ matrix forming elements Cr, Co, W, Mo, Re and Ru diffuse to the horizontal channels in the reverse orientation to increase the width of the $\gamma$ matrix. Under the condition of temperature and stress, the $\gamma'$ rafted structure gradually formed. The formation of regular and perfect $\gamma'$ rafts has a good effect on the stress rupture life of the single-crystal superalloy.

### 4.2. Microstructural Stability of the Alloy

The microstructural stability of high-generation single-crystal superalloys is one of the key technical indexes [21]. The volume fraction and precipitation rate of the TCP phase both declined with the increase in Co content both in the long-term aging experiment and stress rupture properties test. It shows that increasing Co content in the single-crystal superalloys can improve microstructure stability at high temperatures. The TCP phase precipitate of the alloys is ascribed to that of the high-melting-point alloying elements (Re, W, Mo, et al.) in the disordered $\gamma$ matrix which are oversaturated [22]. The equilibrium phase precipitation characteristics of the alloy with different Co content were studied using JMatPro software and the corresponding database. The research result of equilibrium phases of the alloy with different Co contents at 1100 °C is illustrated in Figure 12. There are four equilibrium phases in the alloy: $\gamma$ phase, $\gamma'$ phase, TCP phase and carbide. There is a small change for the TCP phase and MC as the Co content increases from 6% to 12%. However, the amount of $\gamma'$ phase greatly declines and the $\gamma$ matrix greatly increases with the increase in Co content. The three alloys all have the same amount of Re, W, Mo and other $\gamma$ phase formation elements. With the increase in the volume fraction of the $\gamma$ phase, the concentration of these elements in the $\gamma$ matrix can be declined accordingly. So, the supersaturating degree of the high-melting-point alloying elements (Re, W and Mo) within the $\gamma$ matrix will be reduced and the microstructural stability of the alloy can be enhanced as Co content rises.

It is shown that TCP phases precipitated and grew along a fixed direction in all of the specimens in the long-term aging or stress rupture testing. The crystal structures of the TCP phases are extremely complex, and the size of the unit cell is much larger than the lattice of the $\gamma$ and $\gamma'$ phases. A large nucleation barrier serves to prevent the formation of TCP phases in the microstructures of single-crystal superalloys [23]. When TCP phases form in the alloys, they nucleate preferentially on close-packed planes, forming a semicoherent interface, and exhibit distinctive orientation relationships with the parent crystal [24].

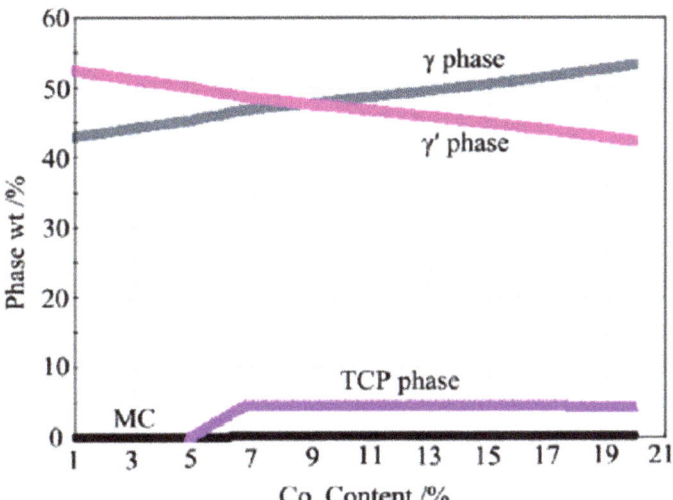

**Figure 12.** The effect of Co content on the equilibrium phases at 1100 °C.

*4.3. Stress Rupture Properties of the Alloy*

The variation of stress rupture properties of the alloy with different Co contents can be attributed to the microstructure changes. The excellent mechanical properties of single-crystal superalloys are mainly attributed to the solution strength of the $\gamma$ phase, the precipitation strengthening of $\gamma'$ phase and $\gamma/\gamma'$ interface strengthening.

Firstly, the atomic radius of Co and Ni is 0.167 nm and 0.162 nm, respectively. The solid solution strengthening effect of the Co element is very small because Co and Ni have very little difference in atomic radius. Co and Ni may be intermiscible indefinitely. Moreover, Co is a $\gamma$ matrix formatting element. The increase in Co content is equivalent to increasing the solvent, diluting the concentration of solution strength elements, such as Mo, W and Re, et al. The solid solution strengthening action of the alloy is reduced as Co content rises.

Secondly, the mechanical property of the alloy is greatly influenced by the $\gamma'$ phase content as considering the precipitation enhancement [9]. Both experiment and phase diagram calculations show that the volume fraction of the $\gamma'$ phase decreases with the increase in Co content, which greatly decreases the precipitation-strengthening effect of the alloy.

Lastly, dislocations and interfaces provide increased diffusion promoting by-passing of the $\gamma'$ phase by climbing at elevated temperatures [25]. The deformation feature of the alloy is the movement of 1/2 [110] dislocations on the octahedral slip systems in the matrix channels [26]. The dislocation networks in the matrix result from the reaction of two sets of dislocations with different Burgers vectors during creep. These dislocations exist in different slip planes. Once these dislocations move to the same slip plane to come across each other, the three-dimensional networks may be formed by the reaction of the dislocation. The denser dislocation networks may be strongly impeded by the subsequent dislocation to cut into the $\gamma'$ precipitate and make the alloy maintain a minimum creep rate [27]. Figure 13 shows the $\gamma/\gamma'$ interfacial dislocation configuration of ruptured samples at 1100 °C/137 MPa. It is shown that the dislocation networks at $\gamma/\gamma'$ interfaces have been clearly formed for the samples. The denser extent of the dislocation network is related to the latlice mifit. The lattice misfit of the alloy is more negative with the decrease in Co content. The dislocation network of the alloy turns sparser and dislocation spacings become bigger with the increase in Co content. Therefore, this is also a reason that the stress rupture lives of the alloy decrease as the Co content rises.

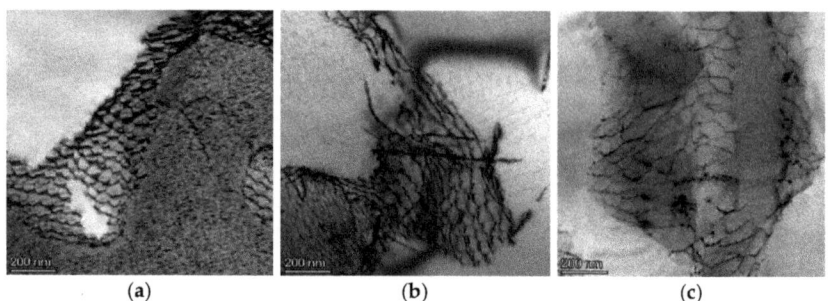

**Figure 13.** TEM images of ruptured specimens of the alloy with different Co content at 1100 °C/137 MPa: (**a**) 6% Co; (**b**) 9% Co and (**c**) 12% Co.

On the other hand, as the Co content rises, the microstructural stability of the alloy is enhanced. It is beneficial to the stress rupture properties of the alloy. There are three main reasons for the negative role of the TCP phase on the stress rupture properties. Firstly, the TCP phase is brittle; it is the site for crack initiation and the easy way for crack propagation during plastic deformation. Secondly, the TCP phase destroys the continuity of the matrix. Finally, the solid solution strengthening elements, such as Re and W, are enriched in the TCP phase, as shown in Table 2, which results in Re and W in the matrix surrounding the TCP phase and decreases the solid solution strengthening of the matrix. Therefore, these lead to an extensive envelope of the $\gamma'$ phase around the TCP phase which may potentially act as a channel for preferential deformation [28].

A combination of beneficial and harmful effects is associated with increasing Co content; the alloy containing 9% Co is the best choice considering the microstructure stability and stress rupture properties.

## 5. Conclusions

(1) The primary dendrite arm spacing of the alloy decreases at first and increases afterwards, and the volume fraction of $\gamma$-$\gamma'$ eutectic decreases with the increase in Co content in the as-cast microstructures.

(2) The size, cubic degree and volume fraction of $\gamma'$ phase of the alloy after standard heat treatment all decrease with the increase in Co content. The microstructural stability of the alloy is enhanced with the increase in Co content. No TCP phase in the alloy with 12% Co precipitate even after aging 1000 h.

(3) The stress rupture lives at two conditions both reduce at different degrees with the increase in Co content. The effect of the Co element on the stress rupture life of the alloy improves with the increase in Co content or test temperature.

(4) The acicular TCP phase appears in the 6% Co alloy and 9% Co alloy in the microstructure of the ruptured specimens with different Co contents. Moreover, the TCP phase content in the 6% Co alloy is much more than that in the 9% Co alloy. No TCP phase is observed in the 12% Co alloy.

(5) The alloy containing 9% Co is the best choice considering the microstructure stability and stress rupture properties.

**Author Contributions:** Experiment conduction, writing—original draft preparation, Z.S.; data and image processing—S.L. All authors have read and agreed to the published version of the manuscript.

**Funding:** This research was funded by the National Science and Technology Major Project (2017-VI-0002-0071).

**Data Availability Statement:** Not applicable.

**Conflicts of Interest:** The authors declare no conflict of interest.

## References

1. Caron, P.; Khan, T. Evolution of Ni-based superalloys for single crystal gas turbine blade applications. *Aerosp. Sci. Technol.* **1999**, *3*, 513–523. [CrossRef]
2. Zheng, Y.R.; Zhang, D.T. *Color Metallographic Investigation of Superalloy and Steels*; National Defence Industry Press: Beijing, Chian, 1999; pp. 6–7.
3. Li, J.R.; Xiong, J.C.; Tang, D.Z. *Advanced High Temperature Structure Materials and Technology*; National Defence Industry Press: Beijing, China, 2012; pp. 19–20.
4. Yang, D.Y.; Jin, T.; Zhao, N.R. The influence of cobalt, tungsten, and titanium on the as-east microstructure of single crystal nickel-base superalloys. *J. Aeronaut. Mater.* **2003**, *23*, 17–20.
5. Yang, D.Y.; Zhang, X.; Jin, T. The Innuence of cobalt, tungsten, and titanium on stress-rupture properties of Nickel-base Single Crystal Superalloy. *Rare Met. Mater. Eng.* **2005**, *34*, 1295–1298.
6. Liu, J.L.; Zhang, J.; Meng, J.; Jia, Y.X.; Jin, T. The effect of Co on the microstructure and stress rupture properties of a single crystal superalloy. *Mater. Res. Innov.* **2014**, *S4*, 414–420. [CrossRef]
7. Wang, W.Z.; Jin, T.; Zhao, N.R.; Wang, Z.H.; Sun, X.F.; Guan, H.R.; Hu, Z.Q. Effect of Cobalt on chemical segregation and solution process in Re-containing single crystal superalloys. In Proceedings of the 2006 Beijing International Materials Week, Beijing, China, 25–30 June 2006; pp. 1978–1981.
8. Zhao, G.Q.; Tian, S.G.; Zhu, X.J. Effect of element Ru on microstructure and creep behaviour of single crystal nickel-based superalloy. *Mater. High Temp.* **2019**, *36*, 132–141. [CrossRef]
9. Shi, Z.X.; Li, J.R.; Liu, S.Z. Effect of long term aging on microstructure and stress rupture properties of a Nickel based single crystal superalloy. *Prog. Nat. Sci. Mater. Int.* **2012**, *22*, 426–432. [CrossRef]
10. Yang, W.C.; Liu, C.; Qu, P.F. Strengthening enhanced by Ru partitioned to $\gamma'$ phases in advanced Nickel-based single crystal superalloys. *Mater. Charact.* **2022**, *186*, 111809. [CrossRef]
11. Erickson, G.L. The development and application of CMSX-10. In *Superalloys*; TMS: Warrendale, PA, USA, 1996; pp. 35–44.
12. Walson, W.S.; O'hara, K.; Ross, E.W.; Pollock, T.M.; Murphy, W.H. RenéN6: Third generation single crystal superalloy. In *Superalloys*; TMS: Warrendale, PA, USA, 1996; pp. 27–34.
13. Walston, S.; Cetel, A.; Mackay, R.; O'hara, K.; Duhl, D.; Deshfield, R. Joint development of a fourth generation single crystal superalloy. In *Superalloys*; TMS: Pennsylvania, PA, USA, 2004; pp. 15–24.
14. Argence, D.; Vernault, C.; Desvallees, Y.; Fournier, D. MC-NG: Generation single crystal superalloy for future aeronautical turbine blades and vanes. In *Superalloys*; TMS: Warrendale, PA, USA, 2000; pp. 829–837.
15. Zhang, J.X.; Murakumo, T.; Koizumi, Y.; Kobayshi, T.; Harada, H. Interfacial Dislocation Networks Strengthening a Fourth-Generation Single-Crystal TMS-138 Superalloy. *Metall. Mater. Trans. A* **2002**, *33*, 3741–3746. [CrossRef]
16. Maciej, Z.; Steffen, N.; Mathias, G. Characterization of $\gamma$ and $\gamma'$ phases in 2nd and 4th generation single crystal Nickel-Base superalloys. *Met. Mater. Int.* **2017**, *23*, 126–131.
17. Shi, Z.X.; Li, J.R.; Liu, S.Z. Effect of Hf on the microstructures and stress rupture properties of DD6 single crystal superalloy. *Rare Met. Mater. Eng.* **2010**, *39*, 1334–1338.
18. Ren, Y.L.; Jin, T.; Guan, H.R.; Hu, Z.Q. The effect of long aging time at high temperature on the structure evolution of $\gamma'$ phase for a Nickel base single crystalline superalloy. *Mater. Mech. Eng.* **2004**, *28*, 10–12. (In Chinese)
19. Lifshitz, M.; Slyozov, V.V. The kinetics of precipitation from supersaturated solid solution. *J. Phys. Chem. Solids* **1961**, *19*, 35–50. [CrossRef]
20. Tian, S.G.; Zhang, J.H.; Zhou, H.H.; Yang, H.C.; Xu, Y.B.; Hu, Z.Q. Aspects of primary creep of a single crystal nickel-base superalloy. *Mater. Sci. Eng. A* **1999**, *262*, 271–280.
21. Yan, H.J.; Tian, S.G.; Zhao, G.Q. Deformation features and affecting factors of a Re/Ru-containing single crystal nickel-based superalloy during creep at elevated temperature. *Mater. Sci. Eng. A* **2019**, *768*, 138437. [CrossRef]
22. Han, Y.F.; Ma, W.Y.; Dong, Z.Q.; Li, S.S.; Gong, S.K. Effect of Ruthenium on microstructure and stress rupture properties of a single crystal Nickel-base superalloy. In *Superalloys*; TMS: Pennsylvania, PA, USA, 2008; pp. 91–97.
23. Neumeier, S.; Pyczak, F.; Goken, M. The influence of Ruthenium and Rhenium on the local properties of the $\gamma$- and $\gamma'$-phase in Nickel-base superalloys and their consequences for alloy behavior. In *Superalloys*; TMS: Pennsylvania, PA, USA, 2008; pp. 109–110.
24. Rae, C.M.F.; Karunaratne, M.S.A.; Small, C.J.; Broomfield, R.W.; Jones, C.N.; Reed, R.C. Topologically close packed phases in an experimental Rhenium-containing single crystal superalloy. In *Superalloys*; TMS: Warrendale, PA, USA, 2000; pp. 767–777.
25. Tian, S.G.; Zhang, J.H.; Zhou, H.H.; Yang, H.C.; Xu, Y.B.; Hu, Z.Q. Formation and role of dislocation networks during high temperature creep of a single crystal nickel–base superalloy. *Mater. Sci. Eng. A* **2000**, *279*, 160–165.
26. Yu, J.J.; Sun, X.F.; Jin, T.; Zhao, N.R.; Guan, H.R.; Hu, Z.Q. High temperature creep and low cycle fatigue of a nickel-base superalloy. *Mater. Sci. Eng. A* **2010**, *527*, 2379–2389. [CrossRef]

27. Tian, S.G.; Zhang, B.S.; Shu, D.L. Creep properties and deformation mechanism of the containing 4.5Re/3.0Ru single crystal nickel-based superalloy at high temperature. *Mater. Sci. Eng. A* **2015**, *643*, 119–126. [CrossRef]
28. Yeh, A.C.; Tin, S. Effect of Ru on the high temperature phase stability of Ni-base single crystal superalloys. *Metall. Mater. Trans. A* **2006**, *37*, 2621–2631. [CrossRef]

**Disclaimer/Publisher's Note:** The statements, opinions and data contained in all publications are solely those of the individual author(s) and contributor(s) and not of MDPI and/or the editor(s). MDPI and/or the editor(s) disclaim responsibility for any injury to people or property resulting from any ideas, methods, instructions or products referred to in the content.

*Article*

# Research on the Microstructure and Properties of a Flux-Cored Wire Gas-Shielded Welded Joint of A710 Low-Alloy High-Strength Steel

Xing Wang, Zhibin Yang * and Lingzhi Du

School of Materials Science and Engineering, Dalian Jiaotong University, Dalian 116028, China
* Correspondence: yangzhibin@djtu.edu.cn

**Abstract:** In this study, a 16 mm thick A710 low-alloy high-strength steel was welded by using flux-cored wire gas-shielded welding with an E81T1-Ni1M flux-cored wire. The microstructure characteristics and mechanical properties of the joints were systematically studied. The results showed that the joint was well formed without obvious welding defects. The center of the weld was mainly needle-like ferrite, the coarse grain area was mainly slat-like and granular bainite, and the fine grain area was mainly ferrite and pearlite. The lowest hardness in the weld area was the weakest area of the joint. The average tensile strength of the joint was 650 MPa, reaching 95% of the base metal; the samples were all fractured in the weld area, and the fracture morphology showed typical plastic fracture characteristics. The low-temperature ($-40\ °C$) impact energy of the joint weld area and the heat-affected zone were 71 J and 253 J; the fracture morphology was characterized by a ductile–brittle mixed fracture, and the ductile area of the specimen fracture in the heat-affected zone was larger. The bending performance was good. Under the specified life of $2 \times 10^6$ cycles; the median fatigue limit and the safety fatigue limit were 520 MPa and 492 MPa, and the fatigue cracks germinated on the surface of the priming weld.

**Keywords:** low-alloy high-strength steel; flux-cored wire; microstructure; mechanical properties; hardness distribution

## 1. Introduction

With the development of industrial technology, traditional steel materials cannot meet the increasingly stringent quality and performance requirements, and low-alloy high-strength steel has attracted more attention. Low-alloy high-strength steel has the advantages of high strength, good weldability, and excellent formability. It can improve the structural strength and toughness while saving materials, which is in line with the national development trend of low carbon and energy savings [1–3]. Welding is one of the key technologies for the application of low-alloy high-strength steel. A solid wire or coating electrode is usually used as its welding material. However, the defects such as an unstable arc, large spatter rate, and slag inclusion during welding limit its application range.

Flux-cored wire is the most promising welding material in the 21st century. It not only has the advantages of a gas-shielded solid wire and flux strip, but it also can obtain good mechanical properties of joints by adjusting the composition and has become more widely used [4]. The flux-cored wire has a high current density, a high thermal efficiency, and a higher deposition efficiency than a solid wire and coating electrode. At the same time, the spray transfer mode is active, and the splash rate is significantly lower than that of a solid wire and coating electrode. The composition of the flux core contains a variety of gas-forming and slag-forming substances. The welding process can realize the combined protection of gas and slag for droplets and molten pools and improve the surface wind resistance [5–7].

Zou et al. [8] used flux-cored wire to weld Q960 low-alloy high-strength steel and studied the effect of aluminum on the microstructure and mechanical properties of Q960 steel weld metal. The results showed that the addition of Al could reduce the content of O and N in the weld, reflecting a good deoxidization and denitrification performance and reducing the porosity content of the weld. When the Al content was about 0.21%, $Al_2O_3$ oxide inclusions were easily formed in the microstructure. They were circular, small, and dispersed, which was conducive to AF (Acicular Ferrite) nucleation, improving the impact energy absorption of the weld metal and the impact toughness of the flux-cored wire. When the Al content was about 1.05%, AlN nitride was formed in the microstructure, the particles were large and polygonal, and microcracks were easily generated between the inclusions and the surrounding matrix, which had an adverse effect on the impact energy absorption of the weld metal. Ilić et al. [9] studied the effect of the welding process on the impact toughness of welded joints of high-strength low-alloy steel. The welding methods included MMA (Manual Metal Arc Welding), MIG (Metal Inert-gas Welding), and MAG (Metal Active Gas Arc Welding). The results showed that when there was a notch in the fusion zone, the parameters were the same except for the weld groove geometry. The experimental value of the fracture energy of the MMA/MAG weld was higher than that of the MIG/MAG weld. In the MMA/MAG welding process, the fracture energy of the specimen with a notch on the side of the root channel was higher than that of the MIG/MAG welding specimen. Therefore, in this regard, the application of MMA/MAG welding could provide better joint characteristics. However, the fracture energy of the MMA/MAG and MIG/MAG weld samples was smaller than that of base metal samples. When there was a notch in the root heat-affected zone, the fracture energy was almost the same. Compared with the MMA/MAG welded joint, the MIG/MAG welded joint sample did not show a termination zone of rapid crack propagation during the test. Kornokar et al. [10] studied the effect of the heat input on the microstructure and mechanical properties of HSLA S500MC gas tungsten arc-welded joints. Six different welding heat inputs were set up in the experiment. The results showed that the weld metal was lath martensite and retained austenite. With the increase in the heat input, the carbides dissolved, the size of the martensite increased, and the content of the retained austenite increased. The coarse grain zone contained a martensite structure, and the fine grain zone was a mixed structure of pearlite and martensite. The maximum tensile strength of the joint was 690 MPa. With the increase in the heat input, the tensile strength tended to decrease. Oktadinata et al. [11] used flux-cored wire to weld SM570-TMC low-alloy high-strength steel. The welding heat inputs of 0.9 KJ/mm and 1.4 KJ/mm were used to study the effect of different heat inputs on the microstructure and impact toughness of the weld. The results showed that the content of acicular ferrite was higher, the microstructure was finer, and the impact toughness was better at a low heat input. Ni et al. [12] welded Q690 high-strength low-alloy steel and studied the microstructure and mechanical properties of narrow-gap GMA (Gas Metal Arc Welding) welded joints with ternary gas protection. The results showed that without preheating before welding, using 80% Ar-10% $CO_2$-10% He ternary protective gas, a complete narrow-gap welded joint of Q690E high strength steel with sufficient side wall penetration was obtained. With the increase in the heat input, the morphology of the bainite ferrite changed from AF to LB (Lath Bainite), resulting in a decrease in the tensile strength and the low-temperature impact toughness of the welded joints. The grain size of the heat-affected zone increased with the increase in the heat input. By changing the welding speed and optimizing the heat input, the tensile strength of the narrow-gap GMA welded joint was improved. When the welding speed was 350 mm/min, the average tensile strength reached 795 MPa. Wen et al. [13] studied the microstructure and mechanical properties of welded joints of ultra-high-strength structural steel under different heat inputs. They thought that the reasonable control of welding heat input would significantly reduce the welding's cold crack defects. At the same time, it was pointed out that when the welding heat input was greater than 7.5 KJ/cm, a twin martensite structure could be avoided, and cold cracks could be effectively avoided. At present, there are relatively few studies on the welding of low-alloy high-strength steel for mining vehicles, and most of the welding materials selected for low-alloy high-strength steel were solid welding wires. There are relatively few studies on the welding of low-alloy high-strength steel for mining vehicles using flux-cored wires [14].

Therefore, based on the principle of equal strength matching, this study used E81T1-Ni1 M flux-cored wire to weld 16 mm thick A710 low-alloy high-strength steel for mining vehicles. The microstructure characteristics, hardness distribution, tensile properties, low-temperature impact properties, bending properties, and fatigue properties of the joints were systematically studied, which provided an experimental basis and theoretical support for the application of flux-cored wire gas-shielded welding to weld low-alloy high-strength steel for mining vehicles.

## 2. Materials and Methods

### 2.1. Materials

The base metal of the experiments was 16 mm thick A710 low-alloy high-strength steel. The chemical composition is shown in Table 1, and the size was 350 mm × 150 mm × 16 mm. Before welding, the surface of the test piece was polished to remove rust and acetone wipe oil. The filler material was E81T1-Ni1M flux-cored wire with a diameter of 1.6 mm, and its chemical composition is shown in Table 2.

Table 1. The chemical composition of the base metal (mass fraction, %).

| Base Metal | C | Mn | Si | P | S | Nb | Ti | Ni | Cu | Cr | Mo | Ni | Al |
|---|---|---|---|---|---|---|---|---|---|---|---|---|---|
| A710 | ≤0.07 | 0.4~0.7 | ≤0.4 | ≤0.025 | ≤0.025 | 0.02~0.05 | ≤0.025 | 0.7~1.4 | 1.0~1.3 | 0.6~0.9 | 0.15~0.25 | 0.7~1.4 | ≤0.05 |

Table 2. The chemical composition of the flux-cored wire (mass fraction, %).

| Flux-Cored Wire | Mn | Si | Cr | Ni | Mo | V | Nb | Al | Cu |
|---|---|---|---|---|---|---|---|---|---|
| E81T1-Ni1M | ≤1.4 | ≤0.8 | ≤0.2 | 0.6~1.2 | ≤0.2 | ≤0.08 | ≤0.05 | ≤2.0 | ≤0.3 |

### 2.2. Methods

The welding experimental was carried out by using a Megmeet ARTSEN PM400FII (MEGMEET, Shenzhen, China) all-digital industrial heavy-duty welding machine. The joint form was a butt joint, the root gap was 6 mm, the groove angle was 30°, the back was forced to form by a liner, the liner thickness was 5 mm, the protective gas was $CO_2$, and the gas flow rate was 20 L/min. The optimized welding process parameters are shown in Table 3.

Table 3. The adopted optimized welding parameters.

| Line | Current (A) | Voltage (V) | Welding Speed (cm/min) | Heat Input (kJ/cm) | Welding Sequence Diagram |
|---|---|---|---|---|---|
| 1 | 250–280 | 26–28 | 20 | 15 | |
| 2 | 250–280 | 26–28 | 19 | 15 | |
| 3–4 | 250–280 | 26–28 | 35 | 8.6 | |
| 5–7 | 250–280 | 26–28 | 42 | 7.6 | |

The macroscopic and microscopic inspection of the weld was carried out according to the GB/T 26955-2011 standard. The metallographic specimen was etched by 4% nitric acid alcohol reagent. The etching time was 4–5 s. The macroscopic morphology and microstructure of the joint were observed by a KEYENCEVHX-1000 (Keyence, Osaka, Japan) video microscope and an OLYMPUS-BX51M (Olympus, Tokyo, Japan) metallographic microscope, respectively. The hardness distribution of the joint was measured by an HV-50A (Huayin, Laizhou, China)

Vickers hardness tester. According to the relevant provisions of the GB/T 4340.1-2009, three paths of 1 mm from the upper surface, 1 mm from the lower surface, and the center of the plate thickness were tested. The loading load was 10 kgf, and the load holding time was 15 s. Tensile and bending tests were performed using a WDW-300E (SUNS, Shenzhen, China) electronic universal testing machine. The tensile test was carried out according to the relevant provisions of the GB/T 2651-2008 and GB/T 25774.1-2010, and the size of the tensile specimen is shown in Figure 1a. The tensile rate was 3 mm/min. The bending test was carried out according to the relevant provisions of the GB/T 2653-2008 and ISO 5173-2000, and the size of the bending specimen is shown in Figure 1b. The bending rate was 1 mm/min. The bending angle of the bending test was 180°, and the diameter of the indenter was 40 mm. The low temperature (−40 °C) impact test was carried out by using a JB-W300A (Time Test, Jinan, China) Microcomputer controlled pendulum impact tester and a DWY-80A (Time Test, Jinan, China) impact test cryostat. The impact performance test was carried out according to the relevant provisions of the GB/T 2650-2008 and GB/T 229-2007. Notches were opened in the weld and the HAZ (Heat-Affected Zone), respectively, and the sampling positions are shown in Figure 1c,d respectively. The impact sample size is shown in Figure 1e. The impact absorption energy was recorded. The fracture morphology of the sample was observed by a ZEISS SUPRA55 (Zeiss, Oberkochen, German) field emission scanning electron microscope, and the fracture characteristics were analyzed. The fatigue test was carried out by using the QBG-200 (Qianbang, Changchun, China) digital high-frequency fatigue testing machine. The fatigue performance test was carried out according to the relevant provisions of the GB/T 13816-1992, and the sample size is shown in Figure 1f. The stress ratio R = 0, the loading frequency was 130–150 Hz, the stress step was 20 MPa, and the maximum number of cycles was $2 \times 10^6$.

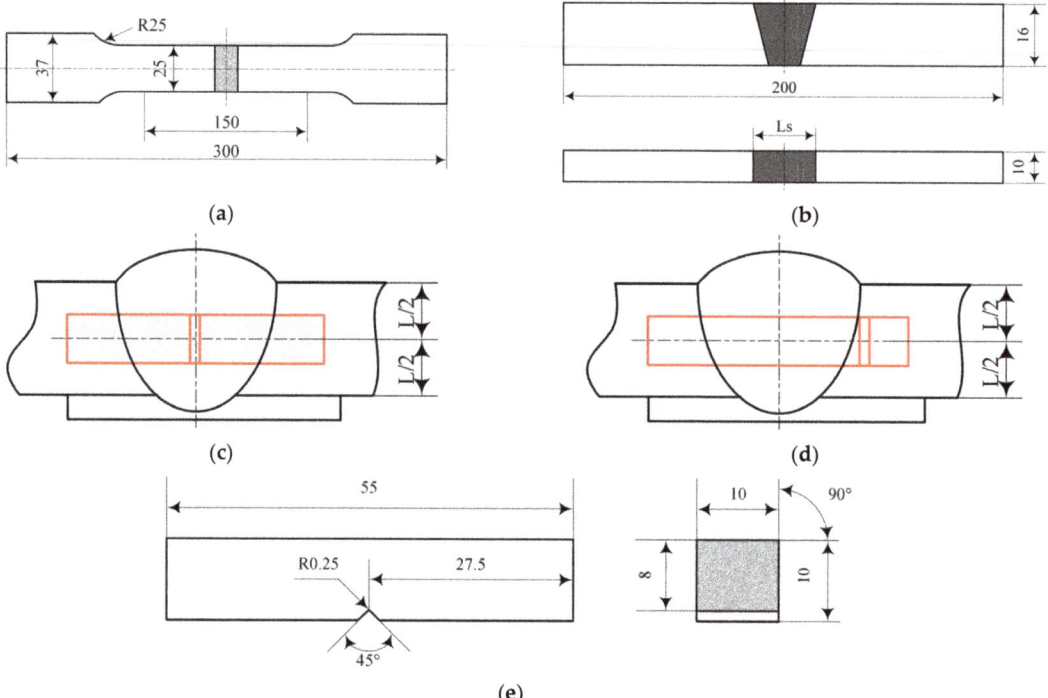

**Figure 1.** *Cont.*

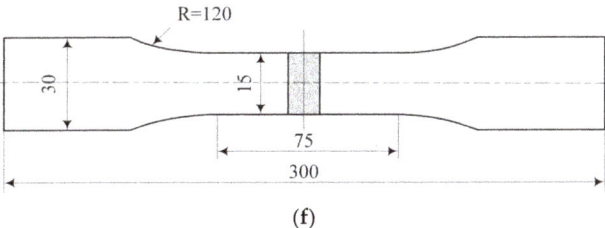

(f)

**Figure 1.** Test specimen size: (**a**) tensile specimen size; (**b**) bending specimen size; (**c**) impact specimen notch in the weld; (**d**) impact specimen notch in the HAZ; (**e**) impact specimen size; (**f**) fatigue specimen size.

## 3. Results and Analysis

### 3.1. Macroscopic and Microstructure Morphology

The macroscopic morphology and microstructure characteristics of the cross section of the joint are shown in Figure 1. The cross section of the joint was well formed, and no obvious welding defects were found, as shown in Figure 2a. The microstructure of the cover area was mainly acicular ferrite, with a small amount of proeutectoid ferrite and side plate ferrite. The acicular ferrite was mainly distributed in the grain boundary, and the direction was disordered. The proeutectoid ferrite and side plate ferrite were mainly distributed at the grain boundary, as shown in Figure 2b. The microstructure of the filling zone was mainly acicular ferrite, lamellar ferrite, and a small amount of granular bainite, as shown in Figure 2c. The microstructure of the backing zone was mainly acicular ferrite, lamellar ferrite, and side lath ferrite, as shown in Figure 2d.

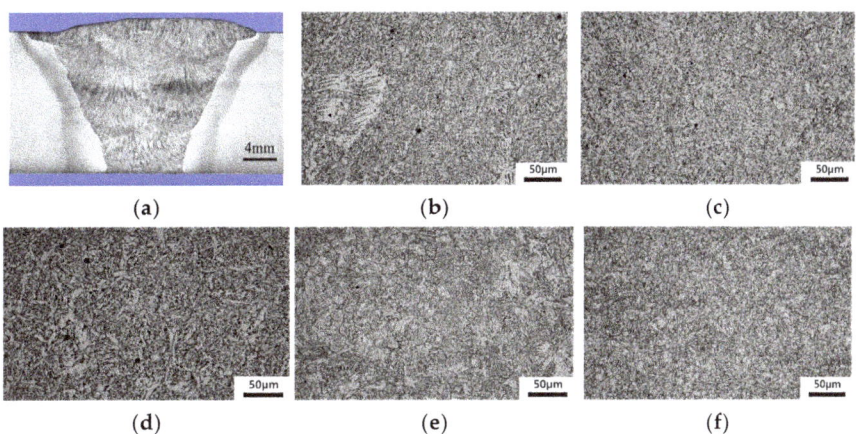

**Figure 2.** The macro- and micromorphology of the joint: (**a**) macromorphology; (**b**) microstructure of the covering surface region; (**c**) microstructure of the filling region; (**d**) microstructure of the backing region; (**e**) microstructure of the coarse-grained region; (**f**) microstructure of the fine-grained region.

The alloying elements in the weld play an important role in the control of the microstructure. The addition of alloying elements in the E81T1-Ni1M flux-cored wire first reduced the formation of proeutectoid ferrite, and the transformation of austenite to ferrite was delayed. During the cooling process of the joint, the content of the intermediate temperature transformation product in the weld increased, thereby generating more acicular ferrite or bainite structure. Acicular ferrite is a thermodynamic nonequilibrium structure, which is essentially an intragranular nucleation of bainite. The difference between the two is that bainite nucleates at the austenite grain boundary, while acicular ferrite nucleates at the nonmetallic inclusions inside the austenite. The E81T1-Ni1M flux-cored wire contains a suitable proportion of Mn and Si content, which can reduce the continuous-cooling phase transition temperature, refine the structure, and significantly increase the content of acicular ferrite. Acicular ferrite is the most desirable microstructure in the weld. It can effectively improve the strength and toughness of the weld. Its content and morphology determine the impact toughness of the weld. A higher amount of randomly oriented fine AF leads to improving the strength and toughness of the material. The side lath ferrite is a conical or serrated ferrite that extends from the grain boundary to the grain. It is essentially a Widmanstätten structure, which will reduce the performance of the joint. The side lath ferrite also belongs to the proeutectoid ferrite, but the formation temperature is lower than that of the grain boundary ferrite. The appearance of side lath ferrite is mainly due to the high welding heat input and the high air-cooling speed after welding [15].

The microstructure of the coarse grain zone was mainly lath bainite and granular bainite, the grain was coarse, and the grain boundary was obvious, as shown in Figure 2e. The microstructure of the fine grain zone was mainly uniform and fine ferrite and pearlite, as shown in Figure 2f.

### 3.2. Hardness Distribution

The hardness distribution characteristics of the joint are shown in Figure 3. The results showed that the average hardness of the base metal was about 245 HV, the average hardness of the weld zone was about 215 HV, and the average hardness of the heat-affected zone was about 230 HV. The hardness values of the weld zone and the heat-affected zone were lower than those of the base metal. The hardness value of the weld zone was the lowest, which was the weakest area of the joint. The main reason for the lowest hardness value in the weld zone was that there were more proeutectoid ferrite and lamellar ferrite in the region, and the multipass welding heat effect increased the grain size and further reduced the hardness value. At the same time, the hardness value of the 1 mm area from the lower surface was slightly higher than that of the central area of the plate thickness and the 1 mm area from the upper surface. This was mainly due to the narrow bottom weld, which led to the fast-cooling rate and high temperature gradient in the welding process. The hardness of the heat-affected zone was lower than that of the base metal mainly because the inappropriate welding thermal cycle led to the formation of coarse lath and granular bainite in the heat-affected zone, resulting in softening of the heat-affected zone.

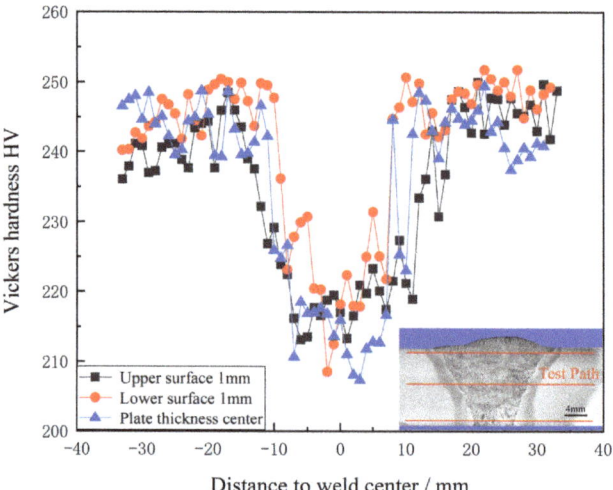

**Figure 3.** The microhardness distribution characteristics of the joint.

## 3.3. Mechanical Properties Analysis

### 3.3.1. Tensile Property

The tensile test results of the joints are shown in Table 4. The test results showed that the average tensile strength of the joint reached 650 MPa, which was about 95% of the tensile strength of the base metal. The average yield strength of the joint was 649 MPa. The average elongation after fracture and the average reduction in the area were about 12% and 43%, respectively. The fracture location and fracture characteristics of the specimen are shown in Figure 4. The tensile specimens were broken in the weld zone, and there was a very obvious necking phenomenon, as shown in Figure 4a. There were no obvious welding defects in the macroscopic fracture, as shown in Figure 4b. The microscopic fracture showed obvious dimple morphology and typical plastic fracture characteristics. There were inclusions of different sizes and quantities at the bottom of the dimples, as shown in Figure 4c. This may be one of the reasons for the close yield strength and tensile strength of the joint. Because the inclusions have a strong influence on the plasticity of the material, there were different degrees of morphological differences and stress concentration in the transverse and longitudinal directions. The cracks competed at the holes of the inclusions, and the cracks first appeared at the positions with a weak bonding force and large stress concentration [16]. This also reduced the plasticity of the joint to a certain extent.

**Table 4.** The results of the tensile test of the joints.

| Sample Number | Tensile Strength/MPa | | Yield Strength/MPa | | Break Elongation/% | | Reduction in Area/% | |
|---|---|---|---|---|---|---|---|---|
| | Single | Average | Single | Average | Single | Average | Single | Average |
| LS-1# | 645 | 650 | 644 | 649 | 12 | 12 | 44 | 43 |
| LS-2# | 655 | | 654 | | 11 | | 41 | |

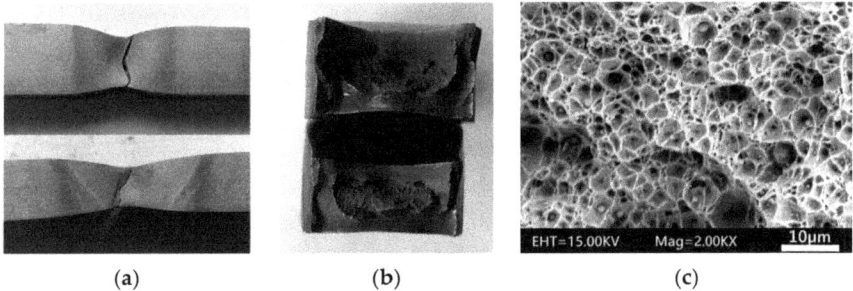

(a) (b) (c)

**Figure 4.** The fracture location and macro- and microfracture morphology of the tensile specimens: (**a**) fracture location; (**b**) macroscopic fracture; (**c**) microscopic fracture.

### 3.3.2. Low-Temperature Impact Properties

The low-temperature ($-40\ ^\circ$C) impact test results of the joint are shown in Table 5, and the V-notches were opened in the weld center and the heat-affected zone, respectively. The results showed that the average low-temperature impact energy of the weld zone and the heat-affected zone was 71 J and 253 J, respectively. The low-temperature impact energy of the weld zone was significantly lower than that of the heat-affected zone.

The macroscopic and microscopic fracture characteristics of the V-notch impact specimens at a low temperature in weld zone and the heat-affected zone are shown in Figures 5 and 6. The test results showed that the proportion of the ductile zone of the specimen fracture was small, and the proportion of the cleavage zone was large, whether in the weld zone or in the heat-affected zone, as shown in Figures 5a and 6a. There were obvious dimple morphology characteristics in the ductile area of the fracture of the weld zone and the heat-affected zone, as shown in Figures 5b and 6b. The cleavage zone of the fracture of the weld zone was mainly characterized by dissociation and quasi-cleavage morphology, as shown in Figure 5c. There were some dimple morphology characteristics in the cleavage zone of the fracture of the heat-affected zone, as shown in Figure 6c. The fracture showed the characteristics of a ductile–brittle mixed fracture as a whole. The area of the fracture ductile zone of the V-notch in the heat-affected zone was obviously larger, and there were some characteristics of the dimple in the cleavage zone, which was also one of the reasons why the low-temperature impact energy of the weld zone was lower than that of the heat-affected zone. The structure determined the performance. Because there were more brittle structures, such as the side plate ferrite and flake ferrite, in the weld zone, the toughness of the weld zone was greatly reduced. The lath bainite in the heat-affected zone increased the energy required for crack propagation due to the staggered distribution, which increased the impact energy of the heat-affected zone.

**Table 5.** The results of the low temperature impact test of the joints.

| Sample Number | Test Temperature /°C | Location of Notch | Low Temperature Impact AKV/J | Average Low Temperature Impact AKV/J |
|---|---|---|---|---|
| CJ-1# |  | Weld | 74 |  |
| CJ-2# |  | Weld | 69 | 71 |
| CJ-3# | −40 | Weld | 69 |  |
| CJ-4# |  | HAZ | 257 |  |
| CJ-5# |  | HAZ | 253 | 253 |
| CJ-6# |  | HAZ | 249 |  |

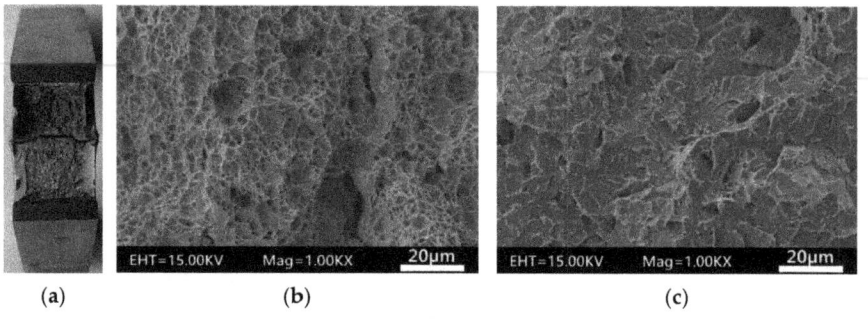

**Figure 5.** The fracture morphology of the impact sample of the weld zone: (**a**) macrofracture; (**b**) ductile region; (**c**) cleavage region.

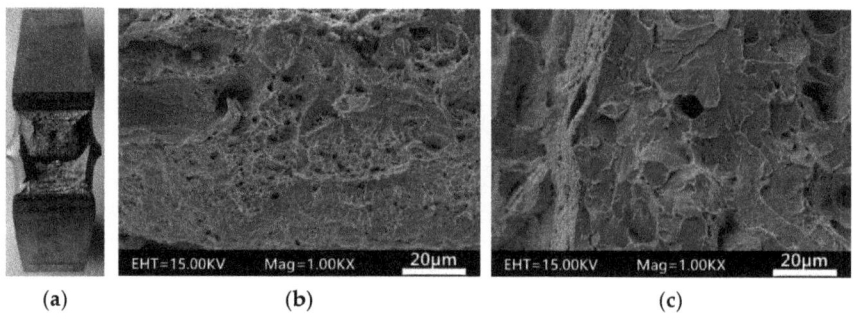

**Figure 6.** The fracture morphology of the impact sample of the heat-affected zone: (**a**) macrofracture; (**b**) ductile region; (**c**) cleavage region.

### 3.3.3. Bending Property

The bending test results of the joints are shown in Table 6. The test results showed that the tensile surface morphology of the bending specimens was good; no obvious cracks were found, and the bending performance of the joints was good.

**Table 6.** The results of the bending test of the joints.

| Sample Number | Pressure Head Diameter | Bending Angle | Bending Results |
|---|---|---|---|
| WQ-1# | | | |
| WQ-2# | 40 mm | 180° | |
| WQ-3# | | | |
| WQ-4# | | | |

### 3.3.4. Fatigue Property

The S-N fatigue curve drawn according to the joint fatigue test results is shown in Figure 7. The median fatigue limit of the 50% survival rate was 520 MPa. The safety fatigue limit of the engineering error $\delta \leq 5\%$, confidence of 95%, and survival rate of 80% was 492 MPa.

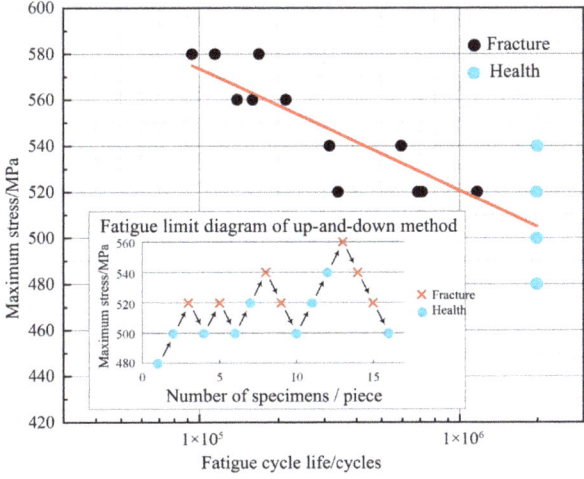

**Figure 7.** The fatigue results and the S-N curve of the fatigue test.

The fracture position and fracture morphology of the fatigue specimen are shown in Figure 8. The test results showed that the fatigue specimen cracked in the weld area, and it was located at the bottom weld bead, as shown in Figure 8a. The crack initiation area and crack propagation zone were clearly observed from the macroscopic fracture, as shown in Figure 8b. The fatigue crack initiated near the surface of the specimen, where there were small size inclusions, a loose structure, and a large stress concentration, as shown in Figure 8c, which significantly reduced the fatigue performance of the joint. There were many lamellar structures in the crack propagation zone, as shown in Figure 8d.

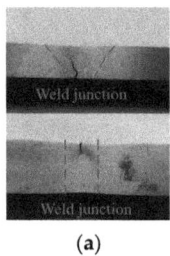  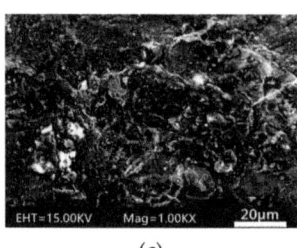 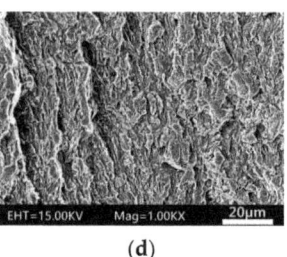

(a) (b) (c) (d)

**Figure 8.** The fracture morphology of the impact sample of the heat-affected zone: (**a**) macrofracture; (**b**) ductile region; (**c**) cleavage region; (**d**) propagation region.

The surface state and microstructure of the specimen are important factors affecting the fatigue strength. Under cyclic loading, the uneven slip of the metal is mainly concentrated on the surface of the specimen, and fatigue cracks are often generated on the surface; so, the surface state of the specimen has a great influence on the fatigue performance. In this study, the surface of the joint fatigue specimen was smooth after finishing; so, the surface state was not the main factor of fracture.

The staggered distribution of acicular ferrite and granular bainite in the weld can inhibit the formation of grain boundary cracks, improve the slip deformation resistance, inhibit the formation and cracking of the cyclic slip bands, increase the grain boundary resistance of crack propagation, and improve the fatigue strength.

The inclusion has a significant effect on the fatigue strength. The stress concentration at the inclusion is large, and it is often the crack initiation area, which greatly reduces the fatigue performance, reduces the inclusion content; reducing the inclusion size can effectively improve the fatigue strength.

## 4. Conclusions

The gas-shielded welding of 16 mm thick A710 low-alloy high-strength steel was carried out by using an E81T1-Ni1M flux-cored wire. The microstructure and mechanical properties of the welded joints were systematically studied. The main conclusions are as follows:

(1) The joint was well formed without obvious welding defects. The microstructure of the cover area was mainly acicular ferrite, with a small amount of proeutectoid ferrite and side plate ferrite. The microstructure of the filling zone was mainly acicular ferrite, flaky ferrite, and a small amount of granular bainite. The microstructure of the bottom zone was mainly acicular ferrite, flaky ferrite, and side plate ferrite. The microstructure of the coarse grain zone was mainly lath bainite and granular bainite with coarse grains. The microstructure of the fine grain zone was mainly uniform and fine ferrite and pearlite. The hardness values of the weld zone and the heat-affected zone were lower than that of the base metal, and the hardness value of the weld zone was the lowest, which was the weakest area of the joint.

(2) The average tensile strength of the joint was 650 MPa, which was about 95% of the tensile strength of the base metal. The specimens all fractured in the weld zone, the necking phenomenon was obvious, and the fracture morphology showed typical plastic fracture characteristics. The tensile surface of the bending specimen was smooth, and no obvious cracks were found.

(3) The average low-temperature ($-40$ °C) impact energy of the weld zone and the heat-affected zone was 71 J and 253 J, respectively. The proportion of the fracture ductile zone was small, and the proportion of the cleavage zone was large. The area of the fracture ductile zone of the heat-affected zone was obviously larger, and the fracture morphology showed the characteristics of a ductile–brittle mixed fracture.

(4) Under the specified life of $2 \times 10^6$ cycles, the median fatigue limit of the 50% survival rate was 520 MPa; the safety fatigue limit of the engineering error $\delta \leq 5\%$, confidence of 95%, and survival rate of 80% was about 492 MPa.

**Author Contributions:** Methodology, Z.Y.; validation, X.W. and L.D.; formal analysis, Z.Y. and X.W.; writing—original draft preparation, X.W.; writing—review and editing, Z.Y.; visualization, Z.Y. All authors have read and agreed to the published version of the manuscript.

**Funding:** This research received no funding.

**Data Availability Statement:** Date sharing not applicable.

**Conflicts of Interest:** The authors declare no conflict of interest.

## References

1. Rodrigues, T.A.; Duarte, V.; Avila, J.A.; Santos, T.G.; Miranda, R.M.; Oliveira, J.P. Wire and arc additive manufacturing of HSLA steel: Effect of thermal cycles on microstructure and mechanical properties. *Addit. Manuf.* **2019**, *27*, 440–450. [CrossRef]
2. Wang, Z.L. Control of microstructure and properties of welded joints of heavy structures of low alloy high strength steels. *Key Eng. Mater.* **2019**, *814*, 171–175. [CrossRef]
3. Berdnikova, O.; Pozniakov, V.; Bernatskyi, A.; Alekseienko, T.; Sydorets, V. Effect of the structure on the mechanical properties and cracking resistance of welded joints of low-alloyed high-strength steels. *Procedia Struct. Integr.* **2019**, *16*, 89–96. [CrossRef]
4. Zeng, H.L.; Wang, C.J.; Yang, X.M.; Wang, X.S.; Liu, R. Automatic welding technologies for long-distance pipelines by use of all-position self-shielded flux cored wires. *Nat. Gas Ind. B* **2014**, *1*, 113–118.
5. Bracarense, A.Q.; Souza, R.; Costa, M.C.M.; Faria, P.E.; Liu, S. Welding current effect on diffusible hydrogen content in flux cored arc weld metal. *J. Braz. Soc. Mech. Sci. Eng.* **2002**, *24*, 278–285. [CrossRef]
6. Li, Z.; Srivatsan, T.S.; Wang, Y.; Zhang, W.; Li, Y. The spectral analysis of different flux-cored wires during arc welding of metals. *Mater. Manuf. Process.* **2012**, *27*, 664–669. [CrossRef]
7. Hayat, F.; Uzun, H. Microstructural and mechanical properties of dual-phase steels welded using GMAW with solid and flux-cored welding wires. *Int. J. Mater. Res.* **2012**, *103*, 828–837. [CrossRef]
8. Zou, Z.; Liu, Z.; Ai, X.; Wu, D. Effect of aluminum on microstructure and mechanical properties of weld metal of Q960 steel. *Crystals* **2022**, *12*, 26. [CrossRef]
9. Ilić, A.; Miletić, I.; Nikolić, R.R.; Marjanović, V.; Ulewicz, R.; Stojanović, B.; Ivanović, L. Analysis of influence of the welding procedure on impact toughness of welded joints of the high-strength low-alloyed steels. *Appl. Sci.* **2020**, *10*, 2205. [CrossRef]
10. Kornokar, K.; Nematzadeh, F.; Mostaan, H.; Sadeghian, A.; Moradi, M.; Waugh, D.G.; Bodaghi, M. Influence of heat input on microstructure and mechanical properties of gas tungsten arc welded HSLA S500MC steel joints. *Metals* **2022**, *12*, 565. [CrossRef]
11. Oktadinata, H.; Winarto, W.; Siradj, E.S. Microstructure and impact toughness of flux-cored arc welded SM570-TMC steel at low and high heat input. *Mater. Sci. Forum* **2020**, *991*, 3–9. [CrossRef]
12. Ni, Z.; Hu, F.; Li, Y.; Lin, S.; Cai, X. Microstructure and mechanical properties of the ternary gas shielded narrow-gap GMA welded joint of high-strength steel. *Crystals* **2022**, *12*, 1566. [CrossRef]
13. Wen, C.; Wang, Z.; Deng, X.; Wang, G.; Misra, R.D.K. Effect of Heat Input on the Microstructure and Mechanical Properties of Low Alloy Ultra-High Strength Structural Steel Welded Joint. *Steel Res. Int.* **2018**, *89*, 1700500. [CrossRef]
14. Ilić, A.; Ivanović, L.; Josifović, D.; Lazić, V.; Živković, J. Effects of welding on mechanical and microstructural characteristics of high-strength low-alloy steel joints. *IOP Conf. Ser. Mater. Sci. Eng.* **2018**, *393*, 012020. [CrossRef]
15. Song, F.; Yin, C.; Hu, F.; Wu, K. Effects of Mn-depleted zone formation on acicular ferrite transformation in weld metals under high heat input welding. *Materials* **2022**, *15*, 8477. [CrossRef] [PubMed]
16. Atkinson, H.V.; Shi, G. Characterization of Inclusions in Clean Steels: A Review Including the Statistics of Extremes Methods. *Prog. Mater. Sci.* **2003**, *48*, 457–520. [CrossRef]

**Disclaimer/Publisher's Note:** The statements, opinions and data contained in all publications are solely those of the individual author(s) and contributor(s) and not of MDPI and/or the editor(s). MDPI and/or the editor(s) disclaim responsibility for any injury to people or property resulting from any ideas, methods, instructions or products referred to in the content.

Article

# Precipitation of Topologically Closed Packed Phases during the Heat-Treatment of Rhenium Containing Single Crystal Ni-Based Superalloys

John Harrison and Paul A. Withey *

School of Metallurgy and Materials, University of Birmingham, Birmingham B15 2TT, UK; jxh389@student.bham.ac.uk
* Correspondence: p.a.withey@bham.ac.uk

**Abstract:** Continual development of nickel-based superalloys for single-crystal turbine applications has pushed their operating temperatures higher and higher, most notably through the addition of rhenium. However, this has left them susceptible to the precipitation of topologically closed packed phases (TCPs), which are widely considered detrimental. Whilst these have long been reported as an end-of-life phenomenon in in-service components, they have more recently been observed during the manufacture of turbine blades. Several rhenium-containing alloys (CMSX-4, CMSX-10K, and CMSX-10N) were cast into single-crystal test bars and studied at different times along their solution heat-treatment process to discern if, when, and where these TCPs precipitated. It was seen that all alloys were susceptible to TCPs at some point along the process, with the higher rhenium-containing alloy CMSX-10N being the most prone. They occurred at the earliest stages of the solution process; this was attributed to aluminium diffusion from the segregated interdendritic regions into the dendrite core, causing the concentration of rhenium into the γ-matrixes until sufficient potential was achieved for TCP precipitation. As the samples became more homogeneous, fewer TCPs were observed; however, in the case of CMSX-10N, this took longer than the typical 24-h solution time used in industry, leading to components entering service with TCPs still present.

**Keywords:** single crystal; rhenium; topologically close-packed phases; solution heat treatment; defects; manufacture

**Citation:** Harrison, J.; Withey, P.A. Precipitation of Topologically Closed Packed Phases during the Heat-Treatment of Rhenium Containing Single Crystal Ni-Based Superalloys. *Crystals* **2023**, *13*, 519. https://doi.org/10.3390/cryst13030519

Academic Editor: Ronghai Wu

Received: 7 March 2023
Revised: 13 March 2023
Accepted: 15 March 2023
Published: 17 March 2023

**Copyright:** © 2023 by the authors. Licensee MDPI, Basel, Switzerland. This article is an open access article distributed under the terms and conditions of the Creative Commons Attribution (CC BY) license (https://creativecommons.org/licenses/by/4.0/).

## 1. Introduction

Due to the service conditions in gas turbines, first for military applications and more and more for commercial applications, turbine blade materials have received enormous amounts of interest for use both in jet engines and industrial gas turbines. Over the decades, these have evolved into highly engineered components made of a single crystal of a metal alloy. The benefits of using single-crystal turbine blades have ranged from increasing the efficiency of the turbines to improving the power-to-weight ratio and increasing the properties of the blades to improve the lifespan. More recent development has focused on improving mechanical properties and microstructural stability at high temperatures [1], as well as cost reduction and improving casting yields [2].

As forming single crystals of this size is not straightforward, difficulties persist in this drive to improve casting yields, as due to a combination of defects such as misaligned grains and freckling, upwards of 30% of produced blades must be rejected as they are unfit for use [3].

Topological close packed phases, simply referred to as TCPs, have also been observed towards the end of the life of turbine blades, are widely considered detrimental to their design life, and are one of the causes of blade removal from service [4]. These are intermetallic phases typically enriched in the elements Cr, Mo, W, and Re. Significant effort has gone into studying the occurrence of TCPs during in-service use, as they appear in increasing quantities within 1000s of hours of operation [5]. These are usually found near the surface of the alloys and are attributed to heightened aluminium content due to the protective coatings applied to the blades [6]. However, little-to-no studies have investigated the appearance of TCPs during manufacture, largely due to their lack of being reported at all, and they have been assumed to be absent from initial manufacture and not an issue until late in the service life. However, certain alloys have shown heightened susceptibility to TCP formation during solution heat treatment [7,8], which has been attributed to the high content of TCP-forming elements such as Cr, Mo, W, and especially Re (Re being particularly likely to enrich due to its propensity to replace nickel-sites during their formation [3]). The presence of TCP phases in as-manufactured components is seen as detrimental to the service life of the blade, but this phenomenon has not previously been investigated for single crystal alloys currently in service; it is simply assumed that they are not present.

Knowing whether there are TCP phases present at the point of entry into use is important for the lifting of the components; however, a comprehensive study of the formation of TCPs at this stage of manufacture has not been undertaken before. This paper investigates the microstructures of common single-crystal alloys (CMSX-4, CMSX-10K, and CMSX-10N), which are typical of second- and third-generation single-crystal superalloys, and determines how susceptible these alloys are to TCP phase formation.

By recreating the production process and observing the diffusion of the composing elements of these alloys throughout the heat-treatment process, primarily through electron microscopy techniques, it was hoped to gain a better and more thorough understanding of when, where, and how these phases formed.

## 2. Materials and Methods

The materials investigated in this paper are the third-generation alloys CMSX-10N and CMSX-10K, and the second-generation alloy CMSX-4, with their compositions (wt.%) shown in Table 1.

**Table 1.** The general composition of the three alloys used [9]. CMSX-4 and CMSX-10 are registered trademarks of Cannon-Muskegon Corporation.

| Alloy | Al | Co | Cr | Ti | Mo | Ta | W | Re | Ni |
|---|---|---|---|---|---|---|---|---|---|
| CMSX-10N | 5.8 | 3.0 | 1.5 | 0.1 | 0.4 | 8.0 | 5.0 | 7.0 | Bal |
| CMSX-10K | 5.7 | 3.0 | 2.0 | 0.2 | 0.4 | 8.0 | 5.0 | 6.0 | Bal |
| CMSX-4 | 5.6 | 9.0 | 6.5 | 1.0 | 0.6 | 6.5 | 6.0 | 3.0 | Bal |

The superalloy castings used in this study were produced at the University of Birmingham School of Materials and Metallurgy using materials manufactured by Cannon Muskegon. For each casting, approximately 1100 g of CMSX-10N was directionally solidified at a temperature of 1550 degrees Celsius in a Retech Single Crystal Furnace with the use of pigtail selectors to induce single crystal growth. This was achieved using investment casting ceramic moulds at a withdrawal rate of 229 mm/hour, each producing two cuboid bars measuring 12 mm in width and depth and approximately 180 mm in length. These were then cut into smaller 20-mm sections for further testing. This was repeated for the CMSX-4 and CMSX-10K samples.

The solution heat treatments were carried out in a type-1 TAV vacuum furnace, using platinum–platinum rhodium thermocouples to monitor the temperature to ±1 degree Celsius. Each test piece was heat treated individually. The furnace was heated to 1361, 1345, and 1310 degrees Celsius for CMSX-10N, CMSX-10K, and CMSX-4, respectively. This was achieved over 90 min at a rate of 15 degrees Celsius per minute to prevent overshooting the solution heat treatment temperature. The samples were then held at these temperatures for times between 0 and 48 h before being quenched in Argon gas.

These samples were then subsequently prepared for SEM and BSE imaging using standard preparation techniques, finishing with a fine polish of OPA (oxide polishing alumina). The samples were then etched with Kaling's Agent 2.

Electron microscopy was carried out using a Jeol6060 microscope with an accelerating voltage of 20 kV and a working distance of 10 mm. EDX maps were taken of dendrite cores and analyzed with INCA software to measure the elemental make-up of the core, while BSE images of the cores were analyzed using ImageJ software to calculate the volume fraction of the TCP σ phases in the cores.

Modelling was carried out using ThermoCalc software. By measuring the varying compositions of the dendritic cores and interdendritic regions throughout the solution process, phase occurrence and volume fractions can be predicted using ThermoCalc and compared to the experimental results. With no available way of imaging these alloys at their solution temperatures, these methods are useful to determine the effect of quenching on the microstructure and what phases would be stable at the heightened temperatures.

## 3. Results

### 3.1. CMSX-4

CMSX-4 was chosen to be used as a baseline comparison of the other two alloys because it has been in use longer than the other alloys and the significant literature and data are available to allow comparison with the other alloys.

BSE images were taken of samples throughout the solution process to see how the compositional makeup of the alloy changed throughout the solution process. EDX was used to quantify these data for use in graphs and modelling.

A segregated structure typical of this alloy can be seen in Figure 1a, where the lighter dendrite cores are significantly enriched in tungsten and rhenium while depleted in tantalum compared to the overall composition presented in Table 1, producing the contrast seen.

By 1 h into the solution process, the dendritic structure is already homogenizing, yet by 8 h, as seen in Figure 1c, TCPs are present toward the centre of the dendrites.

Figure 1d shows after 24 h of the solution process, and in this case, the micrographs suggest an entirely homogenous structure (consisting of evenly distributed ɣ' surrounded with ɣ channels), with no sign of the dendritic structure left. No TCPs were seen throughout the sample after this solution time.

Using ImageJ to calculate a volume fraction of TCPs across the dendrite core region, an 8-h sample was recorded with a 1.15% Vf. This solution time was the only one where TCPs were found to be present. Additionally, the volume fraction is considerably lower than that found in the CMSX-10N samples and only comparable to the CMSX-10Ks' lower recorded values.

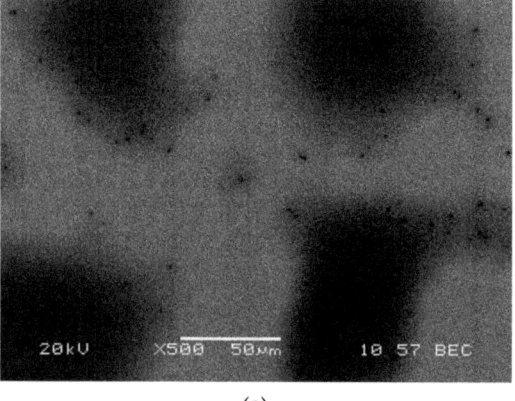
(a)

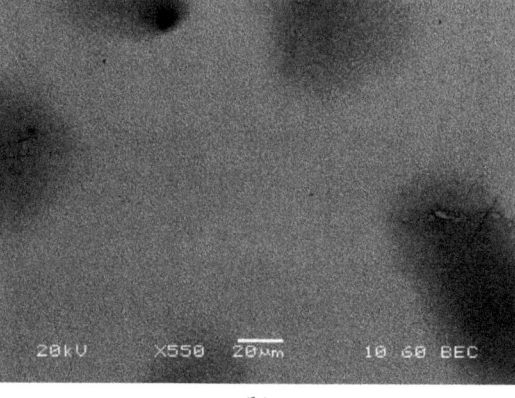
(b)

**Figure 1.** Cont.

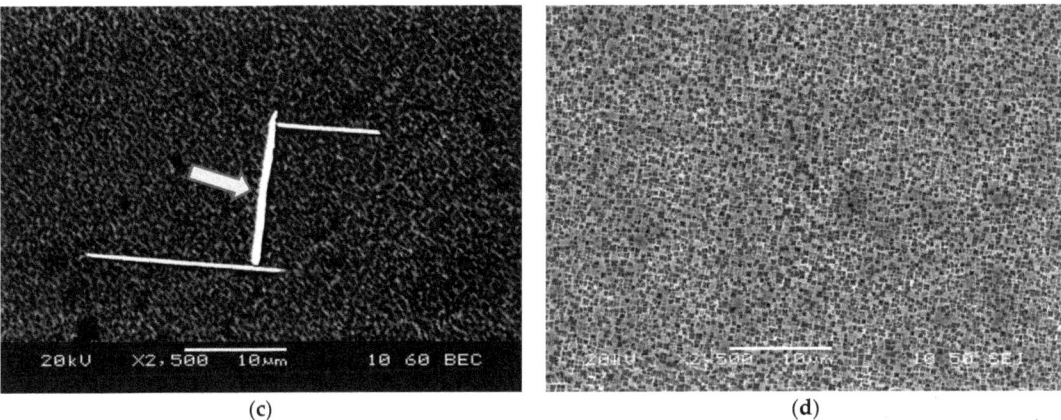

**Figure 1.** The microstructure of the dendritic cores of samples of different solution times: as-cast (**a**), 1 h (**b**), 8 h (**c**), and 24 h (**d**). The TCPs are clearly seen in (**c**) as white linear phases indicated by an arrow.

### 3.2. CMSX-10K

CMSX-10K was chosen for its similar composition to CMSX-10N, the only alloy that has been identified with TCPs after solutions. CMSX-10N was developed from CMSX-10K, and a very similar behaviour was expected; apart from the rhenium content, there are only minor compositional differences between the alloys.

Figure 2 shows micrographs of CMSX-10K through the solution stages. Note that Figure 2b is labelled 0-h; this refers to a sample that underwent the heating regime to the solution temperature of 1345 degrees Celsius before immediate quenching, without holding at the temperature, to study the effects of the heating regime on the microstructure. The lighter regions are the dendrites, where the heavier elements are concentrated. The contrast within the dendrite arms indicates the underlying presence of interdendritic material separating the secondary dendrite arms.

A typical segregated dendritic and interdendritic region can be seen in Figure 2a, where the contrast is more significant compared to Figure 1a, as the dendrite contains significantly higher concentrations of tungsten and rhenium. Note that some of the dendrite arms have a feinter contrast, with some arms possessing a composition similar to that of the dendrite cores, whilst others vary significantly. The gamma–gamma prime structure can be seen in Figure 2b as typical of an as-cast single-crystal material.

Figure 2b shows TCPs already present, with varying morphologies, although typically long, needle-like structures (the images are only 2D, and therefore they could also be plate-like, as the literature would suggest [10,11]), they bridge across the surrounding $\gamma$-$\gamma'$ structure and typically reside in the dendrite core, but in image 2c a small precipitate can be seen in the dendrite arm on the right-hand side of the image.

Figure 2d,e also shows TCPs, which tend to form in small clusters within a single dendrite core. Individual TCPs are also common, but there are also many dendrites with no TCPs, showing large variations in TCP-forming potential between individual dendrites. No TCP phases were observed in the interdendritic regions.

Figure 2f shows a dendrite after 24 h of solutions; here the segregation is now minimal, and while some interdendritic regions can be seen on the edges of the image, a significant amount of homogenization has occurred, and the sample was absent of TCPs throughout.

Figure 3 shows the volume fraction (Vf) of $\sigma$ phase measured from the samples using ImageJ, compared to the value modelled from ThermoCalc using compositional measurements taken via EDX. The model assumed 30 min were available for precipitate formation, and although the quenching process would be much faster than this, the samples were kept in the furnace to prevent the sample from heating up again. Because of this, an overestimation in the model is expected.

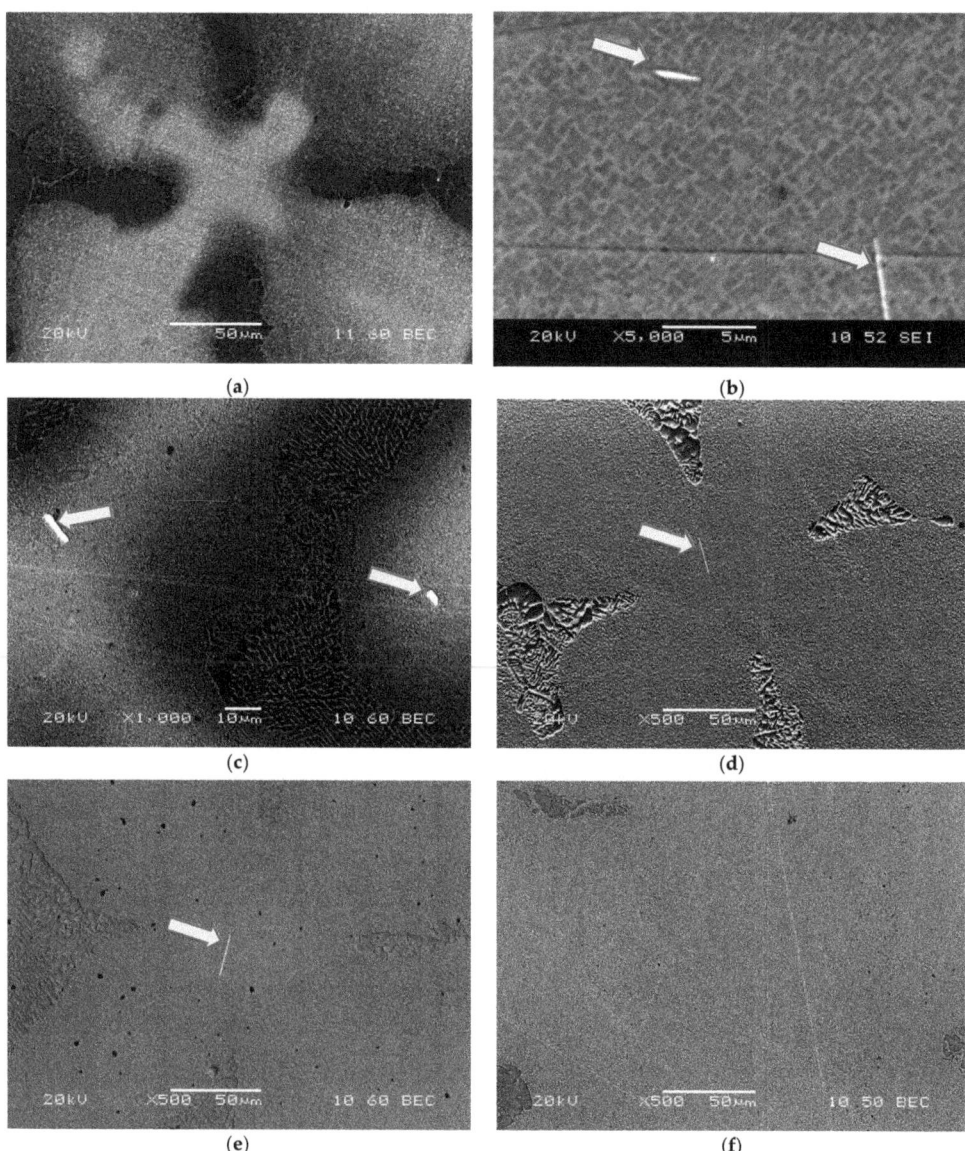

**Figure 2.** The microstructure of the dendritic cores of samples of different solution times: as-cast (**a**), 0 h (**b**), 1 h (**c**), 4 h (**d**), 8 h (**e**), and 24 h (**f**). TCPs are indicated by arrows.

In the as-cast state, the ThermoCalc models suggest there is a sufficient driving force (derived from the composition) for the precipitation of TCPs; however, the sample did not undergo the same 30-min quench but did cool from a molten state. However, no TCPs were seen in situ, and so the circumstances of the cooling during casting appear sufficiently different from cooling from the process of solution to not promote the precipitation of TCPs.

The modelling predictions were very similar to those measured, with the 1 h measured and modelled values varying by less than a percent. However, at 24 h, where none were seen, the modelling still predicted that some would precipitate. Some overestimation in the modelling data was expected, as it was known to overestimate the quenching time so that the results are so similar that one could argue the volume fraction of σ phase is higher than the composition alone could attest to.

Modelled alongside this is the time to the formation of the σ precipitates, also through ThermoCalc in Figure 4. The times refer to the length of time for a 0.1% volume fraction to occur. Immediately it can be seen that the times to formation are very fast, significantly faster than is typically seen from in-service blades [3,5,10]. Their time to formation also increases with time, but still suggests a very rapid formation time at the end of the solution process, despite not being observed in practice.

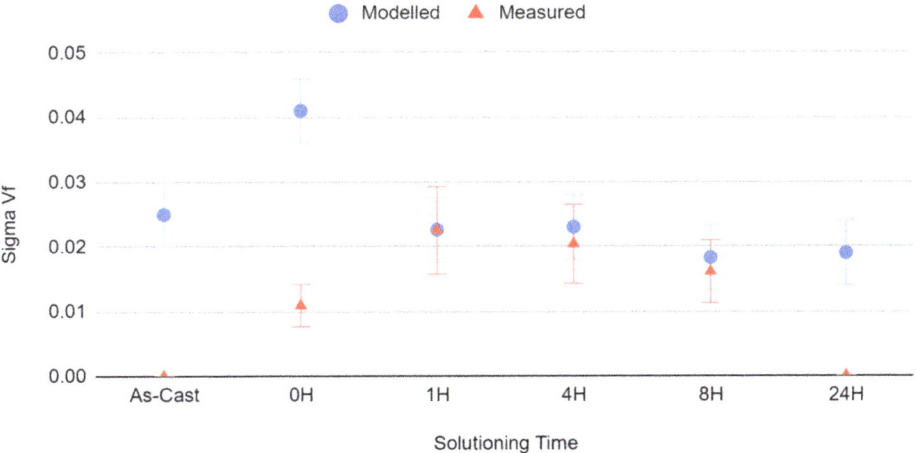

**Figure 3.** The calculated values of σ-phase volume fraction (Vf) using different methods for CMSX-10K.

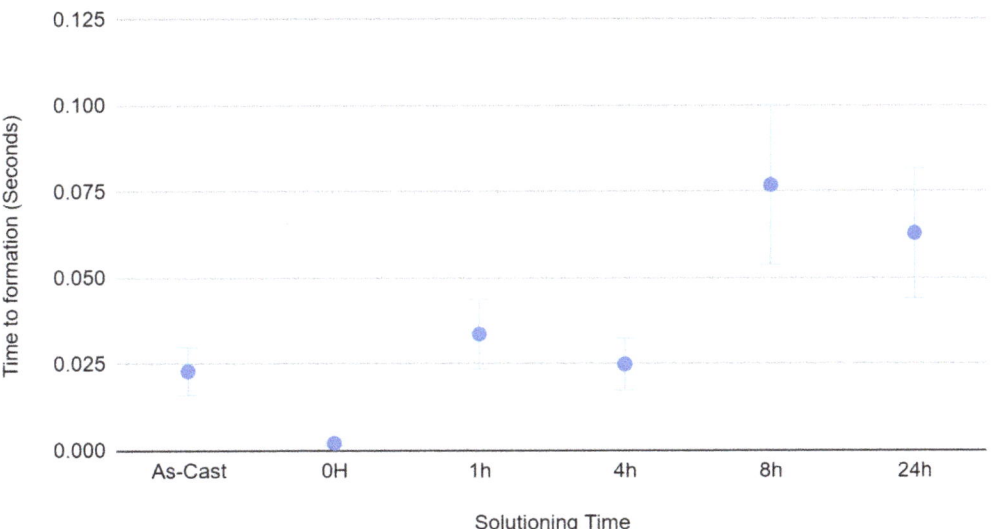

**Figure 4.** The time-to-formation of σ-phase for dendritic core compositions taken from various solution process times for CMSX-10K.

## 3.3. CMSX-10N

CMSX-10N is the primary alloy studied for this work, as it has been reported to contain TCPs after the solution treatment [11]. This previous work also demonstrates that the TCPs seen in these alloys are σ-phase. More tests were carried out on this alloy because of this and its ready availability, as it is a widely used alloy [6,12,13].

Figure 5 shows micrographs taken from each of the solution processes' ages, with many containing TCPs; note that there are also 0 h samples as part of this alloy-specific study. Figure 5a presents a very similar segregated microstructure as Figure 2a; however, Figure 5b lacks any TCPs, and none were found throughout this early stage of the solution process for CMSX-10N. However, the cells in Figure 5c–h all contain TCPs to varying degrees. Figure 5h, showing the microstructure after 24 h of the solution process, shows that TCPs remain, unlike CMSX-10K, and only after 48 h are there no TCPs present (Figure 5i), despite the microstructure not being fully homogenized.

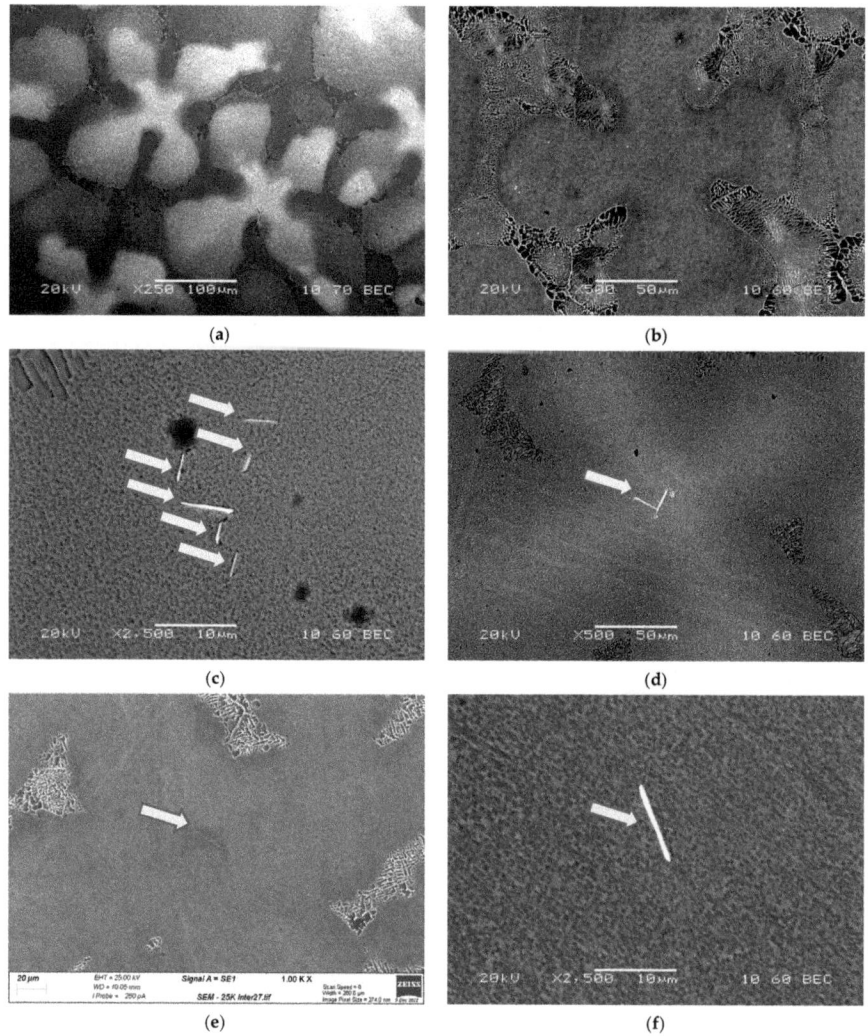

**Figure 5.** *Cont.*

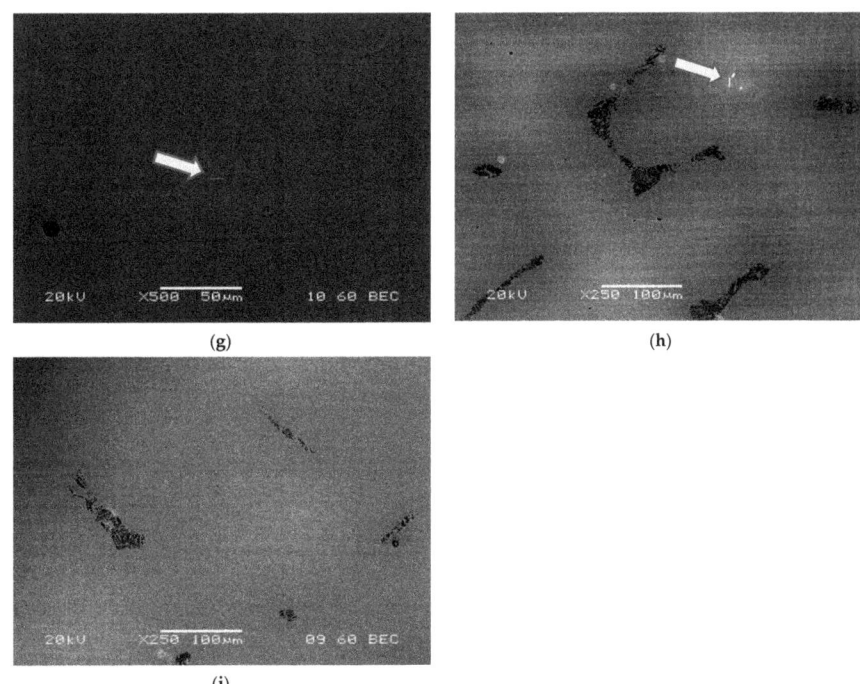

**Figure 5.** The microstructure of the dendritic cores of CMSX-10N samples of solution times: as-cast (**a**), 0 h (**b**), 1 h (**c**), 4 h (**d**), 8 h (**e**), 12 h (**f**), 16 h (**g**), 24 h (**h**), and 48 h (**i**). TCPs are indicated by arrows.

The TCPs present were seen to form either in clusters or as individuals, but almost always within the dendrite core, both as long needle-like precipitates and all smaller circular precipitates; however, these images only present a 2D insight into the material, meaning these needle-like precipitates could be plates and the circular precipitates needles if they were to extend into the depth of the material. This is seen in other works [11]. Some TCP phases were observed in the dendrite arms, but none were seen in the interdendritic regions.

Figure 6 shows the volume fraction (Vf) of σ phase measured from the CMSX-10N samples using ImageJ, compared to the value modelled from ThermoCalc using compositional measurements taken via EDX. The modelled assumptions were the same as those of Figure 3.

The as-cast modelling predicts the occurrence of TCPs, although none were seen, and as with CMSX-10K, the model assumed the quenching behaviour it did not undergo.

An overestimate of TCPs is seen again during the early stages of the solution process, with a very close alignment during the middle section. In contrast to CMSX-10K, the model underpredicted the volume of TCPs at the later stages; however, at 48 h, none were reported in the modelling (a non-zero value is shown to account for errors) nor were any seen during microscopy. Figure 7 shows a segment of a TTT curve for the surrounding temperatures of the solution profile, showing the rapid rate of formation for σ TCP is consistent at all surrounding temperatures.

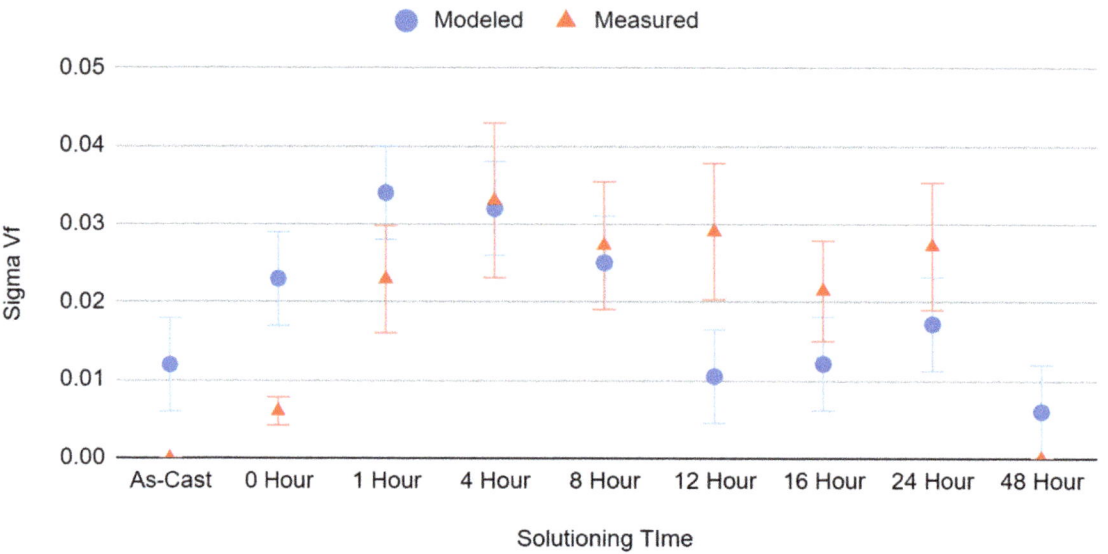

**Figure 6.** The calculated values of σ phase volume fraction (Vf) using different methods for CMSX-10N.

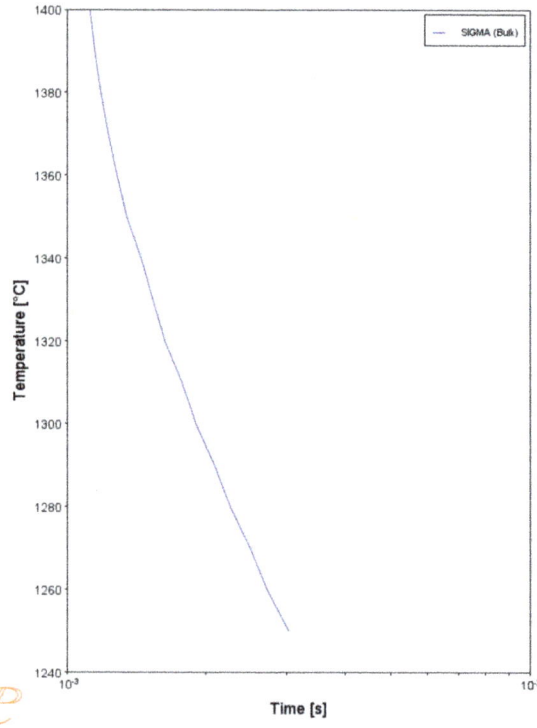

**Figure 7.** The time to the formation of a range of temperatures around the solution range of CMSX-10N.

## 4. Discussion

The immediate observation from the results is the appearance of TCPs at all; whilst there have been some reports of TCPs during the manufacturing process, they are limited. Long et al. [4] reported their occurrence in the as-cast structure, which then decomposed during the solution process; they also reported σ phase being present, along with η, however, Long's σ was nodular, whereas the η was platelet-like. Long did conclude that more σ precipitated from the platelet morphology of the decomposed sites of the nodular TCPs during the solution process.

However, no TCPs were seen in the as-cast state in these samples, and the same phenomenon cannot be used to explain their precipitation during the solution process. However, despite none being seen, there is a possibility they were not observed, and the modelling did suggest the possibility of TCPs, although the modelling regime used did not replicate the solidification process. However, in this study, no platelet η was seen, nor did it appear in the modelling, suggesting that these two occurrences of TCPs are rooted in different mechanisms.

Nader El-Bagoury et al. [14] reported the occurrence of rhenium clusters in the as-cast state, as well as phases that possessed a needle-like morphology; however, neither of these phases was shown by TEM to be TCPs; the needle-like particles were delta phases, found alongside $\gamma''$, and contained Nb, neither of which are observed in the alloys in question here. The rhenium clusters are only a handful of atoms large and are a common occurrence in rhenium-containing alloys [15].

Fuchs suggested that TCPs could occur during the solution process under certain circumstances [7], particularly if the solution process was shortened as it would lead to reduced homogenization. Fuchs reported that rhenium and tungsten diffusion was substantial only above 1340 degrees Celsius. This known behaviour identifies aluminium as the source of TCPs, at least in the 0 h treated sample. At these lower temperatures and shorter time frames, aluminium is the only element to undergo significant diffusion (Cr, Co, and Ni would diffuse at these temperatures too, but segregate relatively evenly on solidification).

As aluminium diffuses from the interdendritic region into the dendrite core, increasing the volume fraction of $\gamma'$, this concentrates the Re and W, which have yet to diffuse, in the $\gamma$ channels. Kim showed the composition of the $\gamma$ and $\gamma'$ is fixed [13], severely limiting its ability to hold the excess rhenium and tungsten in solution. This oversaturation then precipitates out as TCPs, particularly in σ phase. Rae et al. reported that σ is typically the first to precipitate, but as a metastable phase that eventually returns to $\gamma$ or a different type of TCP [5]. However, the ThermoCalc models predicted σ would occur and would remain stable. The kinetics would lend itself to σ forming first, and the enriched concentrations would maintain the stability. This is reinforced by the dropping off of σ volume fraction with increasing solution time, as the Re and W have enough time to diffuse out of the cores into the interdendritic regions, with this reduction in σ volume fraction increasing with time.

The time to the formation of these TCPs is of particular note, as graphs 2 and 4 show precipitation times of less than a second, almost instantaneous, far from the prolonged periods generally assumed for TCPs [16]. These longer-to-form TCPs are generally found at the surface, where there is an increased concentration of Al from the coatings [6], whereas those in this paper are found within the bulk of the material, far from the surface, where the compositional non-equilibrium is the driver.

These data were modelled at the solution temperature, where the TCPs formed and were stable, unlike the metastable σ reported elsewhere [5] and in the absence of $\gamma'$. These TCPs could form during the heat-up cycle, and as Figure 5 shows, the time to the formation of TCPs is small at the surrounding temperatures too. Here the aluminium diffusion into the dendrite core causes an increase in $\gamma'$ before its dissolution into $\gamma$, allowing a concentration of TCP elements in the core and allowing quick precipitation of TCPs that are then stable throughout the solution process, continuing to grow until sufficient rhenium diffusion has occurred. Secondly, the TCPs could form at temperature, despite no $\gamma'$ being present, because the core would still be enriched in rhenium, with an increasing aluminium content in the early stages oversaturating the $\gamma$ matrix, which, as shown by Kim [13], has a fixed composition and must still reject the oversaturation of rhenium causing the precipitation of TCPs. A third option is that the TCPs form on quenching; as $\gamma$-' forms during cooling, the $\gamma$ is supersaturated with rhenium as it is rejected from the $\gamma$-'; from here, the quick formation times of TCPs at these compositions allow for the formation of TCPs on cooling. Whilst this last theory does not reflect the modelling which suggests their stability at the solution temperature, it is hard to distinguish the methods without in situ imaging.

The first of these, where TCPs form on heating, seems the most promising, as the still present $\gamma'$ would still concentrate rhenium into the $\gamma$ matrix, increasing the driving force for TCP precipitation. However, precipitation and growth could still occur at this temperature, albeit to a reduced degree without the $\gamma$-' concentrating effects, and furthermore, precipitation could occur on cooling too. If multiples of these effects are occurring, it would explain the discrepancy between the modelling data and experimental data, as the modelling only assumed at-temperature precipitation.

## 5. Further Work

There is significant precipitation of TCPs during the solution process of rhenium-containing alloys; however, the mechanism by which it occurs is still not fully understood. It is clear that in the early stages of the solution process, the heating cycle has the most significant effect; however, TCP volume fractions continue to increase in the early stages of the solution process. A more thorough investigation of these early times would be necessary to improve our understanding of these mechanisms.

Studying more alloys would also be of use, as rhenium is not the only former of TCPs; furthermore, alloys with different $\gamma'$ fractions would be of interest if this is in fact the prime driver for TCP precipitation via rhenium rejection.

To pinpoint the exact time within the homogenization process that these TCPs form would also be of great use. To do so would require imaging the alloy during the solution process itself, which has numerous practical challenges, due to the extreme temperatures the microscopy equipment would have to be able to operate at.

Furthermore, the length of solution treatment appears vital in the removal of TCPs; there appears to be a cut-off point where TCPs are no longer present, but understanding, if this is a gradual change, or a steeper drop-off would help optimize solution treatments by adjusting the length of solution heat-treatment against minimal TCP precipitation prior to entry into service. These optimal solution times also vary from alloy to alloy, despite only small compositional changes, as each alloy would have a unique window to form TCPs during the solution process. This last step is of high importance, as significant work is currently going into shortening the solution procedure because of its associated high costs during the manufacturing process for turbine blades. However, with certain alloys, these works are likely to run into issues with TCP precipitation.

## 6. Conclusions

TCPs precipitate during solution heat treatment of rhenium-containing alloys, particularly those of high concentration (6%wt or more).

- This occurs even at the earliest stages of the solution process, being present after the heating cycle to the solution window;
- This is most likely driven by aluminium diffusion into the dendrite cores from the interdendritic region before the other elements can homogenize, enriching the $\gamma$ matrix and causing the rejection of the rhenium, potentially from the formation of $\gamma'$, which would further concentrate the rhenium in the $\gamma$ matrix;
- Holding at the solution temperature for sufficient times allows sufficient rhenium diffusion so that TCPs do not form. However, in the case of CMSX-10N, this time frame exceeds the 24 h usually used for its homogenization in a production environment;

- Further work is needed to understand the mechanism behind the initial precipitation of TCPs and how it varies across different alloys.

**Author Contributions:** Conceptualization, J.H. and P.A.W.; methodology, J.H. and P.A.W.; investigation, J.H.; resources, P.A.W.; writing—original draft preparation, J.H.; writing—review and editing, P.A.W. All authors have read and agreed to the published version of the manuscript.

**Funding:** This work was supported by the Engineering and Physical Sciences Research Council [grant number EP/T018518/1] and EPSRC (EPSRC CDT Grant No: EP/L016206/1) in Innovative Metal Processing for financial support.

**Data Availability Statement:** Data is contained within the article.

**Acknowledgments:** The authors would also like to acknowledge the following technical staff at the University of Birmingham for assistance with the experiments: Amy Newell, Grant Holt, Jonathan Davies, and Adrian Caden.

**Conflicts of Interest:** The authors declare no conflict of interest.

## References

1. Sun, F.; Zhang, J.X. Topologically Close-Packed Phase Precipitation in Ni-Based Superalloys. *AMR* **2011**, *320*, 26–32. [CrossRef]
2. Geddes, B.; Leon, H.; Huang, X. *Superalloys: Alloying and Performance*; ASM International: Novelty, OH, USA, 2010.
3. Reed, R.C. *The Superalloys: Fundamentals and Applications*; Imperials College of Science, Technology, and Medicine: London, UK; Cambridge University Press: Cambridge, UK, 2006.
4. Long, F.; Yoo, Y.S.; Jo, C.Y.; Seo, S.M.; Jeong, H.W.; Song, Y.S.; Jin, T.; Hu, Z.Q. Phase Transformation of η and σ Phases in an Experimental Nickel-Based Superalloy. *J. Alloy. Compd.* **2009**, *478*, 181–187. [CrossRef]
5. Rae, C.M.F.; Reed, R.C. The precipitation of topologically close-packed phases in rhenium-containing superalloys. *Acta Mater.* **2001**, *49*, 4113–4125. [CrossRef]
6. Spathara, D.; Sergeev, D.; Kobertz, D.; Müller, M.; Putman, D.; Warnken, N. Thermodynamic study of single crystal, Ni-based superalloys in the γ+γ' two-phase region using Knudsen Effusion Mass Spectrometry, DSC and SEM. *J. Alloy. Compd.* **2021**, *870*, 159295. [CrossRef]
7. Fuchs, G.E. Solution Heat Treatment Response of a Third Generation Single Crystal Ni-Base Superalloy. *J. Mater. Sci. Eng. A* **2001**, *300*, 52–60. [CrossRef]
8. Park, K.; Withey, P. General View of Rhenium-Rich Particles along Defect Grain Boundaries Formed in Nickel-Based Single-Crystal Superalloy Turbine Blades: Formation, Dissolution and Comparison with Other Phases. *Crystals* **2021**, *11*, 1201. [CrossRef]
9. Cannon Muskegon. Available online: https://cannonmuskegon.com/products/vacuum-alloys/ (accessed on 1 January 2023).
10. Darolia, R.; Lahrman, D.; Field, R.D. Formation of Topologically Closed Packed Phases in Nickel Base Single Crystal Superalloys. In Proceedings of the Sixth International Symposium on Superalloys, Seven Springs, Pittsburgh, PA, USA, 18–22 September 1988; pp. 255–264.
11. Kim, K.; Withey, P. Microstructural Investigation of the Formation and development of Topologically Close-Packed Phases in a 3rd generation Nickel-Base Single Crystal Superalloy. *Adv. Eng. Mater.* **2017**, *19*, 1700041. [CrossRef]
12. Warnken, N. Studies on the Solidification Path of Single Crystal Superalloys. *J. Phase Equilib. Diffus.* **2016**, *37*, 100–107. [CrossRef]
13. Park, K.; Withey, P. Compositions of Gamma and Gamma Prime Phases in an As-Cast Nickel-Based Single Crystal Superalloy Turbine Blade. *Crystals* **2022**, *12*, 299. [CrossRef]
14. El-Bagoury, N. Role of Rhenium on Solidification, Microstructure, and Mechanical Properties of Standard Alloy 718. *Metallogr. Microstruct. Anal.* **2012**, *1*, 35–44. [CrossRef]
15. Rao, G.A.; Sirinivas, M.; Sarma, D.S. Effect of solution treatment temperature on microstructure and mechanical properties of hot isostatically pressed superalloy Inconel * 718. *Mater. Sci. Technol.* **2004**, *20*, 1161. [CrossRef]
16. Wilson, A.S. Formation and effect of Topologically close-packed phases in nickel-base superalloys. *Mater. Sci. Technol.* **2017**, *33*, 1108–1118. [CrossRef]

**Disclaimer/Publisher's Note:** The statements, opinions and data contained in all publications are solely those of the individual author(s) and contributor(s) and not of MDPI and/or the editor(s). MDPI and/or the editor(s) disclaim responsibility for any injury to people or property resulting from any ideas, methods, instructions or products referred to in the content.

Article

# Phase Transition of Nb₃Sn during the Heat Treatment of Precursors after Mechanical Alloying

Wanshuo Sun [1,2] and Shunzhong Chen [1,2,*]

[1] Institute of Electrical Engineering, Chinese Academy of Sciences, Beijing 100190, China; sunwanshuo@mail.iee.ac.cn
[2] University of Chinese Academy of Sciences, Beijing 100049, China
* Correspondence: chenshunzhong@mail.iee.ac.cn

**Abstract:** The phase transition process of Nb₃Sn during heat treatment exerts important influences on Nb₃Sn formation and the superconducting characteristics of Nb₃Sn superconductors. A simple method for quickly preparing Nb₃Sn was studied. First, Nb, Sn, and Cu powders were mechanically alloyed to prepare the precursor. Then, the precursor was heat treated at different times to form Nb₃Sn. During the first stage, the morphology and crystal structure of the products were analyzed after different milling times. The results of the transmission electron microscopy showed the poor crystallinity of the products compared with the original materials. During the second stage, heat treatment was performed at different temperatures ranging from room temperature to 1073 K. After treatment, the products were studied via X-ray diffraction analysis to determine how the structure changed with increasing temperature. Only the Nb diffraction peaks in the precursor were observed after high-energy ball milling for more than 3 h. When the heat treatment temperature was above 773 K and heat treatment time was 15 min, Nb₃Sn began to form. When the temperature was above 973 K, some impurities, such as Nb₂O₅, appeared. After 5 h of ball milling, the precursor was heat treated at different times in a vacuum heat treatment furnace. The crystal structure of the product exhibited evident diffraction peaks of Nb₃Sn. The critical temperatures of the samples that were heat treated at different times were between 17 K and 18 K. The magnetic critical current density of the sample versus the applied magnetic field at 4.2 K indicated that the magnetic $J_c$ was approximately 30,000 A/cm².

**Keywords:** Nb₃Sn; superconducting characteristics; phase transition; mechanical alloying

Citation: Sun, W.; Chen, S. Phase Transition of Nb₃Sn during the Heat Treatment of Precursors after Mechanical Alloying. *Crystals* **2023**, *13*, 660. https://doi.org/10.3390/cryst13040660

Academic Editors: Daniel Medyński, Grzegorz Lesiuk and Anna Burduk

Received: 24 February 2023
Revised: 28 March 2023
Accepted: 9 April 2023
Published: 11 April 2023

**Copyright:** © 2023 by the authors. Licensee MDPI, Basel, Switzerland. This article is an open access article distributed under the terms and conditions of the Creative Commons Attribution (CC BY) license (https://creativecommons.org/licenses/by/4.0/).

## 1. Introduction

Since Nb₃Sn was discovered by Matthias in 1954 [1], research on this compound has elicited considerable attention, particularly regarding its preparation and superconductive characteristics. Compared with NbTi, Nb₃Sn has higher critical temperature (above 18 K) and critical magnetic field of around 30 T. Nb₃Sn superconductors have important applications in high magnetic field situations, particularly at above 10 T. Current applications of Nb₃Sn include magnetic resonance imaging [2], nuclear magnetic resonance [3], dipole and quadrupole magnets for particle accelerators, and other areas where high magnetic fields are required, such as the International Thermonuclear Experimental Reactor [4]. Conventional manufacturing methods, such as bronze technology [5,6], internal tin technology [7], and powder-in-tube [8,9] process, require a long period of heat treatment at around 650 °C to allow Sn to react with Nb and form Nb₃Sn superconductors. These methods are time consuming, and their costs are high.

Heat treatment is a key point in the preparation of Nb₃Sn. Heat treatment temperature exerts an important effect on the upper critical field $B_{c2}$, with a higher reaction temperature leading to higher $B_{c2}$ [10]. Many studies on heat treatment have investigated the influences of temperature variation, heat treatment time, temperature gradient, ascending temperature speed, and temperature uniformity on the superconducting characteristics of $Nb_3Sn$ [11]. Various factors that are required to form fine and homogeneous grains of $Nb_3Sn$ were studied through a series of experiments [12,13]. The phase transformations of high $J_c$ $Nb_3Sn$ strands during reaction heat treatment were studied by performing synchrotron X-ray diffraction (XRD) measurements.

The amount of Sn plays an important role in the superconducting properties of $Nb_3Sn$. The critical temperature sharply increases from approximately 6 K to approximately 18 K when the percentage of Sn is increased from below 20% to above 24.5%. For some superconductors in which Sn concentration is insufficient or some regions with a deficient stoichiometric ratio of Sn, the critical temperature of $Nb_3Sn$ may be below 9 K [14]. The upper critical field is also affected by the amount of Sn in accordance with the Werthamer–Helfand–Hohenberg theory. Therefore, increasing the amount of Sn or making the stoichiometry of the $Nb_3Sn$ layer more uniform is important to increase the upper critical field and the critical temperature [4]. The addition of Cu can decrease A15 formation temperature and obtain $Nb_3Sn$ grains with small sizes by limiting grain growth. Thus, higher grain boundary density can be obtained to improve the bulk pinning force [15]. A typical reaction scheme for high-performance $Nb_3Sn$ conductors is increasing the temperature to 483 K and holding this temperature for 100 h, ramping the temperature to 673 K and holding this temperature for 50 h, and then ramping to the final reaction temperature (923–973 K) and holding this temperature for a long period (50–200 h). The heat treatment of the $Nb_3Sn$ bulk is equally important because the bulk can be used to fabricate the $Nb_3Sn$ superconducting joint, which must be heat treated with $Nb_3Sn$ conductors. Reducing the length of the final heat treatment process is significant for $Nb_3Sn$ magnet production.

We report a simple and time-saving synthesis technique for $Nb_3Sn$ that proceeds through a mechanical alloying method, followed by short-term heat treatment. Nb, Sn, and Cu powders were mixed and alloyed through mechanical alloying, and the precursors were synthesized via high-energy ball milling of the as-blended powders. Then, the precursors were heat treated to form $Nb_3Sn$. However, before $Nb_3Sn$ nucleation and growth, other intermetallic phases were synthesized during the large temperature intervals. These phases affected the final reaction in the formation of $Nb_3Sn$. The influence of temperature variation on $Nb_3Sn$ reaction was studied, and changes in crystal structure with temperature were measured to study the change in phase transition during the process. After high-energy ball milling at different times, the precursors were prepared and heat treated at various temperatures ranging from 298 K to 1073 K at 10–50 K intervals.

## 2. Experiment

All original materials used in this work, including Nb, Sn, and Cu, were composed of 99.9% pure powder, which were 5–20 μm in scale. Nb, Sn, and Cu were blended with a molar ratio of 3:1:1. Then, the blended powders were placed in a ball mill tank. The rotational speed of the ball mill was 1725 rpm. The precursors were synthesized after high-energy ball milling. The procedure above was performed in a glove box to prevent the blended powders from oxidizing in a high-temperature and high-pressure environment. The as-blended powders were mechanical alloyed at different times by using a Spex 8000D mixer. The prepared precursors were then heat treated from room temperature to 1073 K. During the process, we studied the changes in the crystal structure of the precursors after the selected temperatures, which was approximately at every 50 K interval from 298 K to 873 K and from 973 K to 1073 K and at every 10 K interval from 903 K to 953 K. The heating rate was 10 K per minute, and sintering holding time was 15 min at each sintering temperature point.

The morphology of the product was recorded using a scanning electron microscope. A transmission electron microscope was used to analyze the crystallinity of the as-prepared samples. The prepared samples were characterized via XRD on an Empyrean X-ray diffractometer. The temperature dependence of magnetizability was measured using a physical property measurement system (PPMS-9).

## 3. Results and Discussion

Figure 1 shows the scanning electron microscopy (SEM) images of the original samples. Figure 2 shows the SEM images and the energy-dispersive X-ray spectroscopy (EDS) mapping images of the products after the milling of Nb–Sn–Cu powders for 5 h. The images indicated that all the elements were homogeneously dispersed. Patankar [16] reported that brittle Nb with a body-centered cubic structure is likely to get fragmented and get coated onto the surface of ductile Sn. Patankar believed that the depth of XRD was insufficient to penetrate the Nb shell. Thus, Sn was not observed from the XRD result. However, the results of the EDS mapping images showed that Sn and Cu were homogeneously dispersed around Nb and did not exhibit the core–shell structure. This phenomenon may be attributed to the amorphization of Sn with increasing milling time. The original material was several microns in size. After mechanical alloying, the dimensions of the particles decreased. Even nanocrystalline materials were formed due to high-energy ball milling.

**Figure 1.** SEM images of original samples: (**a**) Nb, (**b**) Sn, and (**c**) Cu.

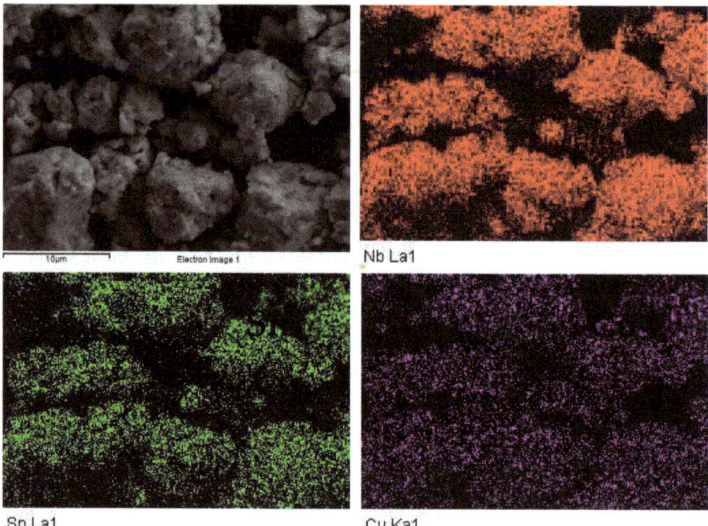

**Figure 2.** SEM image and EDS mapping of the samples after the milling of the Nb–Sn–Cu powder for 5 h.

Figure 3 depicts the elemental composition of the as-blended particles after mechanical alloying, as illustrated in the inset of the figure. From the results of the selected area with a few microns, Nb, Sn, and Cu were included and mixed in stoichiometric ratio, indicating that a new combination of elements was constructed at the microscopic level. This finding exerted a significant effect on the reaction of Nb with Sn to produce $Nb_3Sn$ due to the microscopic stoichiometric ratio and the closer diffusion distance between Nb and Sn in the subsequent heat treatment process.

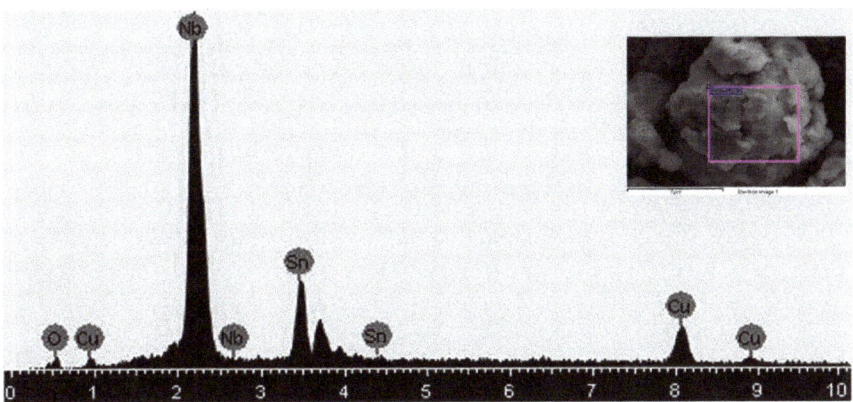

**Figure 3.** EDS analysis of the sample after the milling of the Nb–Sn–Cu powder for 5 h.

Figure 4 shows the typical low-magnification and high-resolution transmission electron microscopy (TEM) micrographs of the samples. The selected area electron diffraction pattern shown in the inset of Figure 4a suggested that the products exhibited poor crystallinity after ball milling for 5 h. One study suggested that Sn transformed into the amorphous phase because Nb and Cu were harder than Sn [17]. Some diffraction spots were observed with halo rings in the figure. The Nb grains were confirmed through the high-magnification micrograph in Figure 4b.

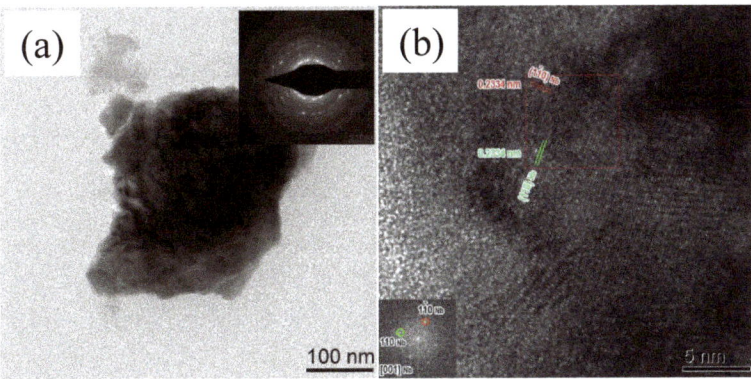

**Figure 4.** TEM micrograph of the sample: (**a**) low-magnification micrograph and (**b**) high-magnification micrograph.

The finding was consistent with the XRD result below. Diffraction intensity became weaker with increasing milling time. This result suggested that on the one hand, the crystalline grain size of products became smaller and even reached nanometer scale, broadening the width of diffraction peaks. On the other hand, the high-energy ball milling introduced many defects into the products. The disorder degree of interfacial atoms increased considerably, reducing diffraction intensity. The high-resolution TEM micrograph showed the interplanar crystal spacing of Nb.

Figure 5 presents the XRD patterns of the particles after mechanical alloying at different times. The results showed that the diffraction peaks of Sn and Cu disappeared when milling time reached 1 h compared with the apparent diffraction peaks of the original as-milled particles. The intensity of peaks decreased with increasing milling time. When ball milling time reached 10 h, the primary diffraction peak of Nb evidently broadened compared with the initial stage. High-energy mechanical alloying caused the continuous cold welding and fracturing of different particles. Grain size decreased with increasing milling time. The results showed that the effects of ball milling on the crystallinity of products became worse with increasing milling time. This effect was also observed from the TEM results of the products. Simultaneously, long-range-order parameters decreased with increasing milling time due to the large number of defects and the high amount of grain boundary energy during high-energy mechanical alloying. Both caused the intensity of peaks to decrease and transform into metastable phases.

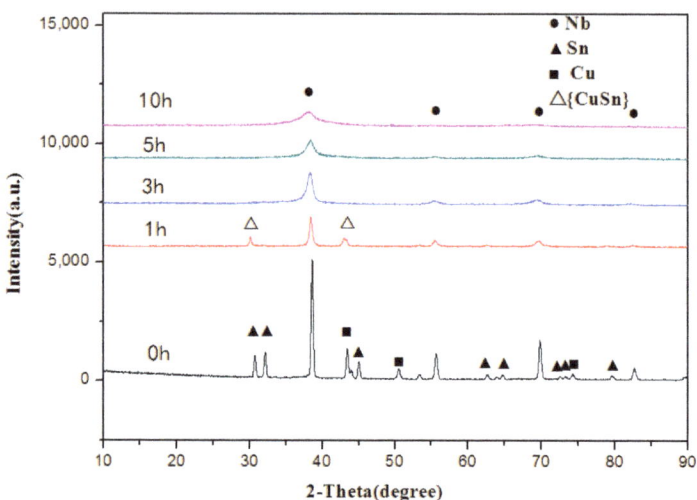

**Figure 5.** XRD patterns of the particles after mechanical alloying at different times.

Figure 6 presents the XRD patterns of the products after different heat treatment temperatures ranging from 298 K to 573 K. These curves represent the crystal structures of the products after each heat treatment at different temperatures. As shown in the figure, the XRD pattern at 298 K presents the measurement result at room temperature, i.e., the crystal structure of the sample after mechanical alloying. Only the Nb diffraction peaks are shown. The Sn and Cu diffraction peaks disappeared after high-energy mechanical alloying.

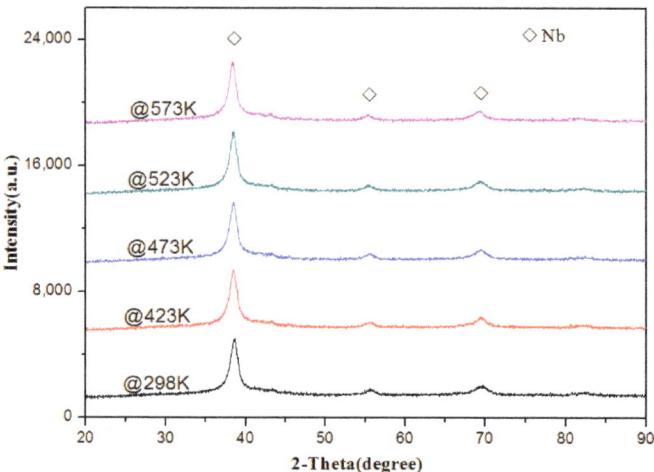

**Figure 6.** XRD patterns of products after heat treatment at different temperatures (from 298 K to 573 K).

After heat treatment from room temperature to 573 K, the XRD patterns showed that the products did not change, and no new diffraction peaks appeared. These results indicated that the crystal structures of the products did not change, and no chemical reaction occurred at below 573 K. The melting point of Sn is below 504.9 K, and Sn will liquify at temperatures above 504.9 K. However, the results obtained after the 573 K heat treatment showed that Sn did not change and turn to liquid, signifying that Sn reacted with the other elements to exist in a new form during high-energy mechanical alloying.

The XRD patterns of the particles after heat treatment from 573 K to 773 K are depicted in Figure 7. The shapes of the peak diffraction were nearly identical from 573 K to 723 K. After heat treatment temperature increased to 773 K, small diffraction peaks of $Nb_3Sn$ appeared. This finding indicated that reactions that produced $Nb_3Sn$ started when the temperature was above 773 K. Simultaneously, the diffraction peaks of Nb began to fade due to the reaction of Nb with Sn or Sn compounds.

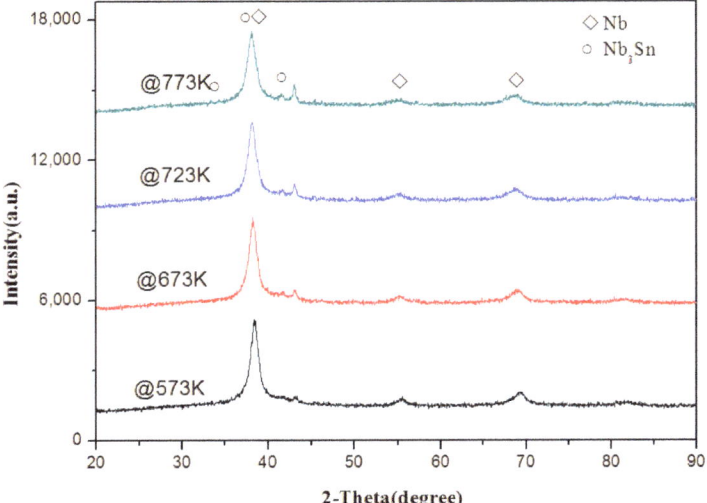

**Figure 7.** XRD patterns of the products after heat treatment at different temperatures (from 573 K to 773 K).

As shown in Figure 8, Nb diffraction peaks gradually attenuated after heat treatment from 773 K to 903 K. Nb diffraction peaks nearly disappeared at temperatures above 903 K. In this process, $Nb_3Sn$ was produced with increasing temperature. After heat treatment at 903 K, $Nb_3Sn$ phase diffraction peaks became highly apparent. Thus, through high-energy mechanical alloying, the as-milled particles should be at least exposed to an environment with a temperature of above 903 K.

To determine the optimum temperature range of heat treatment, temperatures of 903 K, 913 K, and 923 K were selected to study the crystal structure of the products. As shown in Figure 9, the crystal structures were nearly identical at different temperatures, and $Nb_3Sn$ and $NbO_2$ diffraction peaks were observed. $Nb_3Sn$ diffraction peaks exhibited slightly higher intensity with increasing temperature.

As shown in Figure 10, the samples were heat treated at different temperatures of 923 K, 933 K, and 943 K. The results indicated that the locations of the diffraction peaks did not change, and no new structures appeared. Thus, the temperature of 923 K was selected to obtain the final $Nb_3Sn$ product.

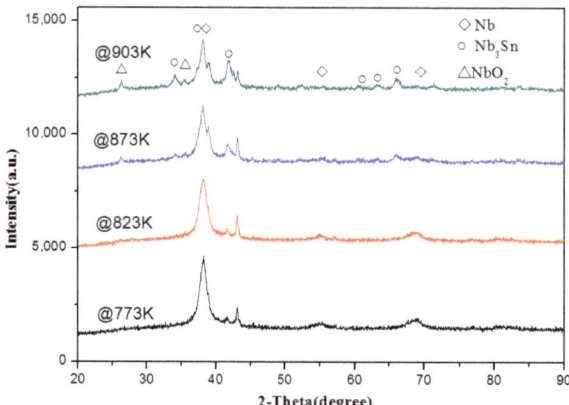

**Figure 8.** XRD patterns of the products after heat treatment at different temperatures (from 773 K to 903 K).

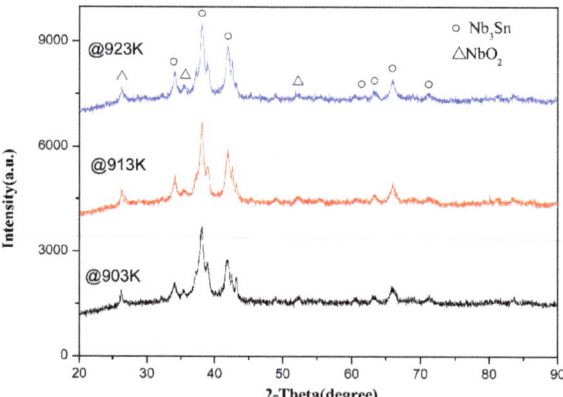

**Figure 9.** XRD patterns of the products after heat treatment at different temperatures (from 903 K to 923 K).

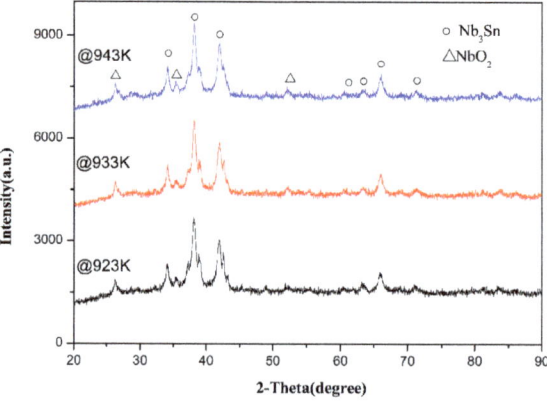

**Figure 10.** XRD patterns of the products after heat treatment at different temperatures (from 923 K to 943 K).

The effect of higher heat treatment temperatures was studied to determine whether the crystal structure of the product changed when exposed to higher temperatures. Figure 11 shows the XRD patterns of the products after heat treatment at different temperatures ranging from 953 K to 1073 K. The results showed that the diffraction peaks of Nb$_3$Sn did not change when exposed to temperatures above 973 K, but other impurities, such as Nb$_2$O$_5$, appeared. Thus, as-milled particles should not be utilized at above 973 K.

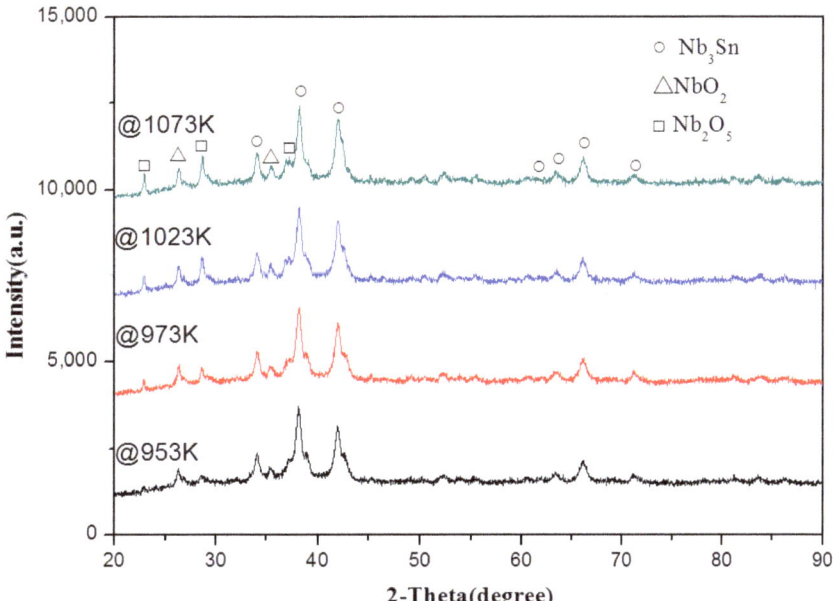

**Figure 11.** XRD patterns of the products after heat treatment at different temperatures (from 953 K to 1073 K).

The key temperature points of Nb$_3$Sn heat treatment were presented to show the change in crystal structure during the heat treatment process. As illustrated in Figure 12, the phase transition process from room temperature to 903 K and then to 1073 K was as follows: the original materials of Nb, Sn, and Cu at 298 K exhibited broadened Nb diffraction peaks at 298 K after mechanical alloying. Then, Nb$_3$Sn and NbO$_2$ diffraction peaks appeared at 903 K. Finally, Nb$_3$Sn, NbO$_2$, and Nb$_2$O$_3$ diffraction peaks appeared at 1073 K. In accordance with the solid-state diffusion reaction, different temperatures correspond to different activation energies of atomic diffusion, affecting the diffusion reaction speed and whether diffusion reaction can occur. In accordance with the result of heat treatment, the as-milled particles should be exposed to temperatures above 903 K. However, when the temperature reached 1073 K, diffraction peaks of impurities appeared.

To maximize the Nb$_3$Sn area, heat treatment time must last for a long time (50–200 h) at high temperature in Nb$_3$Sn formation reaction. The researchers investigated the effects of reaction time (24–150 h) on the superconducting characteristics of Nb$_3$Sn. They determined that a 24 h final reaction reached >90% of the highest obtained J$_c$ [18]. To obtain a short reaction time and good performance, the experimental sample received a final reaction for 24 h at 923 K. With this scheme, we can combine the technologies of Nb$_3$Sn bulk and conductors to fabricate Nb$_3$Sn superconducting joints on magnets in the future.

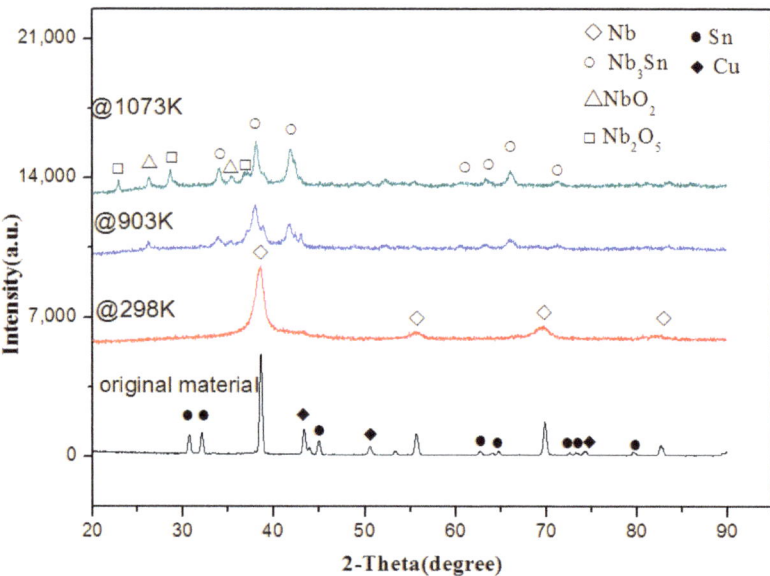

**Figure 12.** XRD patterns of the products after heat treatment at different temperatures (from 298 K to 1073 K).

Figure 13 shows MT curves of the samples that were ball milled for 5 h and heat treated for 24 h and 50 h. The onset $T_c$ of the samples was between 17 K and 18 K. The $T_c$ value was determined by taking the first deviation point from the linearity that signified the transition from the normal state to the superconducting state. As shown in Figure 13, the transitions of all the samples were relatively steep, indicating that the Nb$_3$Sn phase formation qualities of the samples were good. The differences among the prepared samples and between the experiment results and the theoretical value may be attributed to the purity of the products. Critical temperature exhibits a relationship with the concentration of Sn and the stoichiometric ratio of Nb and Sn. After high-energy mechanical alloying, the size of Nb and Sn particles sharply decreased. The distribution of Nb and Sn became more unique, and the solid-state diffusion reaction of Nb and Sn occurred on the basis of a more precise stoichiometric ratio of 3:1. $B_{c2}(0)$ can be calculated using the following equation based on the Werthamer–Helfand–Hohenberg theory [19]:

$$B_{c2}(0) = -0.69 T_c \left.\frac{\partial B_{c2}}{\partial T}\right|_{T_c}$$

where $B_{c2}$ is the upper critical magnetic field, $T_c$ is the critical temperature of a Type II superconductor (Nb$_3$Sn), and 0.69 is the pre-factor constant for the dirty limit [20].

The results led to the conclusion that the $B_{c2}(0)$ of the prepared samples did not appear to depend on the heat treatment time of the samples.

Figure 14 shows the field independence of the critical current density $J_c$ of the heat-treated bulk for 24 h. The heat-treated bulk was about 3 mm in length, 2 mm in width, and 1 mm in thickness. The magnetic $J_c$ was calculated from the magnetization loops based on the critical Bean model [21], as follows:

$$J_c = \frac{\Delta M}{a\left(1 - \frac{a}{3b}\right)},$$

where $a$ and $b$ ($b \geq a$) are the width and length of the sample, respectively.

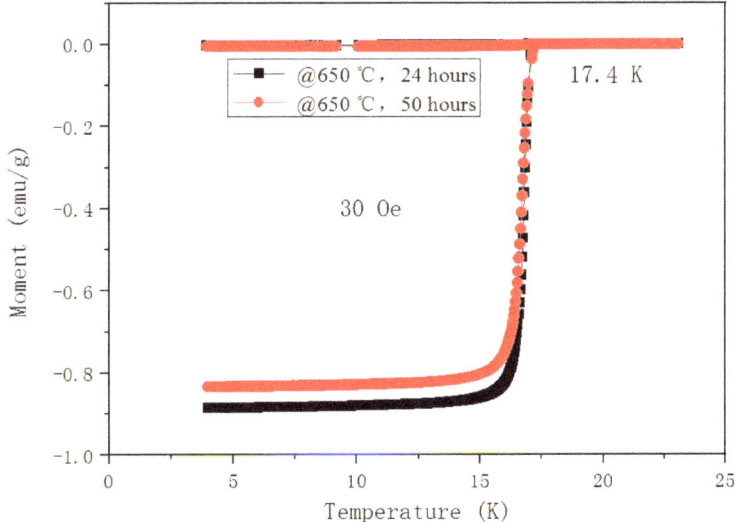

**Figure 13.** MT curve of the samples that were heat treated for 24 h and 50 h.

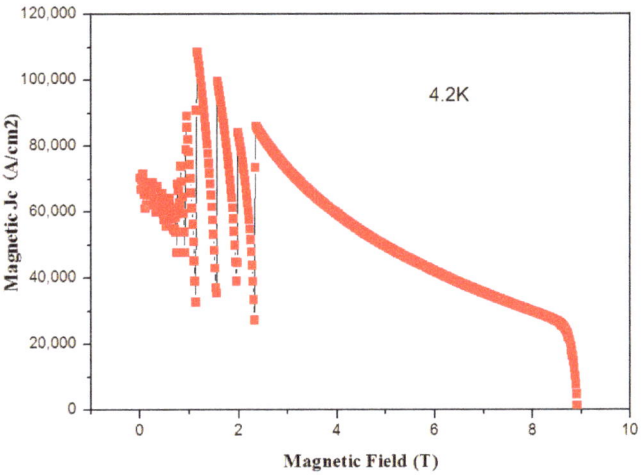

**Figure 14.** Magnetic critical current density of the sample vs. applied magnetic field at 4.2 K.

$J_c$ decreased with increasing magnetic fields. The phenomenon of the curve in the low magnetic field is attributed to the flux jump, which typically occurs in $Nb_3Sn$ materials at low fields in changing magnetic fields. Flux jumps result in the transition between the superconducting and normal states, which has been readily observed during V–H measurements. Compared with the performance of $Nb_3Sn$ in low magnetic field, we are more concerned about the performance of $Nb_3Sn$ in a high magnetic field, because the $Nb_3Sn$ material will contribute more to the high magnetic field region. As shown in the figure, the magnetic $J_c$ was approximately 30,000 A/cm$^2$ under 8 T. Considering the limit of the magnetic field in the PPMS instrument (±9 T), the $J_c$ value under 8 T was used to evaluate the performance of $Nb_3Sn$. The technology can be used for fabricating $Nb_3Sn$ joints in $Nb_3Sn$ magnets. The magnetic intensity in the region of $Nb_3Sn$ joints is lower than that in the center of the magnet. The result can help magnet designers place the $Nb_3Sn$ joint on the location where the joint has no problem to work under such a magnetic field.

## 4. Conclusions

This study presented the process of $Nb_3Sn$ phase transitions during heat treatment. $Nb_3Sn$ was prepared by heat treating the precursors, which were mechanically alloyed first. The size and crystal structure of the products changed after mechanical alloying for some time. The TEM results indicated that the products exhibited poor crystallinity after ball milling, and this finding was consistent with the XRD results. In accordance with the XRD analysis, the diffraction peaks of Cu and Sn became weaker with increasing ball milling time, and only Nb diffraction peaks were observed after 3 h. The precursors were heat treated from room temperature to 1073 K, and simultaneously, the crystal structures were monitored after exposure to different temperatures. When the temperature was above 773 K during heat treatment, $Nb_3Sn$ began to appear. When the temperature was above 973 K, some impurities, such as $Nb_2O_5$, appeared. Thus, the heat treatment temperature of the as-milled particles should not be above 973 K. The critical temperatures of the samples that were heat treated at different times were between 17 K and 18 K. The magnetic critical current density of the sample versus the applied magnetic field at 4.2 K illustrated that magnetic $J_c$ was approximately 30,000 A/cm$^2$ under 8 T. The result can be applied to the fabrication of $Nb_3Sn$ superconducting joints in the future and help magnet designers place the joint in an appropriate location where the joint will not be affected by magnetic intensity.

**Author Contributions:** Conceptualization, W.S. and S.C.; methodology, W.S.; software, W.S.; validation, W.S.; formal analysis, W.S.; investigation, W.S.; resources, W.S. and S.C.; data curation, W.S.; writing—original draft preparation, W.S.; supervision, S.C.; project administration, S.C. All authors have read and agreed to the published version of the manuscript.

**Funding:** This research received no external funding.

**Data Availability Statement:** The authors confirm that the data supporting the findings of this study are available within the article.

**Conflicts of Interest:** The authors declare no conflict of interest.

## References

1. Matthias, B.; Geballe, T.; Geller, S.; Corenzwit, E. Superconductivity of $Nb_3Sn$. *Phys. Rev.* **1954**, *95*, 1435. [CrossRef]
2. Baig, T.; Yao, Z.; Doll, D.; Tomsic, M.; Martens, M. Conduction cooled magnet design for 1.5 T, 3.0 T and 7.0 T MRI systems. *Supercond. Sci. Technol.* **2014**, *27*, 125012. [CrossRef]
3. Iwasa, Y.; Bascuñán, J.; Hahn, S.; Park, D.K. Solid-Cryogen Cooling Technique for Superconducting Magnets of NMR and MRI. *Phys. Procedia* **2012**, *36*, 1348–1353. [CrossRef]
4. Godeke, A.; Hellman, F.; Kate, H.H.J.T.; Mentink, M.G.T. Fundamental origin of the large impact of strain on superconducting $Nb_3Sn$. *Supercond. Sci. Technol.* **2018**, *31*, 105011. [CrossRef]
5. Iwaki, G.; Sato, J.; Inaba, S.; Kikuchi, K. Development of bronze-processed $Nb_3Sn$ superconducting wires for high field magnets. *IEEE Trans. Appl. Supercond.* **2002**, *12*, 1045–1048. [CrossRef]
6. Miyazaki, T.; Kato, H.; Hase, T.; Hamada, M.; Murakami, Y.; Itoh, K.; Kiyoshi, T.; Wada, H. Development of High Sn Content Bronze Processed $Nb_3Sn$ Superconducting Wire for High Field Magnets. *IEEE Trans. Appl. Supercond.* **2004**, *14*, 975–978. [CrossRef]
7. Gregory, E.; Pyon, T. Internal tin $Nb_3Sn$ conductor development for high energy physics applications. *Aip Conf. Proc.* **2002**, *614*, 958–967.
8. Lu, X.F.; Hampshire, D.P. The field, temperature and strain dependence of the critical current density of a powder-in-tube $Nb_3Sn$ superconducting strand. *Supercond. Sci. Technol.* **2010**, *23*, 025002. [CrossRef]
9. Di Michiel, M.; Scheuerlein, C. Phase transformations during the reaction heat treatment of powder-in-tube $Nb_3Sn$ superconductors. *Supercond. Sci. Technol.* **2007**, *20*, 55–58. [CrossRef]
10. Xu, X. A review and prospects for $Nb_3Sn$ superconductor development. *Supercond. Sci. Technol.* **2017**, *30*, 093001. [CrossRef]
11. Di Michiel, M.; Scheuerlein, C. Phase transformations during the reaction heat treatment of $Nb_3Sn$ superconductors. *J. Phys. Conf.* **2010**, *234*, 022032. [CrossRef]
12. Oh, S.; Park, S.H.; Lee, C.; Choi, H.; Kim, K. Heat treatment effect on the strain dependence of the critical current for an internal-tin processed $Nb_3Sn$ strand. *Physical C* **2010**, *470*, 129–133. [CrossRef]
13. Zhang, K.; Zhang, P.X.; Shi, Y.G.; Liu, J.W.; Gao, H.X.; Jia, J.J.; Guo, J.H.; Li, J.F.; Liu, X.H.; Feng, Y. An Investigation Into the Heat Treatment Tolerance of WST $Nb_3Sn$ Strands Produced for Massive Fusion Coils. *IEEE Trans. Appl. Supercond.* **2015**, *25*, 1–6. [CrossRef]

14. Lee, J.; Posen, S.; Mao, Z.; Trenikhina, Y.; He, K.; Hall, D.L.; Liepe, M.; Seidman, D.N. Atomic scale analysis of Nb$_3$Sn on Nb prepared by a vapor-diffusion process for superconducting radiofrequency cavity applications. *arXiv* **2019**, *32*, 024001.
15. Godeke, A. A review of the properties of Nb$_3$Sn and their variation with A15 composition, morphology and strain state. *Supercond. Sci. Technol.* **2006**, *19*, 68–80. [CrossRef]
16. Patankar, S.N.; Froes, F.H. Formation of Nb$_3$Sn Using Mechanically Alloyed Nb—Sn Powder. *Solid State Sci.* **2004**, *6*, 887–890. [CrossRef]
17. Menary, M.A.; Menary, M.A. The Effect of Mechanical Alloying on the Structural and Superconducting Properties of Nb$_3$Sn. 2005, Volume 27, pp. 3–7. Available online: https://www.mendeley.com/catalogue/24a95fbb-8978-3aaa-991e-66b759e6088e/ (accessed on 23 February 2023).
18. Ghosh, A.K.; Sperry, E.A.; Cooley, L.D.; Moodenbaugh, A.M.; Sabatini, R.L.; Wright, J.L. Dynamic stability threshold in high-performance internal-tin Nb$_3$Sn superconductors for high field magnets. *Supercond. Sci. Technol.* **2004**, *18*, 5–8. [CrossRef]
19. Helfand, E.; Werthamer, N.R. Temperature and Purity Dependence of the Superconducting Critical Field, Hc$_2$. *Phys. Rev. Lett.* **1964**, *13*, 686–688. [CrossRef]
20. Rekaby, M.; Awad, R.; Abou-Aly, A.I.; Yousry, M. AC Magnetic Susceptibility of Y$_3$Ba$_5$Cu$_8$O$_{18}$ Substituted by Nd$^{3+}$ and Ca$^{2+}$ Ions. *J. Supercond. Nov. Magn.* **2019**, *32*, 3483–3494. [CrossRef]
21. Bean, C.P. Magnetization of High-Field Superconductors. *Rev. Mod. Phys.* **1964**, *36*, 886–901. [CrossRef]

**Disclaimer/Publisher's Note:** The statements, opinions and data contained in all publications are solely those of the individual author(s) and contributor(s) and not of MDPI and/or the editor(s). MDPI and/or the editor(s) disclaim responsibility for any injury to people or property resulting from any ideas, methods, instructions or products referred to in the content.

Article

# Plasma-Pulsed GMAW Hybrid Welding Process of 6061 Aluminum and Zinc-Coated Steel

Hongchang Zhang [1,2], Wenhu He [1], Huaibei Zheng [3], Jiang Yu [4], Hongtao Zhang [4], Yinan Li [1,*], Jianguo Gao [5] and Zhaofang Su [5]

1. School of Mechanical and Automotive Engineering, Qingdao University of Technology, Qingdao 266520, China
2. School of Rongcheng, Harbin University of Science and Technology, Weihai 264300, China
3. Chengdu Advanced Metal Materials Industrial Technology Research Institute Co., Ltd., Chengdu 610305, China
4. State Key Laboratory of Advanced Welding and Joining, Harbin Institute of Technology, Harbin 150001, China
5. Shandong Classic Group Co., Ltd., Jining 272000, China
* Correspondence: liyinan@qut.edu.cn

**Abstract:** A novel plasma-pulsed GMAW hybrid welding (plasma-GMAW-P) process is proposed for joining 6061 aluminum and zinc-coated steel. The results show that the change in welding heat input has little effect on the microstructure of the joint and the composition of the intermetallic compounds (IMCs) but only changes the thickness of the reaction layer (increased from 5 μm to 12 μm). when the plasma arc current is 20 A and the MIG current is 80 A, the welded joint obtained has the highest tensile-shear force. With the optimal process parameters, the weld strength obtained by filling ER4043 welding wire is the highest, accounting for 65% of the tensile-shear force of the base material. The effect of the plasma arc acting on the joint properties is studied through the microstructure and a tensile-shearing test. The action position of the plasma arc plays a significant role in the Al/steel interface, which directly influences the strength of the welded joints. Regardless of the plasma-GMAW-P style used to obtain the joints, Fe-Al IMCs appear at the interface. When the plasma arc is in front of the welding direction and the GMAW-P arc is in the rear, the tensile-shear force reaches the maximum of 3322 N.

**Keywords:** plasma-GMAW-P; 6061 aluminum; zinc-coated steel; microstructure

## 1. Introduction

Lightweight automobile has gradually become a research hotspot in the automotive field [1–3]. Al alloys were the most popular among the various lightweight alloys due to their features of high mechanical strength, good corrosion resistance, and affordable cost. To ensure quality, it was necessary to produce an Al–steel structure consisting of Al and steel parts [4–6]. Al–steel structures have high temperature resistance, ultra-low temperature resistance, good heat dissipation, and excellent electrical conductivity. They have been widely used in the fields of automobile manufacturing, metallurgy, and aerospace [7]. Due to the differences in their physical and chemical properties, the formation of brittle Fe–Al intermetallic compounds (IMCs) and weld properties appear at the interface and Al side [8–10]. Furthermore, different welding methods, including single fusion welding, laser-arc welding, friction stir welding, etc. [11–13], have been applied to join Al and steel to improve the interfacial reaction and optimize the thickness and distribution of IMCs [14–16]. However, these methods have stricter requirements for welding heat input [17], joint shape, and assembly conditions, which are difficult to meet given the need for rapid manufacturing in the industrial field [18–20].

Hybrid welding has been considered as one of the effective ways to improve welding quality and efficiency since it was proposed [21,22], and has gradually become a research hotspot in the field of welding [23,24]. Hybrid heat source is an effective method to solve practical problems, which conventional fusion welding could not [25,26]. In this study, 6061 aluminum and zinc-coated steel were innovatively welded with a plasma-pulsed GMAW (Plasma-GMAW-P) hybrid welding process. The plasma arc current, the MIG current and filler metal, and the different hybrid welding types were investigated. The aim of this work was to provide a new method for achieving efficient and high-quality welding of dissimilar metals.

## 2. Materials and Methods

### 2.1. Welding Test System and Test Material

Figure 1 shows the physical and schematic diagram of the plasma-GMAW-P hybrid welding system. The system is mainly composed of the following parts: 1, welding robot; 2, self-designed plasma-GMAW-P hybrid welding gun; 3, MIG power supply; 4, PAW power supply; 5, magnetron power supply; 6, cooling system. The welding robot was a Shanghai Xinshida SRC2.4(SRC2.4) (Xinshida Robotics Co., Ltd., Shanghai, China). The models of PAW power supply, MIG power supply, and water chiller were a DC pulse LHM8-300A (Chengdu Electric Welding Machine Research Institute, Chengdu, China), Panasonic YD-400GE (Panasonic Welding Systems (Tangshan) Co., Ltd., Tangshan, China), and CW-6000 (Guangzhou Teyu Electromechanical Co., Ltd., Guangzhou, China), respectively.

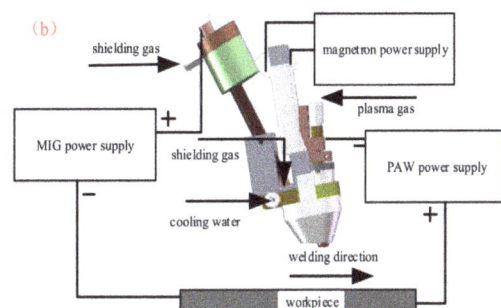

**Figure 1.** The physical and schematic diagram of plasma-GMAW-P hybrid welding system, (**a**) the physical diagram of plasma-GMAW-P hybrid welding system, and (**b**) the schematic diagram of plasma-GMAW-P hybrid welding system.

The zinc-coated steel was made of Q235 and the thickness of the zinc-coat was 11 microns. The whole materials were cut into pieces of 150 mm × 80 mm × 2 mm. The surface oxide film of the base materials was removed by steel brush and then cleaned with alcohol and acetone prior to the welding process.

### 2.2. Test Design

In order to explore the optimal process parameters for plasma-GMAW-P hybrid welding, welding tests were conducted using hybrid process parameters of different styles, as shown in Table 1. The schematic diagram of the assembled plasma-GMAW-P hybrid welding system is depicted in Figure 2. The Al alloy was placed on the top of the zinc-coated steel with an overlapping width of 10 mm. In the plasma-GMAW-P welding process, the angle between the plasma torch and the GMAW torch was 45°, and the distance between the sources was 7 mm. Plasma welding current, welding speed, nozzle height, plasma shielding gas flow rate, shielding gas flow rate, GMAW-P welding current, and magnetic current were recorded in detail. When the nozzle height was constant at 5 mm, the welding speed was 5 mm/s. Pure argon was used as a plasma shielding gas with a 3 L/min flow rate. The shielding gas, composed of argon (80 vol%) and carbon dioxide (20 vol%), was chosen at a volume flow rate of 12 L/min in all experiments. Moreover, to investigate the effect of the plasma welding arc on the joint, different styles plasma-GMAW-P hybrid welding were applied (Figure 3). Plasma welding current and GMAW-P welding current were 20 A and 80 A, respectively. The filler metal used in the plasma-GMAW-P welding process with different styles was ER4043 with a diameter of 1.2 mm. The chemical composition of the different welding wires is shown in Table 2.

**Table 1.** The hybrid process parameters of different styles.

| Parameter Set | Plasma Arc Current | MIG Current | Filler Metal |
|---|---|---|---|
| #1 | 40, 50, 60 | 60 | ER4043 |
| #2 | 30, 40, 50, 60 | 70 | ER4043 |
| #3 | 10, 20, 30, 40 | 80 | ER4043 |
| #4 | 20 | 80 | ER4047, ER1070, ER5356 |

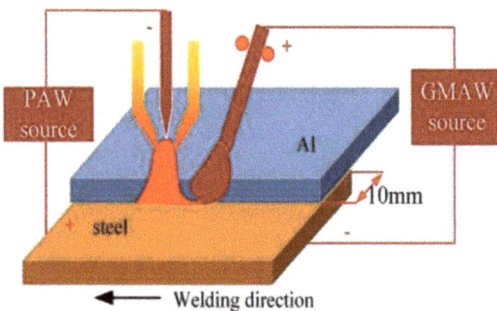

**Figure 2.** Schematic diagram of plasma−GMAW−P hybrid welding.

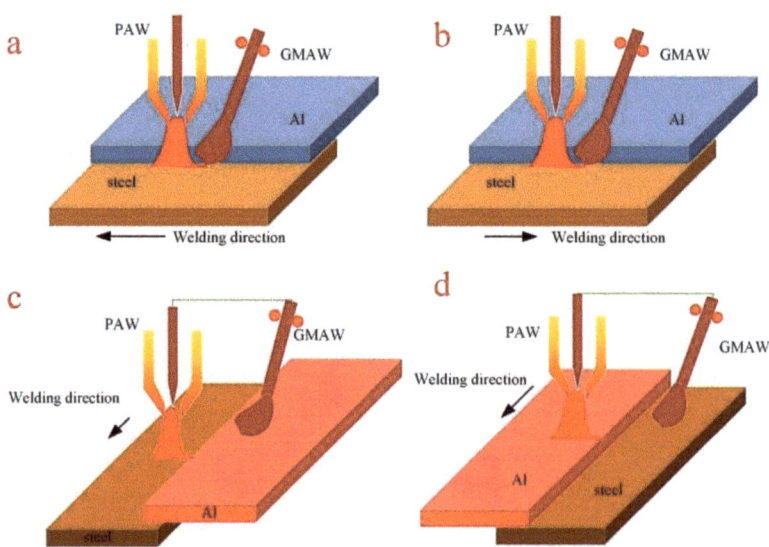

**Figure 3.** Schematic diagram of plasma−GMAW−P hybrid welding with different types: (**a**) model 1 (plasma arc was in front and GMAW arc was in rear), (**b**) model 2 (GMAW arc was in front and plasma arc was in rear), (**c**) model 3 (plasma arc was located on steel side and GMAW arc was located on Al side), and (**d**) model 4 (plasma arc was located on Al side, and GMAW arc was located on steel side).

Table 2. The chemical compositions of different welding wires (wt.%).

| Welding Wire | Si | Cu | Fe | Mn | Ti | Zn | Mg | Al |
|---|---|---|---|---|---|---|---|---|
| ER4043 | 4.50–6.00 | 0.30 | 0.80 | 0.05 | 0.20 | 0.10 | 0.05 | Bal. |
| ER4047 | 11.0–13.0 | 0.30 | 0.80 | 0.15 | - | 0.20 | 0.10 | Bal. |
| ER1070 | 0.3 | - | 0.3 | - | - | - | - | Bal. |
| ER5356 | 0.25 | 0.10 | 0.40 | 0.05–0.20 | 0.06–0.20 | 0.10 | 4.50–5.50 | Bal. |

The specimen dimensions were 12 mm × 10 mm, and the specimens' surfaces were brushed with a buffing machine and then cleaned with acetone and alcohol, in turn. All the workpieces were polished to remove the surface oxidation film before welding and then cleaned with alcohol and acetone. All the metallographic specimens and tensile specimens were cut out from the weld joints by wire cutting after welding. After the metallographic specimens were ground, polished, and then etched, the metallographic test was conducted through an Olympus-DSX510 optical microscope and scanning electron microscope (SEM). A 110 mm × 10 mm overlap joint was removed by wire cutting as a tensile sample. Three sets of tensile specimens were selected for each set of parameters, and the average was taken as the final test results. The tensile tests were carried out with an electronic tensile testing machine (CSS-44400) with a travel speed of 0.5 mm/min.

## 3. Results and Discussion

### 3.1. Effects of Plasma Arc Current and MIG Current on Microstructure and Mechanical Properties of Welded Joints

Figure 4 shows the weld formation and cross-sectional microstructure of #1 under different plasma arc currents when the MIG current was 60 A. As can be seen from Figure 4, the welded joint could be divided into a zinc-rich zone, an interface zone, and a weld zone. As the plasma arc current increased, the bead width decreased. This is because the increase in heat input accelerates the evaporation of the zinc coating. The wetting and spreading effect of zinc coating on aluminum is weakened.

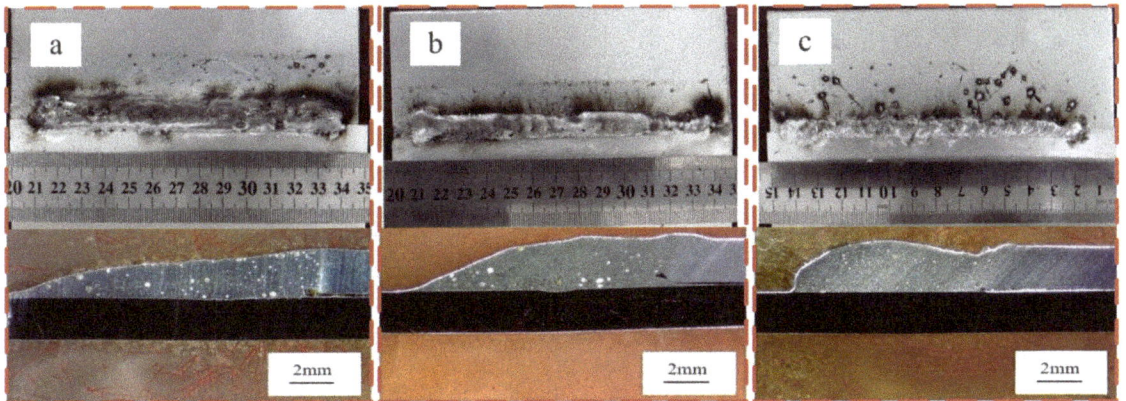

**Figure 4.** The weld formation and the cross-sectional microstructure with different plasma arc currents when the MIG current is 60 A. (**a**) Plasma arc current is 40 A; (**b**) plasma arc current is 50 A; (**c**) plasma arc current is 60 A.

The corresponding tensile-shear force and the fracture surface of the welded joint are shown in Figures 5 and 6, respectively. It can be seen in Figures 5 and 6 that with an increase in plasma arc current, the tensile-shear force of the welded joint increased, and the fracture basically occurred near the fusion line. This is because the increase in plasma current enhances the preheating effect, resulting in a slower rate of temperature increase. This change refines the grain size and improves the tensile-shear force.

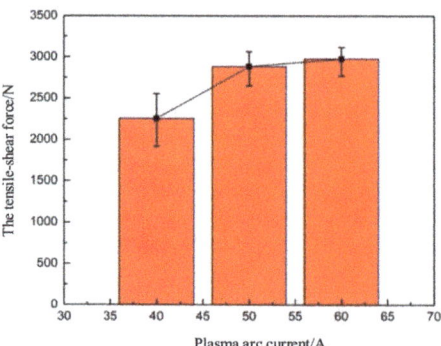

**Figure 5.** The tensile-shear force of the welded joint with different plasma arc currents when the MIG current is 60 A.

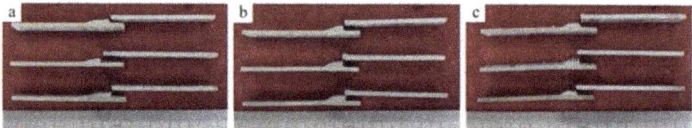

**Figure 6.** The fracture surface of the welded joint with different plasma arc currents when the MIG current is 60 A. (**a**) Plasma arc current is 40 A; (**b**) plasma arc current is 50 A; (**c**) plasma arc current is 60 A.

Figure 7 shows the weld formation and optical microstructure of the welded joints of #2 with different plasma arc currents when the MIG current is was 70 A. Whether the plasma arc current is too small or too large, the welded joints showed porosity defects. The corresponding tensile-shear force and the fracture surface of the welded joint are shown in Figures 8 and 9, respectively. As can be seen from the figure, with the increase in the plasma arc current, the tensile-shear force of the welded joint first increased and then decreased. The maximum value occurred when the plasma arc current was 50 A, and the maximum tensile-shear force was close to 3000 N. This was due to the increase in plasma arc current, which improved the wetting and spreading effect of aluminum on the steel surface. However, the weak spot of the weld was the heat-affected zone near the fusion line.

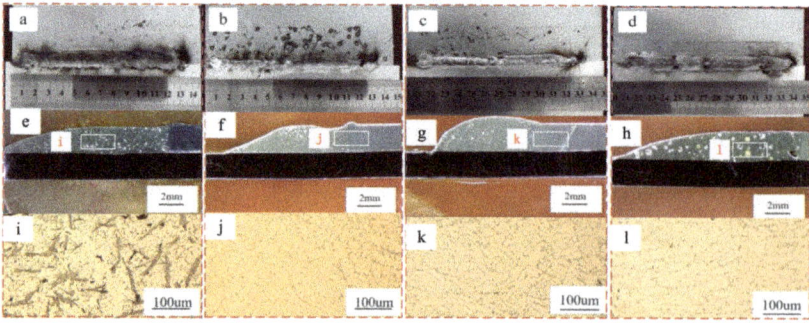

**Figure 7.** The weld formation and optical microstructure of the welded joints with different plasma arc currents when the MIG current is 70 A. (**a,e,i**) plasma arc current is 30 A; (**b,f,j**) plasma arc current is 40 A; (**c,g,k**) plasma arc current is 50 A; (**d,h,l**) plasma arc current is 60 A.

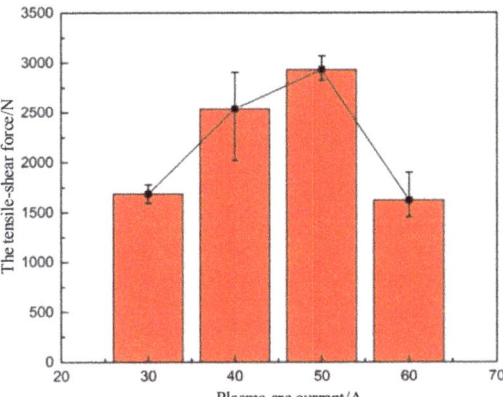

**Figure 8.** The tensile-shear force of the welded joint by plasma-GMAW-P with different plasma arc currents when the MIG current is 70 A.

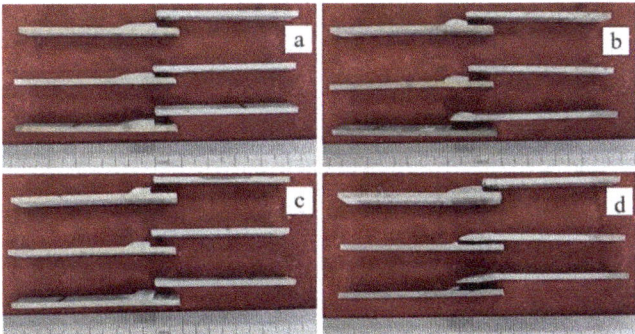

**Figure 9.** The fracture surface of the welded joint by plasma-GMAW-P with different plasma arc currents when the MIG current is 70 A. (**a**) Plasma arc current is 30 A; (**b**) plasma arc current is 40 A; (**c**) plasma arc current is 50 A; (**d**) plasma arc current is 60 A.

At the same time, Figure 7 shows that the melting zone was mainly composed of α-Al and Al-Si eutectic structure. As the plasma arc current increased, the grain size increased. When the plasma arc current was 40 A and 50 A, there were fewer porosity and crack defects in the joint. This was because the increase in the plasma arc current allowed more time for the gas in the weld pool to escape, especially hydrogen. This improved the porosity defects of the weld. However, excessive plasma arc current could accelerate the cooling speed of the weld, which increased the brittleness of the weld and decreased the tensile-shear force, as shown in Figure 8. Figure 9a,c shows that all fractures were located near the weld fusion line, while Figure 9b,d shows that some fractures were located on the boundary line.

Figure 10 shows the weld formation and cross-sectional microstructure of #3 with different plasma arc currents when the MIG current was 80 A. When the plasma arc current was 40 A, the bead width was larger than that of the other groups. The corresponding tensile-shear force and the fracture surface of the welded joint are shown in Figures 11 and 12, respectively. As can be seen in Figure 11, with the increase in the plasma arc current, the tensile-shear force of the weld first increased and then decreased. When the plasma arc current was 20 A, the maximum tensile-shear force occurred, which was the maximum value for all groups of samples, accounting for 65% of the tensile-shear force of the base material. Except for that shown in Figure 12c, all fractures occurred near the weld fusion line.

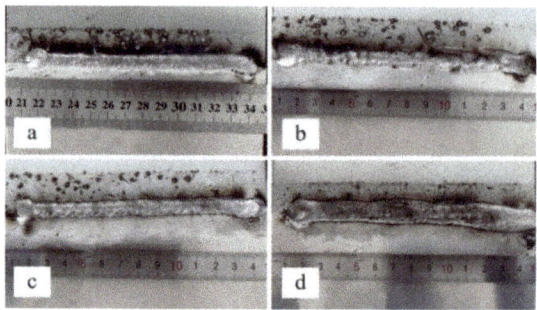

**Figure 10.** The weld formation of the welded joints with different plasma arc currents when the MIG current is 80 A. (**a**) Plasma arc current is 10 A; (**b**) plasma arc current is 20 A; (**c**) plasma arc current is 30 A; (**d**) plasma arc current is 40 A.

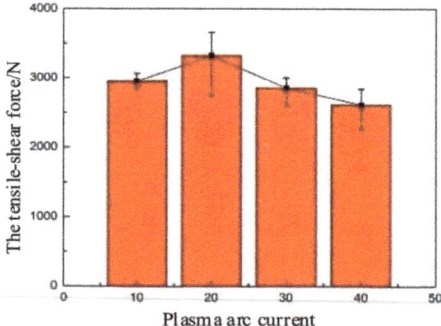

**Figure 11.** The tensile-shear force of the welded joint.

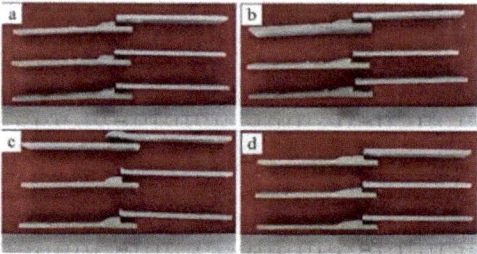

**Figure 12.** The fracture surface of the welded joint. (**a**) Plasma arc current is 10 A; (**b**) plasma arc current is 20 A; (**c**) plasma arc current is 30 A; (**d**) plasma arc current is 40 A.

The optical microstructure of the welded joints is shown in Figure 13. The melting zone mainly had an α-Al and Al-Si eutectic structure composition. With the increase in the plasma arc current, the grain size increased, but the microstructure morphology changed little. When the plasma arc current was 20 A and 30 A, there were fewer defects such as pores and cracks in the joint. Meanwhile, the microstructure of the joint interface area is shown in Figure 13. The reaction layer in the interface region was composed of $Fe_2Al_5$ that was dense near the steel side and $FeAl_3$ that grew needle-like toward the aluminum side. Welding heat input had no significant impact on the type and morphology of interfacial compounds but only determined the thickness of the reaction layer. As the plasma arc current increased, the thickness of the reaction layer increased from 5 μm to 12 μm.

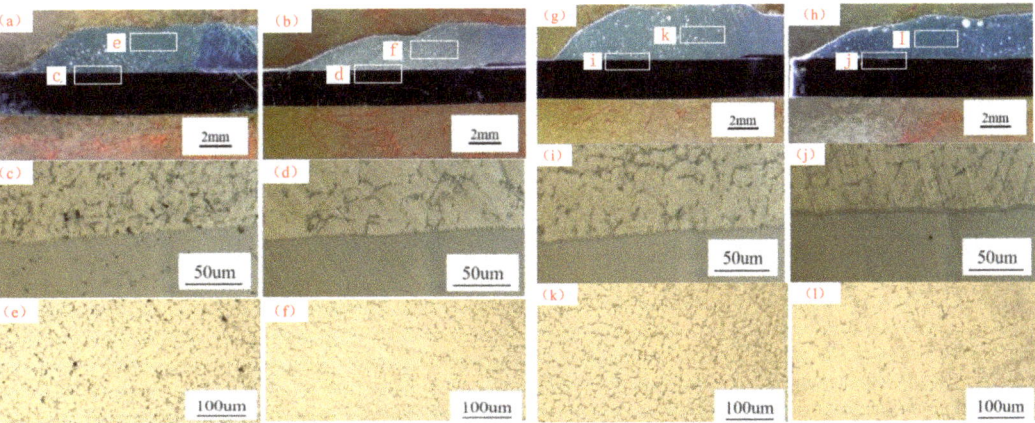

**Figure 13.** Optical microstructure of the welded joints. (**a,c,e**) plasma arc current is 10 A; (**b,d,f**) plasma arc current is 20 A; (**g,i,k**) plasma arc current is 30 A; (**h,j,l**) plasma arc current is 40 A.

Suitable welding heat input can obtain good performance of the welded joints. According to tests #1–#3, when the plasma arc current was 20 A and the MIG current was 80 A, the welded joint obtained had the highest tensile-shear force.

*3.2. Effect of Filler Metal on Microstructure and Mechanical Properties of Welded Joints*

With the optimal welding process parameters, four different types of welding wires, ER4043, ER4047, ER1070, and ER5356, were selected for cladding, and good weld formation was obtained. The bead width obtained by filling ER4043 and ER4047 is larger. This is because the Si element in the filler metal promotes the wetting and spreading of aluminum on the steel surface, increasing the bead width. According to the XRD results, the obtained joints all contained the $Fe_{0.905}Si_{0.905}$ phase. The welded joints filled with different metals were observed by SEM and analyzed using EDS. The results are shown in Figure 14 and Tables 3–6.

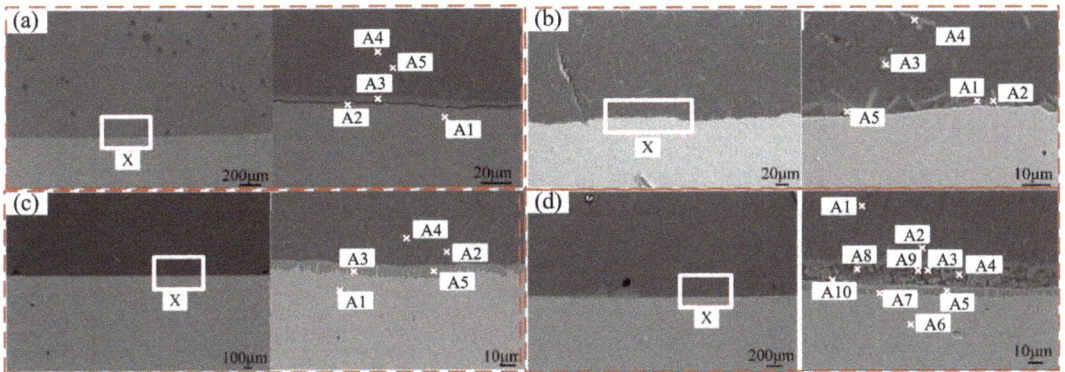

**Figure 14.** The SEM microstructure of the overall image and the enlargement image of zone X with different types of welding wires. (**a**) ER4043; (**b**) ER4047; (**c**) ER1070; (**d**) ER5356.

Table 3. The EDS results of joints filled with ER4043.

| Elements (at.%) | Fe | Al | Si. | Cu | O |
|---|---|---|---|---|---|
| A1 | 98.48 | 0.20 | 0.26 | 1.06 | - |
| A2 | 38.90 | 57.91 | 2.37 | 0.82 | - |
| A3 | 23.65 | 71.31 | 4.40 | 0.64 | - |
| A4 | 3.57 | 88.94 | 6.28 | 0.28 | - |
| A5 | 0.32 | 98.10 | - | 0.24 | 0.24 |

Table 4. The EDS results of joints filled with ER4047.

| Elements (at.%) | Fe | Al | Si. | Cu | Zn |
|---|---|---|---|---|---|
| A1 | 10.95 | 72.87 | 15.89 | 0.29 | - |
| A2 | 9.72 | 66.60 | 23.41 | 0.27 | - |
| A3 | 9.89 | 66.60 | 22.95 | 7.33 | - |
| A4 | 8.98 | 78.24 | 11.43 | - | 1.35 |
| A5 | 13.26 | 63.86 | 21.36 | - | 1.52 |

Table 5. The EDS results of joints filled with ER1070.

| Elements (at.%) | Fe | Al | C | Au | O | Mg |
|---|---|---|---|---|---|---|
| A1 | 76.95 | - | 20.86 | 1.16 | - | - |
| A2 | 1.97 | 97.48 | - | - | - | - |
| A3 | 78.02 | - | 20.49 | 0.70 | - | - |
| A4 | 0.82 | 89.75 | 6.95 | - | 0.75 | 0.98 |
| A5 | 77.63 | - | 21.37 | - | - | - |

Table 6. The EDS results of joints filled with ER5356.

| Elements (at.%) | Fe | Al | C | Mg | O | Zn |
|---|---|---|---|---|---|---|
| A1 | 2.13 | 71.04 | 20.64 | 3.84 | 1.24 | - |
| A2 | - | 76.37 | 17.41 | 3.85 | 1.71 | - |
| A3 | - | 29.93 | 8.29 | 1.20 | 51.95 | 6.23 |
| A4 | 0.68 | 64.36 | 16.86 | 1.36 | 6.45 | 10.28 |
| A5 | 67.17 | 11.93 | 20.23 | - | - | - |
| A6 | 74.49 | - | 24.62 | - | - | - |
| A7 | 74.30 | - | 23.41 | - | - | - |
| A8 | 0.51 | 33.41 | 18.86 | 2.84 | 37.05 | 6.04 |
| A9 | 0.71 | 36.88 | 10.43 | 2.44 | 38.97 | 8.75 |
| A10 | 22.31 | 57.95 | 19.74 | - | - | - |

It can be seen in Figure 14 that the thickness of the intermetallic compound in the weld interface region obtained by filling ER4043 was 7 to 9 μm, and the weld pore diameter was 30~100 μm. According to the XRD results, the main IMC was $Al_{0.5}Fe_3Si_{0.5}$. The melting zone located on the upper side of the interface zone mainly had of an α-Al and Al-Si eutectic structure composition. The base metal structure of galvanized steel is composed of α ferrite and fine carbide particles. The IMCs in the weld interface region obtained by filling ER4047 were acicular and diffuse toward the aluminum side. According to the XRD results, the main IMC was also $Al_{0.5}Fe_3Si_{0.5}$.

The thickness of the reaction layer in the weld interface area obtained by filling ER1070 was 10–12 μm. It extended toward the steel side. The content of C near the reaction layer was large, and the element C in the galvanized steel diffused here, while the content of C in the melting zone above the reaction layer was very low. There were many welding joint defects obtained by filling ER5356 welding wire. Al, C, and O compounds with large grain sizes appeared on the aluminum side above the reaction layer.

The tensile-shear force and the fracture surface of the welded joints filled with different metals are shown in Figures 15 and 16. The tensile-shear force obtained by filling ER4043 welding wire was the highest followed by that of ER4047, while ER5356 had the lowest force. They accounted for 65%, 56%, 46%, and 27% of the tensile-shear force of the base metal, respectively. The tensile-shear force of the weld filled with ER4047 welding wire was less than ER4043. This was because too much Si element in the ER4047 welding wire entered the molten pool during the welding process, increasing the viscosity of the weld pool. It increased the difficulty of gas escaping from the weld pool, resulting in porosity defects in the welded joint.

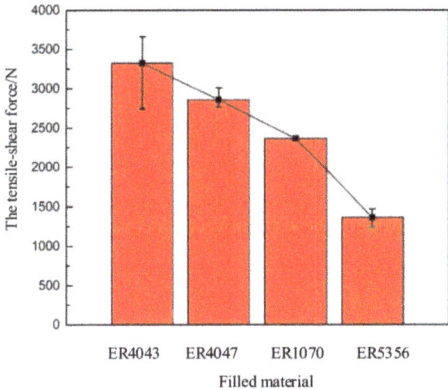

**Figure 15.** The tensile-shear force of the welded joints filled with different metals.

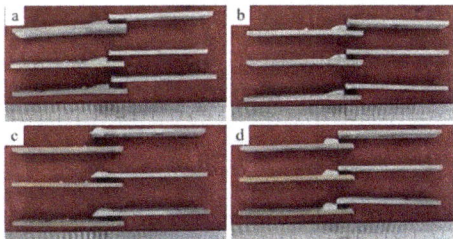

**Figure 16.** The fracture surface of the welded joints filled with different metals. (**a**) ER4043; (**b**) ER4047; (**c**) ER1070; (**d**) ER5356.

Figure 16 shows the fracture surface of a welded joint, where the fracture of the welded joint obtained by filling ER4043, ER4047, and ER5356 welding wires was located near the fusion line. However, the fracture of the welded joint filled with ER1070 welding wire was located at the interface. According to EDS analysis, there were high-C compounds near the interface region of the joint filled with ER1070 welding wire, which was the reason for the fracture of the sample at the interface.

### 3.3. The Effect of the Plasma Welding Arc on the Joint

The macromorphology of the welded joints obtained by plasma-GMAW-P hybrid welding with different styles is shown in Figure 17. As shown in Figure 17, the welding spatters were generated during the whole welding process but were located on the zinc-coated steel side. However, there was no effective connection between 6061 aluminum and zinc-coated steel in Figure 17b. Furthermore, the cross-section of the welded samples and the tensile-shear specimens could not be obtained. This is because model 2 did not take advantage of plasma arc preheating and MIG arc filling. The microstructure and joint properties via plasma-GMAW-P are compared with those of the other three styles below.

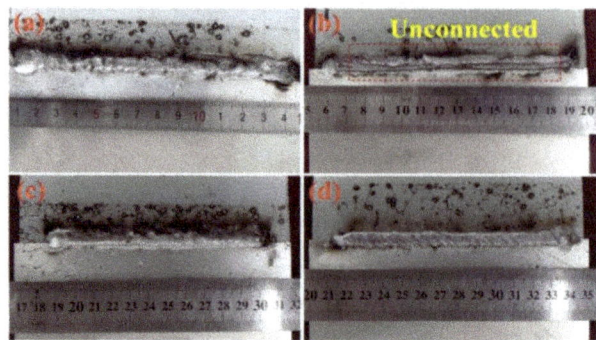

**Figure 17.** Surface appearances of plasma-GMAW-P with different styles. (**a**) model 1; (**b**) model 2; (**c**) model 3; (**d**) model 4.

Figure 18 depicts the optical microstructure of the joints produced by plasma-GMAW-P with different styles. As shown in Figure 18a–c, the height of the fusion zone changed in different ways. The height of the fusion zone of the joints welded via plasma-GMAW-P with model 4 reached the maximum, but the wetting angle of the joints welded via plasma-GMAW-P with model 1 was the minimum. This was attributed to the plasma arc located on the Al side, where most of the generated heat melted the Al alloy. Furthermore, in comparison with Figure 18g–i, the microstructure of the fusion zone was mainly composed of an α-Al and Al-Si eutectic structure. The grain size of the welded joints in Figure 18h is smaller than that of the other two styles. This was attributed to the plasma arc mainly acting on the steel side in this model and the reduced heat distributed to the Al alloy side, resulting in grain refinement in the fusion zone. As shown in Figure 18d–f, there was no obvious difference in the interface structure.

**Figure 18.** Optical microstructure of the joints welded via plasma-GMAW-P with different styles. (**a,d,g**) model 1; (**b,e,h**) model 3; (**c,f,i**) model 4.

The SEM images of the interface structure of the joints welded via Plasma-GMAW-P with different styles are shown in Figures 19–21. As shown in Figure 19, the interface did not form a valid connection between the Al alloy and the zinc-coated steel via plasma-GMAW-P with model 1. The pores with different sizes appeared in the fusion zone. According to the EDS results, $FeAl_2$ and $FeAl_3$ IMCs were formed at interface.

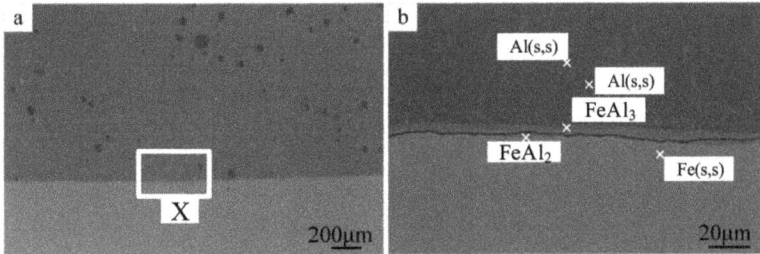

**Figure 19.** The SEM microstructure of joints via plasma-GMAW-P with model 1. (**a**) The overall image; (**b**) the enlargement image of zone X.

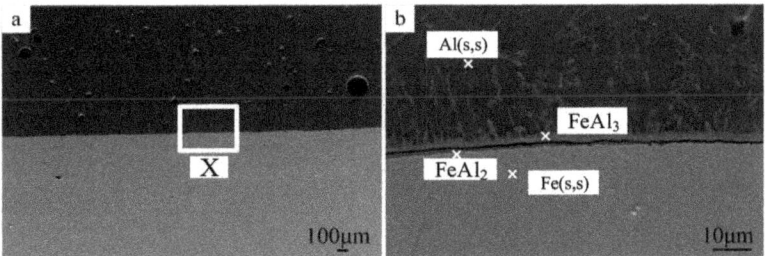

**Figure 20.** The SEM microstructure of joints formed via plasma-GMAW-P with model 3. (**a**) The overall image; (**b**) the enlargement image of zone X.

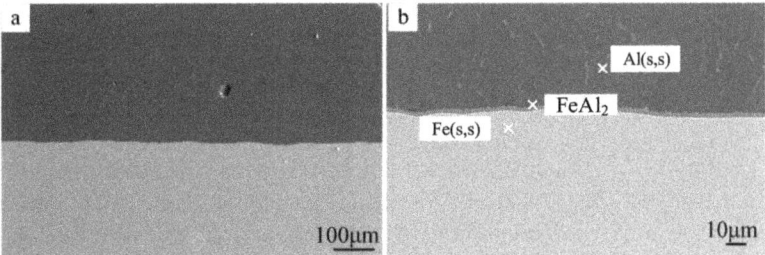

**Figure 21.** The SEM microstructure of joints formed via plasma-GMAW-P with model 4. (**a**) The overall image; (**b**) the enlargement image of zone X.

Figure 20 shows the SEM image of the interface structure obtained via plasma-GMAW-P with model 3. As shown in Figure 20b, a transverse crack was formed at the interface, which did not achieve a valid connection. An obvious reaction layer was formed on the Al side. This proved that the Fe elements and Al elements diffused each other. According to the EDS results, $FeAl_2$ and $FeAl_3$ IMCs were formed at the interface.

Figure 21 shows the SEM image of the interface structure via plasma-GMAW-P with model 4. As shown in Figure 21b, an obvious reaction layer was formed at the interface, which achieved a valid connection. According to the EDS results, $FeAl_2$ IMC was formed at the interface.

The tensile-shear force and the fracture surface of the welded joints by plasm-GMAW-P with different styles are shown in Figures 22 and 23. As shown, the tensile-shear force of the joint made by plasma-GMAW-P with model 1 reached a maximum of 3322 N, and a fracture occurred at the weld seam. Moreover, the fracture of the joints welded by plasma-GMAW-P with model 3 occurred at the weld seam. However, the fracture of the joints welded by plasma-GMAW-P with model 4 occurred at the interface, which reached the minimum tensile-shear force of 2236 N, attributed to the IMCs' thickness.

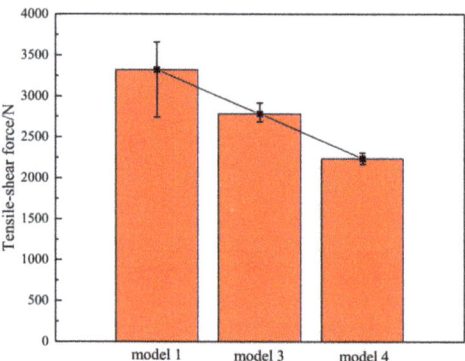

**Figure 22.** The tensile-shear force of the joint welded by plasma-GMAW-P with different styles.

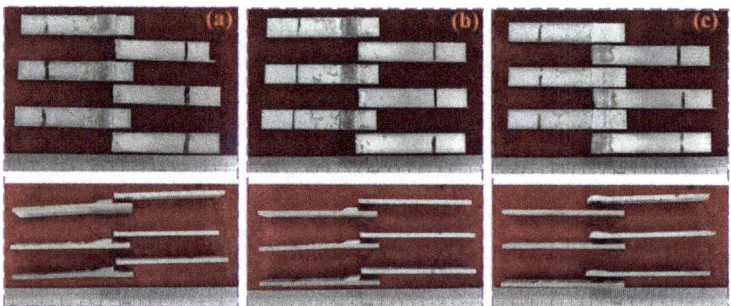

**Figure 23.** The fracture surface of the joint welded by plasma-GMAW-P with different styles. (**a**) model 1; (**b**) model 3; (**c**) model 4.

## 4. Conclusions

Novel plasma-GMAW-P hybrid welding was applied to join 6061 aluminum and zinc-coated steel. The following conclusions were obtained:

1. According to the microstructure characteristics, the welded joint can be divided into a zinc-rich zone, an interface zone, and a weld zone. When the heat input is approximately equal, the combination of the lower plasma arc current and the higher MIG current can achieve a higher tensile-shear force and a good weld microstructure. When the plasma arc current is 20 A and the MIG current is 80 A, the specimen with the highest tensile-shear force is obtained, reaching 65% of the 6061 aluminum base material.
2. The welding test conducted with the optimal process parameters showed that the weld strength obtained by filling ER4043 welding wire was the highest, followed by that obtained with ER4047, ER1070, and ER5356, accounting for 65%, 56%, 46%, and 27% of the tensile-shear force of the base material, respectively. The joint samples obtained by filling only ER1070 welding wires were fractured at the boundary line, while the rest of the samples were fractured near the fusion line. According to the EDS results, there were high-C compounds near the interface region of the ER1070 welding wire sample, which was the reason for the fracture at the interface. In the ER4047 welding wire joint, excessive Si elements entered the weld pool, increasing the viscosity of the weld pool and the difficulty of gas escape and resulted in more porosity defects. This also led to a lower tensile-shear force of the joint with ER4047 than ER4043.
3. The action position of the plasma arc played a significant role in the Al/steel interface, which directly influenced the strength of the welded joints. Regardless of the style of plasma-GMAW-P used to obtain the joints, Fe-Al IMCs appeared at the interface.

When the plasma arc was in front of the welding direction and GMAW-P arc was in the rear, the tensile-shear force reached a maximum of 3322 N.

**Author Contributions:** Conceptualization, H.Z. (Hongchang Zhang) and H.Z. (Hongtao Zhang); methodology, H.Z. (Hongchang Zhang); software, J.G.; validation, W.H. and Z.S.; formal analysis, W.H.; investigation, H.Z. (Hongtao Zhang); resources, J.Y.; data curation, J.G.; writing—original draft preparation, H.Z. (Hongchang Zhang); writing—review and editing, Y.L.; visualization, W.H.; supervision, Y.L.; project administration, H.Z. (Hongtao Zhang); funding acquisition, H.Z. (Hongtao Zhang) and Y.L. All authors have read and agreed to the published version of the manuscript.

**Funding:** This work was sponsored by the National Natural Science Foundation of China (No. U22B20127), the National Natural Science Foundation of China (No. 52175305), and Jining City Global List of Major Projects (No. 2022JBZP004).

**Data Availability Statement:** Not applicable.

**Conflicts of Interest:** The authors declare no conflict of interest.

## References

1. Zhou, L.; Luo, L.Y.; Tan, C.W.; Li, Z.Y.; Song, X.G.; Zhao, H.Y.; Huang, Y.X.; Feng, J.C. Effect of Welding Speed on Microstructural Evolution and Mechanical Properties of Laser Welded-Brazed Al/Brass Dissimilar Joints. *Opt. Laser Technol.* **2018**, *98*, 234–246. [CrossRef]
2. Lu, Y.; Sage, D.D.; Fink, C.; Zhang, W. Dissimilar Metal Joining of Aluminium to Zinc-Coated Steel by Ultrasonic plus Resistance Spot Welding–Microstructure and Mechanical Properties. *Sci. Technol. Weld. Join.* **2020**, *25*, 218–227. [CrossRef]
3. Lu, Y.; Walker, L.; Kimchi, M.; Zhang, W. Microstructure and Strength of Ultrasonic Plus Resistance Spot Welded Aluminum Alloy to Coated Press Hardened Boron Steel. *Met. Mater. Trans. A* **2020**, *51*, 93–98. [CrossRef]
4. Li, Y.; Geng, S.; Zhu, Z.; Wang, Y.; Mi, G.; Jiang, P. Effects of Heat Source Configuration on the Welding Process and Joint Formation in Ultra-High Power Laser-MAG Hybrid Welding. *J. Manuf. Process.* **2022**, *77*, 40–53. [CrossRef]
5. Kaushik, P.; Dwivedi, D.K. Effect of Tool Geometry in Dissimilar Al-Steel Friction Stir Welding. *J. Manuf. Process.* **2021**, *68*, 198–208. [CrossRef]
6. Zhang, G.; Su, W.; Zhang, J.; Wei, Z. Friction Stir Brazing: A Novel Process for Fabricating Al/Steel Layered Composite and for Dissimilar Joining of Al to Steel. *Met. Mater Trans. A* **2011**, *42*, 2850–2861. [CrossRef]
7. Ge, Y.; Xia, Y. Mechanical Characterization of a Steel-Aluminum Clinched Joint under Impact Loading. *Thin-Walled Struct.* **2020**, *151*, 106759. [CrossRef]
8. Su, Y.; Hua, X.; Wu, Y. Effect of Input Current Modes on Intermetallic Layer and Mechanical Property of Aluminum–Steel Lap Joint Obtained by Gas Metal Arc Welding. *Mater. Sci. Eng. A* **2013**, *578*, 340–345. [CrossRef]
9. Lu, Y.; Mayton, E.; Song, H.; Kimchi, M.; Zhang, W. Dissimilar Metal Joining of Aluminum to Steel by Ultrasonic plus Resistance Spot Welding—Microstructure and Mechanical Properties. *Mater. Des.* **2019**, *165*, 107585. [CrossRef]
10. Yazdipour, A.; Heidarzadeh, A. Dissimilar Butt Friction Stir Welding of Al 5083-H321 and 316L Stainless Steel Alloys. *Int. J. Adv. Manuf. Technol.* **2016**, *87*, 3105–3112. [CrossRef]
11. Su, Y.; Hua, X.; Wu, Y. Influence of Alloy Elements on Microstructure and Mechanical Property of Aluminum–Steel Lap Joint Made by Gas Metal Arc Welding. *J. Mater. Process. Technol.* **2014**, *214*, 750–755. [CrossRef]
12. Wan, L.; Huang, Y. Microstructure and Mechanical Properties of Al/Steel Friction Stir Lap Weld. *Metals* **2017**, *7*, 542. [CrossRef]
13. Xia, H.; Tao, W.; Li, L.; Tan, C.; Zhang, K.; Ma, N. Effect of Laser Beam Models on Laser Welding–Brazing Al to Steel. *Opt. Laser Technol.* **2020**, *122*, 105845. [CrossRef]
14. Zhang, G.; Chen, M.; Shi, Y.; Huang, J.; Yang, F. Analysis and Modeling of the Growth of Intermetallic Compounds in Aluminum–Steel Joints. *RSC Adv.* **2017**, *7*, 37797–37805. [CrossRef]
15. Das, A.; Shome, M.; Goecke, S.-F.; De, A. Numerical Modelling of Gas Metal Arc Joining of Aluminium Alloy and Galvanised Steels in Lap Joint Configuration. *Sci. Technol. Weld. Join.* **2016**, *21*, 303–309. [CrossRef]
16. Yu, F.; Wei, Y.; Ji, Y.; Chen, L.-Q. Phase Field Modeling of Solidification Microstructure Evolution during Welding. *J. Mater. Process. Technol.* **2018**, *255*, 285–293. [CrossRef]
17. Miranda-Pérez, A.F.; Rodríguez-Vargas, B.R.; Calliari, I.; Pezzato, L. Corrosion Resistance of GMAW Duplex Stainless Steels Welds. *Materials* **2023**, *16*, 1847. [CrossRef]
18. Figner, G.; Vallant, R.; Weinberger, T.; Enzinger, N.; Schröttner, H.; Paśič, H. Friction Stir Spot Welds between Aluminium and Steel Automotive Sheets: Influence of Welding Parameters on Mechanical Properties and Microstructure. *Weld World* **2009**, *53*, R13–R23. [CrossRef]
19. Yang, M.; Ma, H.; Shen, Z.; Chen, D.; Deng, Y. Microstructure and Mechanical Properties of Al-Fe Meshing Bonding Interfaces Manufactured by Explosive Welding. *Trans. Nonferrous Met. Soc. China* **2019**, *29*, 680–691. [CrossRef]
20. Fujii, H.T.; Goto, Y.; Sato, Y.S.; Kokawa, H. Microstructure and Lap Shear Strength of the Weld Interface in Ultrasonic Welding of Al Alloy to Stainless Steel. *Scr. Mater.* **2016**, *116*, 135–138. [CrossRef]

21. Kanemaru, S.; Sasaki, T.; Sato, T.; Era, T.; Tanaka, M. Study for the Mechanism of TIG-MIG Hybrid Welding Process. *Weld World* **2015**, *59*, 261–268. [CrossRef]
22. Chen, J.; Wu, C.S.; Chen, M.A. Improvement of Welding Heat Source Models for TIG-MIG Hybrid Welding Process. *J. Manuf. Process.* **2014**, *16*, 485–493. [CrossRef]
23. Kanemaru, S.K.; Sasaki, T.; Sato, T.; Mishima, H.; Tashiro, S.; Tanaka, M. Study for the Arc Phenomena of TIG-MIG Hybrid Welding Process by 3D Numerical Analysis Model. *Querterly J. Jpn. Weld. Soc.* **2012**, *30*, 323–330. [CrossRef]
24. Cai, D.T.; Han, S.G.; Zheng, S.D.; Yan, D.J.; Luo, J.Q.; Liu, X.L.; Luo, Z.Y. Plasma-MIG Hybrid Welding Process of 5083 Marine Aluminum Alloy. In Proceedings of the Materials Science Forum, Salt Lake City, UT, USA, 23–27 October 2016.
25. Ding, M.; Liu, S.S.; Zheng, Y.; Wang, Y.C.; Li, H.; Xing, W.Q.; Yu, X.Y.; Dong, P. TIG–MIG Hybrid Welding of Ferritic Stainless Steels and Magnesium Alloys with Cu Interlayer of Different Thickness. *Mater. Des.* **2015**, *88*, 375–383. [CrossRef]
26. Jiang, H.; Liao, Y.; Gao, S.; Li, G.; Cui, J. Comparative Study on Joining Quality of Electromagnetic Driven Self-Piecing Riveting, Adhesive and Hybrid Joints for Al/Steel Structure. *Thin-Walled Struct.* **2021**, *164*, 107903. [CrossRef]

**Disclaimer/Publisher's Note:** The statements, opinions and data contained in all publications are solely those of the individual author(s) and contributor(s) and not of MDPI and/or the editor(s). MDPI and/or the editor(s) disclaim responsibility for any injury to people or property resulting from any ideas, methods, instructions or products referred to in the content.

*Article*

# Friction and Wear Behavior of NM500 Wear-Resistant Steel in Different Environmental Media

Guobo Wang [1,2], Hao Zhao [1,2], Yu Zhang [1,2], Jie Wang [1,2], Guanghui Zhao [1,2] and Lifeng Ma [1,2,*]

[1] School of Mechanical Engineering, Taiyuan University of Science and Technology, Taiyuan 030024, China; wgb0506@tyust.edu.cn (G.W.); zhaohaotg@163.com (H.Z.); zhangyu20230410@163.com (Y.Z.); wangjie0102022@163.com (J.W.); zgh030024@163.com (G.Z.)

[2] Shanxi Provincial Key Laboratory of Metallurgical Device Design Theory and Technology, Taiyuan 030024, China

* Correspondence: mlf_zgtyust@163.com

**Abstract:** The study aims to investigate the influence of environmental media on the friction and wear behavior of low-alloy wear-resistant steels and to provide practical references for their application. This article conducted sliding wear tests on NM500 wear-resistant steel under different loads under air atmosphere, deionized water, and 3.5 wt% NaCl solution conditions. Someone quantitatively measured the friction coefficient and wear amount of each friction pair. The present study employed scanning electron microscopy, energy dispersive spectroscopy, and a white light interference three-dimensional surface profiler to analyze the surface structure, cross-sectional morphology, element distribution, and wear mechanism of the wear scars under various experimental conditions. The results show that: In deionized water, NM500 has the best wear resistance, while the dry state is the worst. The lubricating and cooling effect of the liquid, as well as the corrosive effect of the NaCl solution, play an essential role in the wear behavior of NM500. Under dry friction conditions, the wear mechanism of NM500 is principally adhesive wear, fatigue wear, and oxidation wear. In the case of wear testing in deionized water, the researchers characterized the dominant wear mechanism as adhesive wear in conjunction with fatigue wear and abrasive wear. In contrast, when they carried out the wear testing in NaCl solution, the wear mechanism was primarily driven by corrosion wear and adhesive wear, with only a minor contribution from fatigue wear.

**Keywords:** NM500; wear mechanism; friction and wear; environmental media

**Citation:** Wang, G.; Zhao, H.; Zhang, Y.; Wang, J.; Zhao, G.; Ma, L. Friction and Wear Behavior of NM500 Wear-Resistant Steel in Different Environmental Media. *Crystals* **2023**, *13*, 770. https://doi.org/10.3390/cryst13050770

Received: 10 April 2023
Revised: 27 April 2023
Accepted: 4 May 2023
Published: 5 May 2023

**Copyright:** © 2023 by the authors. Licensee MDPI, Basel, Switzerland. This article is an open access article distributed under the terms and conditions of the Creative Commons Attribution (CC BY) license (https://creativecommons.org/licenses/by/4.0/).

## 1. Introduction

As industries rapidly develop, the demand for wear-resistant materials in friction and wear also increases. Hence, it is crucial to consider using products made from such materials to minimize material consumption [1]. In recent years, industrial machinery has increasingly stringent requirements for the performance of wear-resistant materials, so low-alloy high-strength wear-resistant steel with good wear-resistance properties has gradually become the focus of the application. The relevant market demand is also growing [2].

Compared with high manganese steel and high chromium cast iron, as a widely used low alloy high strength wear-resistant steel, NM500 has the advantages of low cost and easy processing [3–6]. It corresponds to the AR500 grade under the ASTM standard. Currently, the research on low alloy wear-resistant steel is more focused on the effects of different processing techniques on its tribological properties [7–9]. Wen et al. [10] studied the impact of the heat treatment process on the friction properties of low alloy wear-resistant steel. The experiment indicated that the best wear resistance was obtained when steel was tempered at 200 °C. Huang et al. [11] and Deng et al. [12] studied the friction properties of TiC particle-reinforced low alloy wear-resistant steel under different wear conditions. The experiments have shown that the wear resistance of TiC-particle-reinforced steels was proportional to the density of the TiC particles. Huang et al. [13] studied the three-body abrasive wear performance of TiC-reinforced low alloy wear-resistant steel prepared by different processes. Liu et al. [14] found that the wear resistance of low alloy martensite steels with additional Ti content will increase linearly. Kostryzhev et al. [15] found that Ti alloying enhances the friction properties of wear-resistant steel. The fracture development at the particle-matrix interface was located to govern the wear mechanism in the studied steels.

However, in practical engineering applications, the wear mechanism often varies due to differences in the working conditions encountered by materials [16–25]. Xu et al. [26] compared the wear mechanism of C17200 copper under dry conditions and 3.5% NaCl solution and found that oxidative wear disappeared in the NaCl solution. Zhang et al. [27] found that the wear mechanism of ZrO2 thin films in liquid environments increased more abrasive wear compared to dry conditions. Tribological properties are not intrinsic to the material, but the influence of the environmental medium on them is also significant [28]. Zhao et al. [29] compared the performance of stainless steel in terms of wear under various environmental conditions. They found that dry conditions have the most severe wear compared to liquid environments, and the deformation of the subsurface layer of stainless steel is the highest. It is found that the microwear performance of SAF 2507 super duplex stainless steel is related to the lubricating medium. Aqueous solutions can effectively improve the wear performance of materials. Artificial seawater has a better lubrication effect than deionized water [30]. Environmental media has a significant impact on the tribological properties of steel. In practical application, it is inevitable for wear-resistant steel to encounter diverse ecological press, such as outdoor environments characterized by high humidity levels, seawater environments, and others. Consequently, investigating the wear behavior of low alloy wear-resistant steel under varying environmental conditions holds great significance.

Therefore, this paper studied the friction and wear behavior of NM500 wear-resistant steel under a series of normal loads in different environmental media, explored the effect of ecological media on wear resistance, and analyzed the corresponding wear mechanism. At the same time, it will provide a reference for selecting and applying low-alloy wear-resistant steel materials in different environmental media.

## 2. Experimental Procedure

### 2.1. Material and Sample Preparation

The material used in the experiment is the quenched and tempered NM500 steel made by Baosteel, with a large amount of martensite structure. Its tensile strength is 1500 MPa, elongation is 8%, and Brinell hardness is 503HBW. The NM500 steel plate was processed into 20 mm × 15 mm × 8 mm rectangular blocks by Wire Electrical Discharge Machining. The composition of NM500 steel is given in Table 1. Considering the need to reduce the error caused by the inconsistency of the surface condition, the surface of the specimen was sanded with sandpaper and then polished before the experiment. Finally, the sample was cleaned with anhydrous ethanol and an ultrasonic cleaning machine and dried with a blower.

**Table 1.** Chemical composition of NM500 (wt%).

| C | Si | Mn | P | S | Cr | Ni | Mo | Ti | B | ALs |
|---|----|----|---|---|----|----|----|----|---|-----|
| 0.38 | 0.70 | 1.70 | 0.020 | 0.010 | 1.20 | 1.00 | 0.65 | 0.050 | 0.00045 | 0.010 |

### 2.2. Friction and Wear Tests

The test was performed with an American Rtec MFT 5000 tester, using a 6.35 mm diameter silicon nitride ceramic ball as the friction substrate. The hardness of silicon nitride ceramic balls is approximately 1530 HV. The testing frequency is 1 Hz. The experimental loads were selected as 50 N, 100 N, and 150 N. The friction was carried out in a linear reciprocating motion mode, as illustrated by Figure 1a. The test time was 30 min, and the reciprocating stroke was 5 mm.

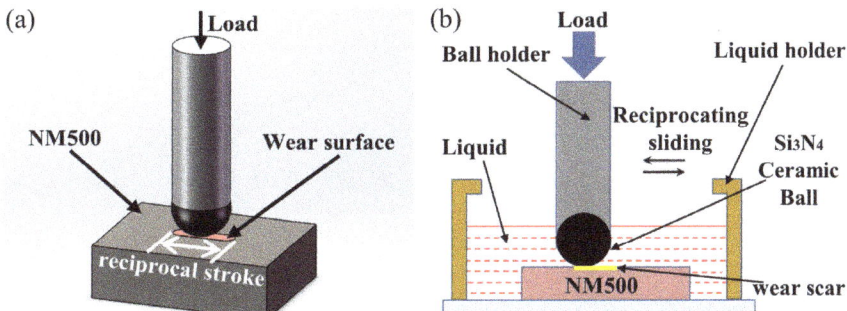

**Figure 1.** Schematic illustration of friction and wear test: (**a**) reciprocating sliding, (**b**) liquid condition.

Dry friction testing began with the stable clamping of the sample to the test platform. In the liquid medium environment, after the piece was installed, a certain amount of liquid medium was needed to inject into the liquid pool (the same amount of liquid medium was injected in each test). At the same time, the sample and the ceramic ball were always guaranteed to be immersed in the liquid, as shown in Figure 1b. After each test, the liquid medium shall be drained, and the liquid pool and test bench shall be thoroughly cleaned to avoid impacting the subsequent test results.

*2.3. Analysis Methods*

The three-dimensional topography of the worn surface was measured by white light interference 3D surface profiler (Rtec, American). The wear volume was determined using Gwyddion image analysis software (free and open-source). We performed each calculation to minimize the error and considered the average value of the final result. The wear rate was calculated by the following formula shown in Equation (1) [31]:

$$W = V/(F \cdot S) \tag{1}$$

where $W$ is the wear rate (mm$^3$/N·m), $V$ is the wear volume (mm$^3$), $F$ is the applied load (N), and $S$ is the total sliding distance (m).

The morphology, chemical composition, wear surface, and cross-section of NM500 wear-resistant steel were analyzed using a scanning electron microscope (SEM, ZEISS 300, Jena, Germany) equipped with an energy dispersive spectrometer (EDS). Before characterization, a nickel protective layer was electrodeposited on the surface of the wear mark in advance. The thickness of the nickel protective layer is about 30 μm. Then, the longitudinal section of the wear mark parallel to the sliding direction was obtained by wire cutting. The specimen sections were ground, polished, and then etched using a 4% nitric acid alcohol solution.

## 3. Results and Discussion

*3.1. Friction Coefficient*

Figure 2a shows the variation curve of the friction coefficient of NM500 with sliding time under different loads under dry friction conditions. It has prominent stage change characteristics, roughly divided into three stages: initial running-in stage, climbing stage, and stable wear stage. In the initial running-in stage, the micro-convex body on the friction surface contacts first, causing adhesion and plastic deformation. Because the contact area is small, the contact stress is significant, and the micro convex body is severely worn, the friction coefficient rises sharply. When the friction coefficient reaches the peak, the surface becomes relatively smooth. Additionally, some abrasive particles act similarly to ball bearings to reduce friction, decreasing the friction coefficient. This friction coefficient curve rising to the peak and then falling is one of the most common friction coefficient curves in adhesive wear experiments. During the climbing stage, the effective contact area between the surface increases as the wear progresses, leading to an increase in wear particle generation and a consequent intensification of abrasive particle wear. Additionally, the generation of frictional heat induces an oxidation reaction, resulting in a micro-softening of the surface—the adhesion ability of the friction surface increases. The friction coefficient stops decreasing and starts climbing instead. When the generation and overflow of the debris reach a comparatively balanced state, under the overall impact of the softening effect of friction heat, the protection effect of oxidized surface, and the hardening work effect caused by plastic deformation, the friction curve enters a stable wear stage.

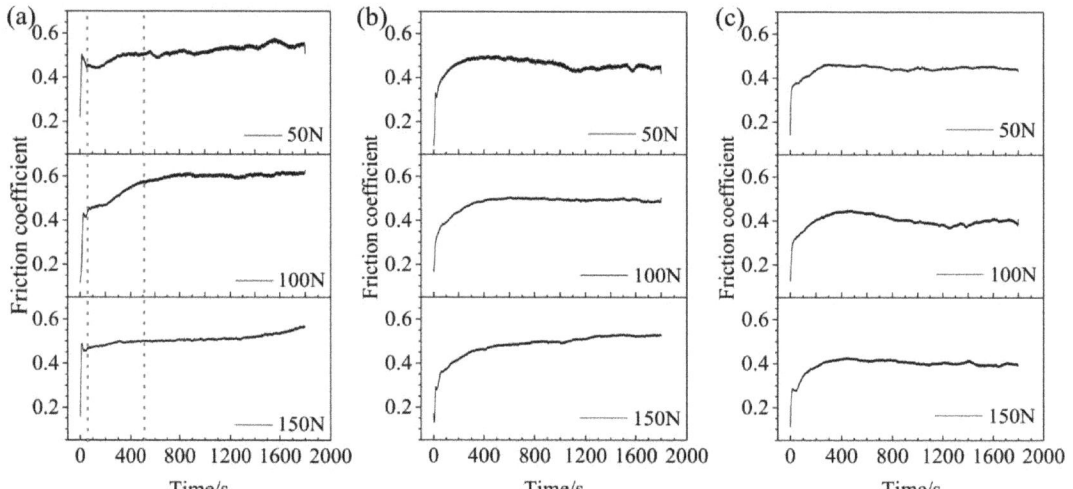

**Figure 2.** Friction coefficient of the NM500 wear-resistant steel at 50, 100, and 150 N in different conditions: (**a**) dry condition, (**b**) deionized water, (**c**) and NaCl solution.

Figure 2b,c show the variation curves of the friction coefficient with sliding time for NM500 wear-resistant steel in the liquid medium environment under different loads. Compared with the curve under dry friction conditions, it is smoother, and the peak value of the curve is smaller or has no noticeable peak change. The alteration in the contact state between the ball and surface due to the presence of a liquid medium causes a lubricating effect. At the same time, the liquid reduces the effect of frictional heat, mitigates adhesive wear, leads to a lower coefficient of friction, and weakens the peak variation.

Under dry friction, the friction coefficient tends to rise and then fall as the average load increases. When the load is small, the contact area is small and abrasive particles gather in the contact space. The oxidation of the surface and the abrasive particles are explicit in friction reduction, so the friction coefficient is small. When the load is 100 N, the oxidation and plastic deformation of the material deepens, and the contact area increases, promoting the generation of cracks and abrasive particles, and the wear increases. The surface becomes rough, and the friction coefficient increases. When the load is 150 N, the increase in load promotes the generation of friction heat, and the temperature of the surface rises, prompting the formation of the oxide film to accelerate so that the oxide film coverage protects the material surface. Due to the lubricating effect of the oxide film, the friction coefficient decreases.

In deionized water, the friction resistance decreases because of the liquid's lubricating influence. The friction coefficient is smaller than that underneath dry friction, and it tends to increase with the increase in average load. Abrasive particles are easily cleaned by liquid, making the friction surface smoother. At the same time, the cooling effect of the liquid can effectively reduce the impact of frictional heat and reduce the occurrence of adhesion so that the friction coefficient is reduced. Furthermore, the liquid medium can inhibit the oxidation reaction and weaken the oxidation protection of the surface. With the increase of load, the contact area increases, the formation of the liquid lubricating film becomes difficult, and the lubrication effect of deionized water gradually decreases; with the increase of load, there is a tendency for the friction coefficient to increase.

In NaCl solution, the friction coefficient is smaller than that in deionized water, and its variation trend decreases gradually with the load increase. This is because NaCl solution contains active chloride ions that are easy to cause pitting corrosion, which can promote the electrochemical reaction on the surface of the material to form a layer of corrosion product film that is easy to shear, play the role of lubrication protection, and further reduce the friction coefficient. Under the double influence of the lubrication protection and corrosion of NaCl solution, the adhesion ability of the material surface is further weakened. During sliding, the ratio of load growth is greater than the product of the ratio of shear force and contact area increase, so the findings suggest that an increase in load results in a propensity for a reduction in the friction coefficient.

### 3.2. Surface Profiles, Wear Volume, and Wear Rate

Figure 3 shows the three-dimensional morphology of wear marks under different conditions when the load is 100 N and the cross-section curve of wear marks. Under the dry friction condition, the cross-sectional curves of the wear marks are "V" shaped, while in the liquid medium, they are "U" and "W" shaped. When experiencing dry friction, the abrasion marks exhibit their greatest width and depth, and the width and depth of the abrasion marks in NaCl solution are the second largest. In deionized water, the breadth and depth of the abrasion marks are the smallest. Based on the above different macroscopic morphologies, it is generally accepted that the environmental medium significantly affects the wear mechanism.

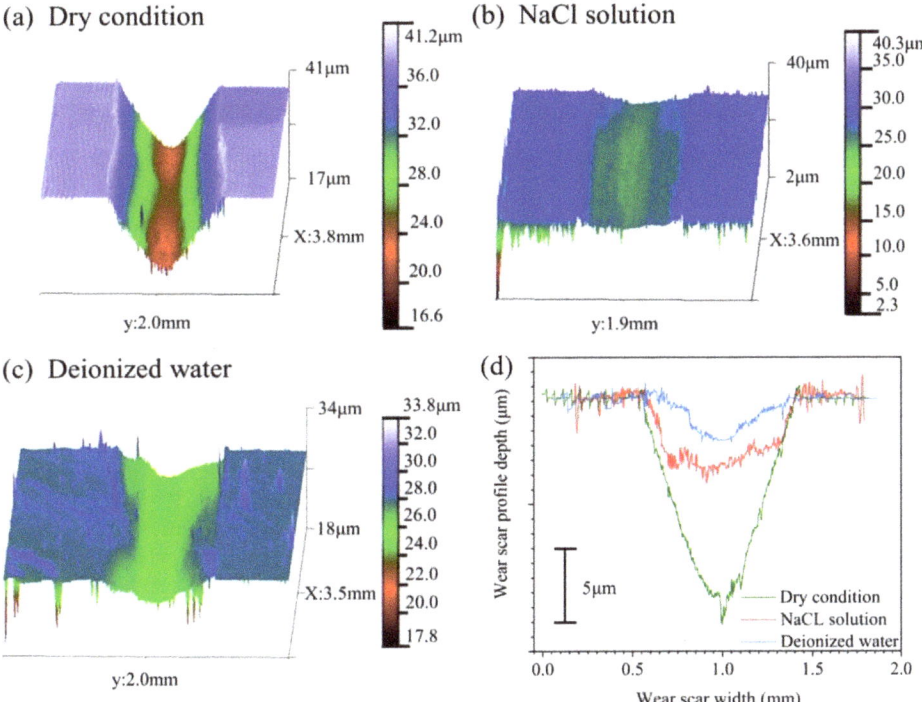

**Figure 3.** 3D morphology (**a**–**c**) and wear scar cross-section curve (**d**) of the wear surfaces generated in different conditions (100 N).

Figure 4a,b show the variation of wear volume and wear rate with load for NM500 in different environmental media, respectively. Wear-resistant steel's wear volume and wear rate are significantly higher under dry friction conditions than in liquid media. In contrast, the wear volume and wear rate in NaCl solution are always more significant than those in deionized water. For example, when the load is 50 N, the wear volume in the air is about 5.17 times the wear volume when exposed to a NaCl solution and seven times the wear volume in deionized water. The wear rate of dry friction is about 5.14 times the wear rate in NaCl solution and seven times in deionized water. The reason for this is the lubricating and cooling properties of the liquid medium, which reduces the surface wear. In addition, the corrosion caused by NaCl solution is relatively weak, and the resulting wear leads to a product film that exhibits robust corrosion resistance and lubricating properties. Therefore, the wear of NM500 is far less severe than its wear in the air. The increase in dislocation density and surface defects is attributed to wear-induced effects on the material surface, the physicochemical activity increases, and there is an increase in the efficiency of corrosion; the surface of the corroded material is loose and porous, which is easily rubbed off by abrasives or cleaned off by liquids, thus aggravating the wear [32].

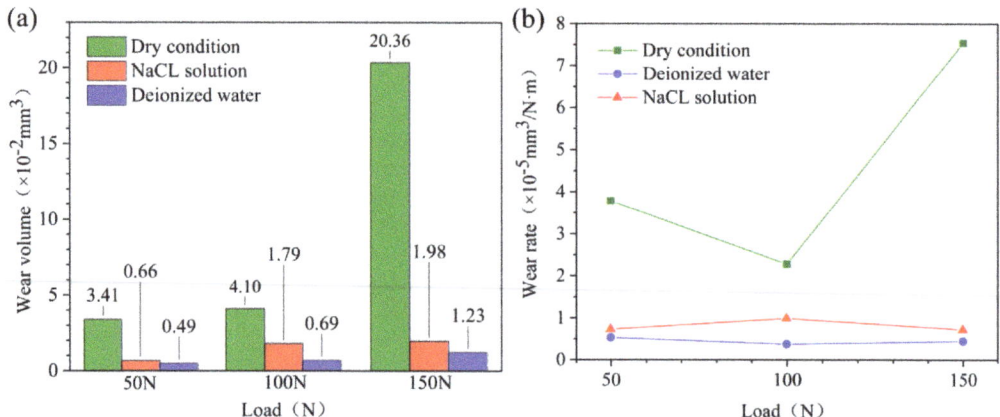

**Figure 4.** Wear results of the NM500 wear−resistant steel at various loads with the corresponding environment: (**a**) wear volume and (**b**) wear rate.

Within the same environmental medium, the wear volume increases with the load. The wear rate exhibits a trend of first decreasing and then increasing with the increase in load under both dry friction and deionized water conditions. According to the classical Archard's law of wear, the amount of wear is inversely proportional to the hardness of the softer material in the friction pair [33–36]. Therefore, at a load of 100 N, the work-hardening effect caused by plastic deformation continuously increases the surface hardness of NM500. The wear volume of dry friction increases only slightly by 20.2%, and the wear rate decreases by $1.51 \times 10^{-5}$ mm$^3$/(N·m); the wear volume in deionized water increases by 40.8%, while the wear rate decreases by $0.16 \times 10^{-5}$ mm$^3$/(N·m). At the load of 150 N, the wear surface softens, and adhesive wear intensifies under the influence of the temperature distribution in wear-resistant steel, which is characterized by the surface temperature and the temperature gradient in the depth direction. So the wear volume under dry friction conditions showed a significant increase of 396.6%, and the wear rate increased by $5.26 \times 10^{-5}$ mm$^3$/(N·m); the wear volume in deionized water increased by 78.3%, and the wear rate increased by $0.07 \times 10^{-5}$ mm$^3$/(N·m). The wear rate of dry friction reaches a maximum value of $7.54 \times 10^{-5}$ mm$^3$/(N·m) at 150 N, and the wear rate reaches a minimum value of $0.38 \times 10^{-5}$ mm$^3$/(N·m) in deionized water at 100 N.

In NaCl solution, the wear rate increases first and then decreases with the load increase. When the load increases from 50 N to 100 N, the wear volume increases significantly by 171.2%, and the wear rate increases by $0.26 \times 10^{-5}$ mm$^3$/(N·m) due to the promotion of wear by the corrosive effect of the solution. When the load increases to 150 N, the wear volume only increases by 10.6%, and the wear rate decreases by $0.26 \times 10^{-5}$ mm$^3$/(N·m) due to the protection of the corrosion product film. The results show that under the action of 100 N load, the wear rate of wear-resistant steel in deionized water reaches the minimum value of $0.38 \times 10^{-5}$ mm$^3$/(N·m), with the best wear resistance. Thus, it seems that the environmental medium has a more significant influence on the wear resistance of NM500 wear-resistant steel.

### 3.3. Morphologies of the Wear Surfaces

Figure 5a–c show the wear surface morphology of NM500 under dry friction conditions. It can be observed that there are peeling pits, cracks, and dark areas of oxidation on each surface. Figure 6 shows the element distribution of the wear surface under dry friction. As the observed data suggests, there is a considerable deposition of oxygen and silicon on the worn surface. Additionally, the distribution of oxygen elements coincides with the dark area in the electronic image, indicating oxidation reactions on the wear surface. The distribution of oxygen elements decreases and then increases with increasing load, indicating the presence of oxidative wear; silicon elements suggest the presence of material migration between the wear-resistant steel and the Si3N4 ceramic ball, which is a typical characteristic of adhesive wear.

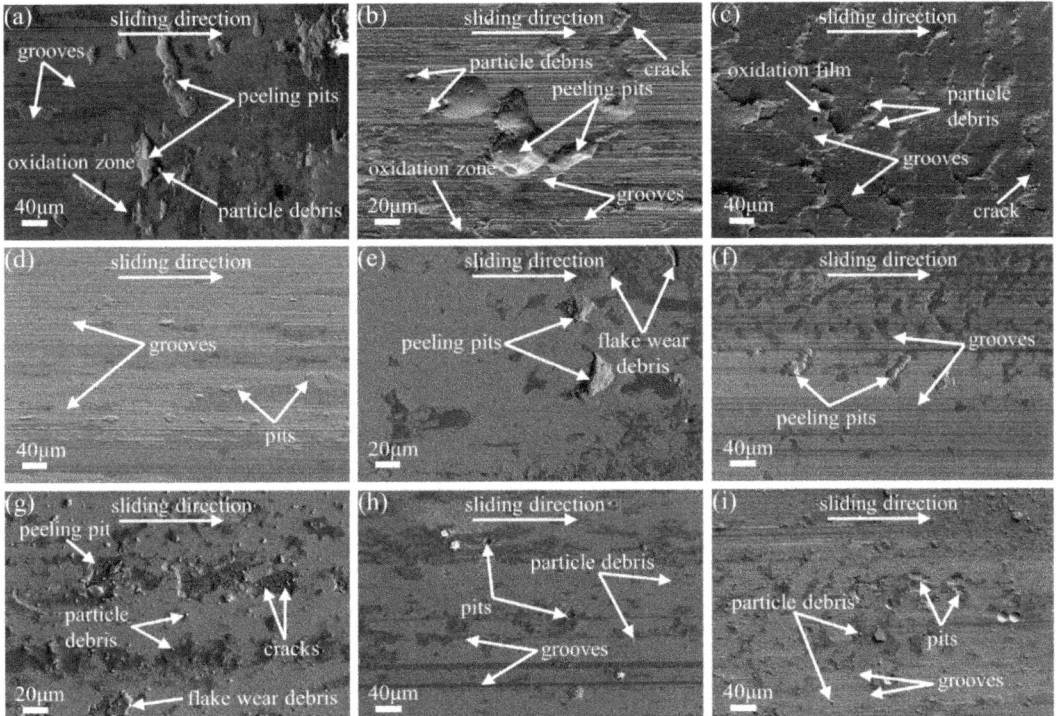

**Figure 5.** Scanning electron microscopy images of the wear surfaces: (**a**) dry condition-50 N, (**b**) dry condition-100 N, (**c**) dry condition-150 N, (**d**) deionized water-50 N, (**e**) deionized water-100 N, (**f**) deionized water-150 N, (**g**) NaCl solution-50 N, (**h**) NaCl solution-100 N, and (**i**) NaCl solution-150 N.

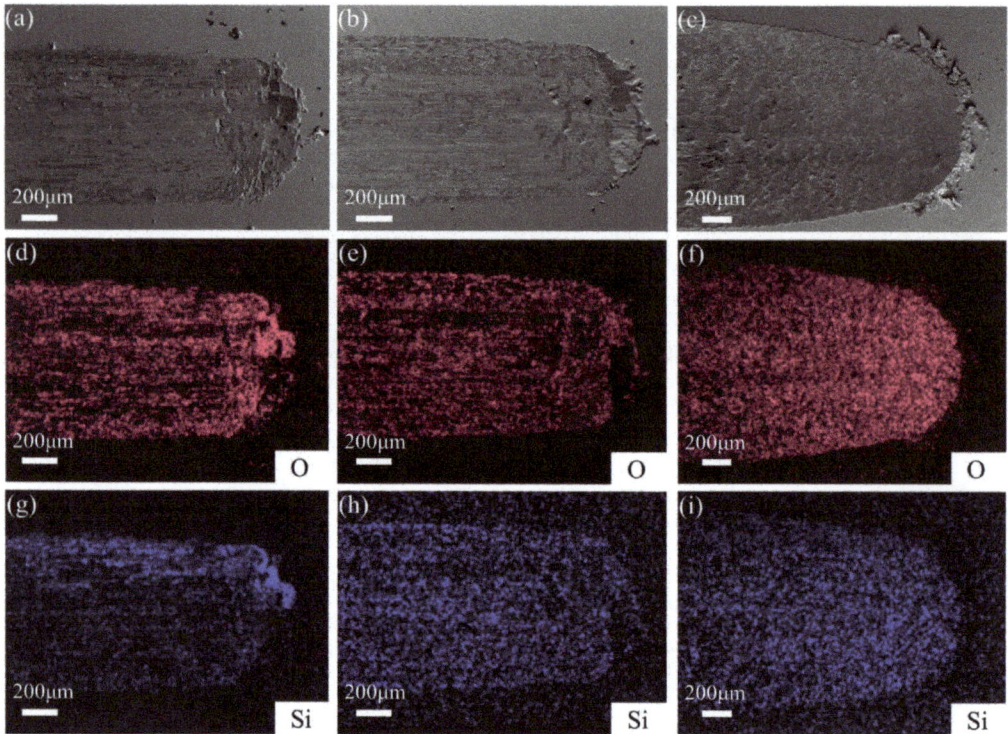

**Figure 6.** Elemental mapping of O, Si on the wear surfaces of the NM500 wear-resistant steel in dry conditions for 50 N (**a–c**), 100 N (**d–f**), and 150 N (**g–i**).

In Figure 5a, peeling pits with different depths and areas are distributed on the surface, which reflects the fatigue wear mechanism. The delicate and shallow grooves parallel to the slippery direction indicate the presence of weaker abrasive wear. With the load increase, the freshly exposed surface resulting from the detachment of the oxidized area expands accordingly. The oxidative wear accelerates, decreasing the spread of oxygen elements in Figure 6. Particle debris around the peeling pit increase, and abrasive wear increases, which makes the grooves in Figure 5b obvious, and the surface roughness increases. When the load is 150 N, the wear surface in Figure 5c is covered with a layer of scale-like oxide film formed by the rolling and fusion of flake-peeling wear debris under the load. Under alternating contact stress, some lamella edges become warped, and apparent cracks are produced. The surface undergoes three-body abrasive wear as abrasive particles disperse and take part in the process, which makes groove marks more noticeable. The main wear mechanisms underneath dry friction are fatigue, adhesive, and oxidation wear.

Figure 5d–i show the wear surface morphology in the liquid medium, which is smoother than the wear surface of dry friction. The liquid's ability to lubricate and cool enhances the anti-fatigue and anti-adhesion ability of the character. Therefore, the size and number of peeling pits on the material surface are significantly reduced, the degree of plastic deformation is reduced, and the parallel grooves caused by abrasive wear are shallower and finer than those under dry friction. Figure 7 shows the element distribution of the wear surface under liquid media. The wear surface is covered with a certain amount of silicon elements, and the migration between materials indicates the presence of adhesive wear.

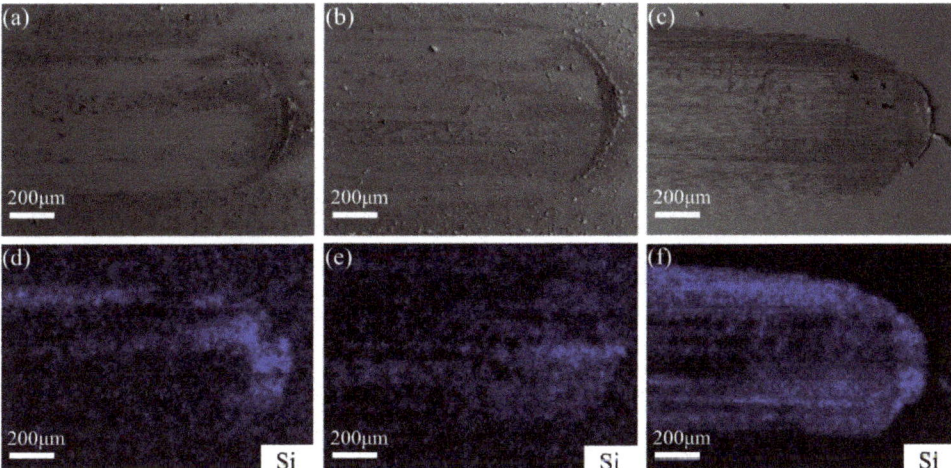

**Figure 7.** Elemental mapping of Si on the wear surfaces of the NM500 wear-resistant steel in liquid condition: (**a**) NaCl solution-50 N, (**b**) NaCl solution-100 N, (**c**) deionized water-150 N, (**d**) NaCl solution-50 N, (**e**) NaCl solution-100 N, and (**f**) deionized water-150 N.

In addition to adhesive wear, the wear mechanisms in deionized water also include fatigue wear and abrasive wear. In Figure 5d, the surface has scattered pitting pits and small shallow peeling pits, a typical feature of fatigue wear. During the wear process, fine and external groove marks were left by the wear debris, indicating the existence of abrasive wear. In Figure 5e, the surface shows a reduction in the peeling pits, reducing fatigue wear. The groove marks become inconspicuous. The weakening of abrasive wear has occurred. When we increase the load to 150 N, under the softening effect caused by frictional heat, the surface's adhesive and abrasive wear in Figure 5f is aggravated. Therefore, the parallel groove widens and becomes more profound, the surface roughness increases, the friction coefficient and wear rate rise, and the peeling pits decrease.

In NaCl solution, the wear mechanism includes adhesive wear, fatigue wear, and corrosion wear. As shown in Figure 5, the wear surface in NaCl solution is the smoothest compared with other environmental conditions. The wear surface in Figure 5g is scattered with flaky and granular abrasive debris, and shallow corrosion peeling pits and cracks mark the existence of fatigue wear and corrosion wear. The crack network in the peeling pit is caused by the corrosion of the solution, which accelerates the crack propagation. With the increase of load, the fatigue wear is weakened under the dual influence of the hardening work effect and lubrication protection, and the dimples and breaks on the wear surface in Figure 5h are reduced. At the same time, the abrasive chips are changed into fine particles by repeated grinding and corrosion, which polishes the surface and accelerates the wear of NM500. Moreover, the particle wear debris leaves extremely fine and shallow groove marks, the surface roughness decreases, and the friction coefficient decreases. When the load is increased to 150 N, there are fewer pits and cracks on the wear surface in Figure 5i, and the parallel grooves become broader and more profound. At the same time, much wear debris is gathered and compacted under load to form a layer of corrosion product film, and the friction coefficient and wear rate decrease significantly.

### 3.4. Cross-Sectional Morphologies of the Wear and Tear Surfaces

Figure 8 shows the cross-sectional morphology of wear-resistant steel's wear and tear surface under entirely different conditions. It is seen from the image that the morphology of the cross-section presents a layered distribution. The cross-section is composed of three parts: the reactants of the abrasive chips and environmental media are subject to a combination of mechanical mixing, chemical and thermal effects, resulting in the formation of a mechanically mixed layer (MML) [37]; the plastic deformation layer (PDL) formed due to the plastic strain caused by shear and positive force applied on the area of contact where friction occurs; and the whole substrate part.

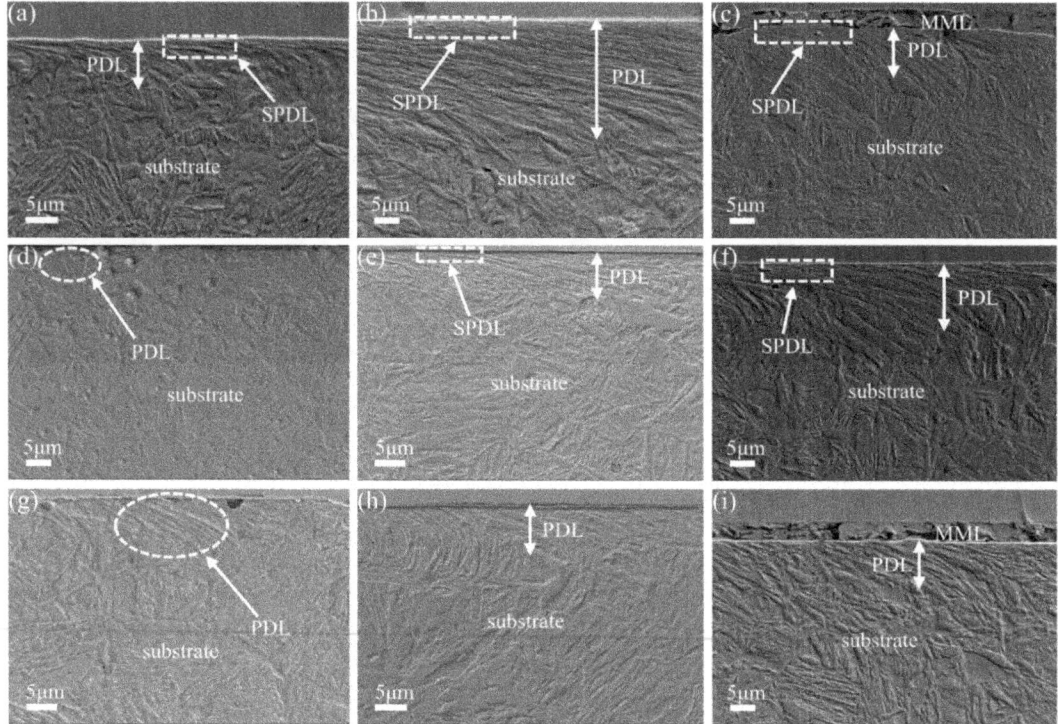

**Figure 8.** Cross-sectional morphology of the wear surfaces of the NM500 steel under different conditions: (**a**) dry condition-50 N, (**b**) dry condition-100 N, (**c**) dry condition-150 N, (**d**) deionized water-50 N, (**e**) deionized water-100 N, (**f**) deionized water-150 N, (**g**) NaCl solution-50 N, (**h**) NaCl solution-100 N, and (**i**) NaCl solution-150 N. (MML: mechanically mixed layer; PDL: plastic deformation layer; SPDL: severe plastic deformation layer).

Work hardening and grain refinement caused by plastic deformation can increase the surface hardness of materials and reduce the wear rate. Martensite that when subjected to significant plastic deformation, the material near the surface will experience elongation and become highly refined, forming a filamentous structure strain-induced surface layer that is narrowly arranged and in alignment with the worn surface [38]. The layer is also known as the layer of severe plastic deformation (SPDL layer), which can enhance the materials' resistance to wear.

In the process of sliding wear, the formation of a mechanically mixed layer is related to the structural composition of the material, wear average load, strain degree, and other factors [38]. Not all conditions can produce a mechanically mixed layer. In this study, a mechanically mixed layer with a thickness of about 3.2~6.2 μm composed of abrasive chips and oxides is found in Figure 8c. A mechanically mixed layer with a thickness of about 3.6 μm composed of corrosion products is found in Figure 8i. Under other conditions, the mechanically mixed layer can hardly be observed. The mechanically mixed layer can play the role of lubrication protection.

Figure 9 shows the variation of NM500 plastic deformation layer depth with load in different environmental media. Under the condition of dry friction, the variation of plastic deformation layer depth is similar to that of the friction coefficient, which shows a trend of first rising and then declining. As the load increases from 50 N to 100 N, the degree of plastic deformation deepens, and the depth of the plastic deformation layer increases by 17.7 µm. The depth of the SPDL increases from 1.7 µm in Figure 8a to 3.1 µm in Figure 8b. The surface hardness of the material increases. When the load is 150 N, a mechanically mixed layer is formed on the wear surface, which reduces the degree of plastic deformation, so the depth of the plastic deformation layer decreases by 15.2 µm. The depth of the SPDL in Figure 8c decreases to 2.3 µm.

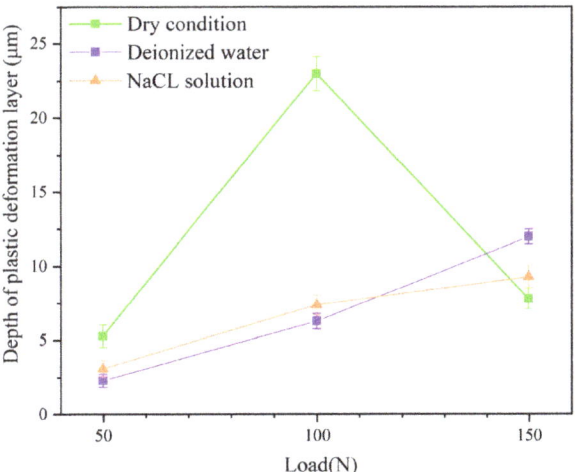

**Figure 9.** Variation of plastic deformation layer depth of the NM500 wear-resistant steel at various loads with the corresponding environment.

In the liquid medium, the depth of the PDL gradually deepens with the increase of load. When there is no mechanically mixed layer, the thickness of the plastic deformation layer in a liquid medium is smaller than that in air. For example, at the load of 50 N, the plastic deformation layer depth of dry friction is about 1.71 times that of NaCl solution and 2.3 times that of deionized water. This results from the liquid's ability to lubricate and provide cooling, which weakens the adhesion ability of the friction surface and the influence of the temperature gradient along the depth direction. When the load is 150 N, the mechanically mixed layer appears under the dry friction condition, and the depth of the plastic deformation layer decreases significantly. The depth of the plastic deformation layer in the liquid medium exceeds that under the dry friction condition.

In deionized water, the plastic deformation of the contact surface layer is not apparent when the load is small, and the plastic deformation region can be observed only at the local location in Figure 8d. When the shipment is 100 N, the depth of the plastic deformation layer increases by 4µm. Meanwhile, SPDL with a depth of 1 µm appears in Figure 8e, further reducing the wear rate. When the load is increased to 150 N, the plastic deformation layer depth increases by 5.7 µm due to the influence of surface friction heat, and the growth rate of the depth value increases by 42.5%. In Figure 8f, the depth of the SPDL increases to 2 µm, indicating significant wear of the surface.

As shown in Figure 8, the NaCl solution has no obvious SPDL. This could be attributed to the combined impact of corrosion and abrasion, working in tandem, which accelerates the loss of materials and prevents the formation of SPDL on the surface. Therefore, the depths of plastic deformation layers in NaCl solution are all greater than those in deionized water when no mechanically mixed layer appears. At 50 N load, as in deionized water, only local plastic deformation can be observed in Figure 8g. With the increase in burden, the depth of the plastic deformation layer increases by 4.3 µm. The growth rate of depth value is similar to that of deionized water. When the load reaches 150 N, a mechanical mixing layer is formed on the surface, and the depth of the plastic deformation layer only increases by 1.9 μm. The growth rate of depth value decreases by 55.8%.

In the air, when the wear is light, adhesive wear is the primary mechanism, with fatigue wear and oxidation wear also being present; when the wear reaches a critical level, an oxide film is formed on the wear surface, and the predominant wear mechanism comprises of adhesive and fatigue wear. At the same time, some degree of oxidation wear is also observed. A liquid medium readily generates a lubricating film that weakens surface adhesion and thermal softening effects, impeding oxidation reactions. In deionized water, the wear mechanism is characterized by adhesive wear with some degree of fatigue and abrasive wear. In NaCl solution, adhesive and corrosion wear is the predominant mechanisms with some accompanying fatigue wear.

## 4. Conclusions

The researchers systematically investigated the friction and wear behavior of NM500 steel under sliding wear tests in air atmosphere, deionized water, and 3.5wt% NaCl solution conditions. The following conclusions can be drawn:

1. Under dry friction conditions, the friction coefficient and wear rate of NM500 are much higher than those under other conditions. The maximum friction coefficient of 0.6 can be obtained at 100 N, and the maximum wear rate of $7.54 \times 10^{-5}$ mm$^3$/(N·m) is received at 150 N. In the liquid medium environment, NM500 wear-resistant steel in NaCl solution has the lowest friction coefficient and obtains the minimum value of 0.39 at 150 N; in deionized water, wear-resistant steel has the lowest wear rate and brings the minimum value of $0.38 \times 10^{-5}$ mm$^3$/(N·m) at 100 N. Therefore, the wear resistance of NM500 steel is the best in deionized water (100 N) and the worst in dry friction;

2. Under dry friction conditions, the wear mechanism of NM500 steel is mainly adhesive wear, fatigue wear, and oxidation wear. The wear process in deionized water is dominated by adhesive wear as the primary mechanism, accompanied by some degree of fatigue wear and abrasive wear as secondary mechanisms. The wear mechanism prevailing in the NaCl solution is predominantly ascribed to corrosion and adhesive wear, with a small amount of fatigue wear;

3. When there is no mechanically mixed layer, the magnitude of the plastic deformation layer's thickness in dry friction is about 2~3 times that in the liquid environment under the same load. This is because the lubrication and cooling action of liquid affects the work hardening and the surface's tendency to undergo thermal softening and also causes the reduction of the friction coefficient and wear rate. In addition, the corrosion of the NaCl solution is the main reason for the lowest friction coefficient and higher wear rate of wear-resistant steel.

**Author Contributions:** Conceptualization, G.W.; analysis, G.W.; investigation, H.Z.; data curation, Y.Z.; writing—original draft preparation, G.W.; writing—review and editing, G.W. and G.Z.; supervision, J.W. and Y.Z.; funding acquisition, L.M. and G.Z. All authors have read and agreed to the published version of the manuscript.

**Funding:** This work was supported by the National Natural Science Foundation of China (U1910213), the Fundamental Research Program of Shanxi Province (20210302123207 and 20210302124009), Scientific and Technological Innovation Programs of Higher Education Institutions in Shanxi (2021L292), Taiyuan University of Science and Technology Scientific Research Initial Funding (20212026).

**Data Availability Statement:** Not applicable.

**Acknowledgments:** The authors would like to thank Shanxi Provincial Key Laboratory of Metallurgical Device Design Theory and Technology for providing financial support for this research.

**Conflicts of Interest:** The authors declare no conflict of interest.

## References

1. Wei, S.; Xu, L. Review on research progress of steel and iron wear-resistant materials. *Acta Metall. Sin.* **2020**, *56*, 523–538.
2. Deng, X.; Wang, Z.; Han, Y.; Zhao, H.; Wang, G. Microstructure and Abrasive Wear Behavior of Medium Carbon Low Alloy Martensitic Abrasion Resistant Steel. *J. Iron Steel Res. Int.* **2014**, *21*, 98–103. [CrossRef]
3. Xue, H.; Zhang, Y.; Zhu, M.; Yin, X.; Zhang, W.; Liu, S. The size effect of martensite laths and precipitates on high strength wear-resistant steels. *Mater. Res. Express.* **2021**, *8*, 126528. [CrossRef]
4. Konat, Ł.; Zemlik, M.; Jasiński, R.; Grygier, D. Austenite Grain Growth Analysis in a Welded Joint of High-Strength Martensitic Abrasion-Resistant Steel Hardox 450. *Materials* **2021**, *14*, 2850. [CrossRef] [PubMed]
5. Hietala, M.; Ali, M.; Khosravifard, A.; Keskitalo, M.; Järvenpää, A.; Hamada, A. Optimization of the tensile-shear strength of laser-welded lap joints of ultra-high strength abrasion resistance steel. *JMR&T* **2021**, *11*, 1434–1442.
6. Kaijalainen, A.; Hautamäki, I.; Kesti, V.; Pikkarainen, T.; Tervo, H.; Mehtonen, S.; Porter, D.; Kömi, J. The Influence of Microstructure on the Bendability of Direct Quenched Wear Resistant Steel. *Steel Res. Int.* **2019**, *90*, 1900059. [CrossRef]
7. Damião, C.A.; Alcarria, G.C.; Teles, V.C.; de Mello, J.D.B.; da Silva, W.M., Jr. Influence of metallurgical texture on the abrasive wear of hot-rolled wear resistant carbon steels. *Wear* **2019**, *426–427*, 101–111. [CrossRef]
8. Bratkovsky, E.V.; Shapovalov, A.N.; Dema, R.R.; Kharchenko, M.V.; Platov, S.I.; Rubanik, V.V. Evaluation Method of Impact and Abrasive Steel Resistance. *J. Frict. Wear* **2019**, *40*, 133–138. [CrossRef]
9. Naseema, S.; Gaurav, B.; Das, G.; Srivatsava, V.C.; Rao, K.S.; Krishna, K.G. TEM Analysis of Precipitates in a Low Alloy Wear Resistant Steel and Correlating with its Corrosion Behaviour. *Adv. Mater. Res.* **2018**, *1148*, 77–81. [CrossRef]
10. Erding, W.; Song, R.; Xiong, W. Effect of Tempering Temperature on Microstructures and Wear Behavior of a 500 HB Grade Wear-Resistant Steel. *Metals* **2019**, *9*, 45.
11. Huang, L.; Deng, X.; Li, C.; Jia, Y.; Wang, Q.; Wang, Z. Effect of TiC particles on three-body abrasive wear behaviour of low alloy abrasion-resistant steel. *Wear* **2019**, *434–435*, 202971. [CrossRef]
12. Deng, X.; Huang, L.; Wang, Q.; Fu, T.; Wang, Z. Three-body abrasion wear resistance of TiC-reinforced low-alloy abrasion-resistant martensitic steel under dry and wet sand conditions. *Wear* **2020**, *452–453*, 203310. [CrossRef]
13. Huang, L.; Deng, X.; Wang, Q.; Wang, Z. Microstructure, Mechanical Properties and Wear Resistance of Low Alloy Abrasion Resistant Martensitic Steel Reinforced with TiC Particles. *ISIJ Int.* **2020**, *60*, 2586–2595. [CrossRef]
14. Liu, L.; Liang, X.; Liu, J.; Sun, X. Precipitation Process of TiC in Low Alloy Martensitic Steel and Its Effect on Wear Resistance. *ISIJ Int.* **2020**, *60*, 168–174. [CrossRef]
15. Kostryzhev, A.G.; Killmore, C.R.; Yu, D.; Pereloma, E.V. Martensitic wear resistant steels alloyed with titanium. *Wear* **2020**, *446–447*, 203203. [CrossRef]
16. Ayyagari, A.; Barthelemy, C.; Gwalani, B.; Banerjee, R.; Scharf, T.W.; Mukherjee, S. Reciprocating sliding wear behavior of high entropy alloys in dry and marine environments. *Mater. Chem. Phys.* **2018**, *210*, 162–169. [CrossRef]
17. Wieczorek, A.N.; Jonczy, I.; Bała, P.; Stankiewicz, K.; Staszuk, M. Testing the Wear Mechanisms of the Components of Machines Used in Fossil Energy Resource Extraction. *Energies* **2021**, *14*, 2125. [CrossRef]
18. Lindroos, M.; Apostol, M.; Kuokkala, V.T.; Laukkanen, A.; Valtonen, K.; Holmberg, K.; Oja, O. Experimental study on the behavior of wear resistant steels under high velocity single particle impacts. *Int. J. Impact Eng.* **2015**, *78*, 114–127. [CrossRef]
19. Rastegar, V.; Karimi, A. Surface and subsurface deformation of wear-resistant steels exposed to impact wear. *J. Mater. Eng. Perform.* **2014**, *23*, 927–936. [CrossRef]
20. Valtonen, K.; Keltamäki, K.; Kuokkala, V.T. High-stress abrasion of wear resistant steels in the cutting edges of loader buckets. *Tribol. Int.* **2018**, *119*, 707–720. [CrossRef]
21. Ojala, N.; Valtonen, K.; Antikainen, A.; Kemppainen, A.; Minkkinen, J.; Oja, O.; Kuokkala, V.-T. Wear performance of quenched wear resistant steels in abrasive slurry erosion. *Wear* **2016**, *354–355*, 21–31. [CrossRef]
22. Magnol, R.V.; Gatti, T.; Romero, M.C.; Sinatora, A.; Scandian, C. Liquid media effect on the abrasion response of WC/Co hardmetal with different cobalt percent. *Wear* **2021**, *477*, 203815. [CrossRef]
23. Wieczorek, A.N.; Jonczy, I.; Filipowicz, K.; Kuczaj, M.; Pawlikowski, A.; Łukowiec, D.; Staszuk, M.; Gerle, A. Study of the Impact of Coals and Claystones on Wear-Resistant Steels. *Materials* **2023**, *16*, 2136. [CrossRef] [PubMed]
24. Laukkanen, A.; Lindgren, M.; Andersson, T.; Pinomaa, T.; Lindroos, M. Development and validation of coupled erosion-corrosion model for wear resistant steels in environments with varying pH. *Tribol. Int.* **2020**, *151*, 106534. [CrossRef]
25. Kalácska, Á.; De Baets, P.; Hamouda, H.B.; Theuwissen, K.; Sukumaran, J. Tribological investigation of abrasion resistant steels with martensitic and retained austenitic microstructure in single- and multi–asperity contact. *Wear* **2021**, *482–483*, 203980. [CrossRef]
26. Xu, Z.; Huang, Z.; Wang, Y.; Lin, C.; Xu, X. Friction and Wear Behavior of C17200 Copper-Beryllium Alloy in Dry and Wet Environments. *J. Mater Eng. Perform.* **2021**, *30*, 7542–7551. [CrossRef]
27. Zhang, Z.; Ji, G.; Shi, Z. Tribological properties of ZrO2 nanofilms coated on stainless steel in a 5% NaCl solution, distilled water and a dry environment. *Surf. Coat. Technol.* **2018**, *350*, 128–135. [CrossRef]
28. Liu, Y.; Ma, S.; Gao, M.C.; Zhang, C.; Zhang, T.; Yang, H.; Wang, Z.; Qiao, J. Tribological properties of AlCrCuFeNi 2 high-entropy alloy in different conditions. *Metall. Mater. Trans. A* **2016**, *47*, 3312–3321. [CrossRef]
29. Zhao, G.; Li, J.; Zhang, R.; Li, H.; Li, J.; Ma, L. Wear behavior of copper containing antibacterial stainless steel in different environmental media and EBSD analysis of its sub surface structure. *Mater. Charact.* **2023**, *197*, 112690. [CrossRef]

30. Wang, M.; Wang, Y.; Wang, J.; Fan, N.; Yan, F. Effect of Heat Treatment Temperature and Lubricating Conditions on the Fretting Wear Behavior of SAF 2507 Super Duplex Stainless Steel. *ASME J. Tribol.* **2019**, *141*, 101601. [CrossRef]
31. Greer, A.L.; Rutherford, K.L.; Hutchings, I.M. Wear resistance of amorphous alloys and related materials. *Int. Mater. Rev.* **2002**, *47*, 87–112. [CrossRef]
32. Li, X.; Zhou, Y.; Cao, H.; Li, Y.; Wang, L.; Wang, S. Wear behavior and mechanism of H13 steel in different environmental media. *J. Mater. Eng. Perform.* **2016**, *25*, 4134–4144. [CrossRef]
33. Varenberg, M. Adjusting for running-in: Extension of the Archard wear equation. *Tribol. Lett.* **2022**, *70*, 59. [CrossRef]
34. Yilmaz, N.G.; Goktan, R.M.; Onargan, T. Correlative relations between three-body abrasion wear resistance and petrographic properties of selected granites used as floor coverings. *Wear* **2016**, *372–373*, 197–207. [CrossRef]
35. Verbeek, H.J. Tribological systems and wear factors. *Wear* **1979**, *56*, 81–92. [CrossRef]
36. Archard, J.F. Contact and rubbing of flat surfaces. *J. Appl. Sci.* **1953**, *24*, 981–988. [CrossRef]
37. Kapoor, A.; Franklin, F.J. Tribological layers and the wear of ductile materials. *Wear* **2000**, *245*, 204–215. [CrossRef]
38. Cao, Y.; Yin, C.; Liang, Y.; Tang, S. Lowering the coefficient of martensite steel by forming a self-lubricating layer in dry sliding wear. *Mater. Res. Express* **2019**, *6*, 055024. [CrossRef]

**Disclaimer/Publisher's Note:** The statements, opinions and data contained in all publications are solely those of the individual author(s) and contributor(s) and not of MDPI and/or the editor(s). MDPI and/or the editor(s) disclaim responsibility for any injury to people or property resulting from any ideas, methods, instructions or products referred to in the content.

# Thermal-Mechanical Fatigue Behavior and Life Assessment of Single Crystal Nickel-Based Superalloy

Juan Cao [1], Fulei Jing [2] and Junjie Yang [3,*]

[1] Aero Engine Corporation of China, Beijing 100097, China
[2] Aero Engine Academy of China, Aero Engine Corporation of China, Beijing 101304, China; jingfulei@163.com
[3] Institute for Aero Engine, Tsinghua University, Beijing 100084, China
* Correspondence: yangjunjie@tsinghua.edu.cn

**Abstract:** Thermal-mechanical fatigue (TMF) tests and isothermal fatigue (IF) tests were conducted using thin-walled tubular specimens under strain-controlled conditions. The results of TMF tests showed a strong correlation between mechanical behavior and temperature cycling. Under different phases of temperature and mechanical loading, the hysteresis loop and mean stress of the single crystal superalloy showed noticeable variations between the stress-controlled and strain-controlled conditions. In the strain-controlled TMF test, temperature cycling led to stress asymmetry and additional damage, resulting in a significantly lower TMF life compared to IF life at the maximum temperature. Moreover, the OP TMF life is generally lower than that of the IP TMF at the same strain amplitude. The Walker viscoplastic constitutive model based on slip systems was used to analyze the TMF mechanical behavior of the single crystal superalloy, and the change trends of the maximum Schmid stress, the maximum slip shear strain rate, and the slip shear strain range were analyzed, and their relationship with the TMF life was investigated. Finally, a TMF life prediction model independent of the loading mode and phase was constructed based on meso-mechanical damage parameters. The predicted TMF lives for different load control modes and phases fell within the twofold dispersion band.

**Keywords:** single crystal superalloy; thermal-mechanical fatigue; strain-controlled; cyclic stress–strain relationship; life assessment; meso parameters; viscoplastic constitutive model

## 1. Introduction

Air-cooled turbine blades in single crystal nickel-based superalloy are essential components in advanced aero engines. However, their life assessment remains a challenging issue due to thermal-mechanical fatigue, a typical failure mode caused by cyclical thermal and mechanical loads during operation [1,2]. In different regions of air-cooled blades, the phases between temperature and stress/strain vary significantly, which affects their fatigue life. For example, when an engine starts, the initial temperature rise rate on the outer surface of air-cooled blades exposed to the high-temperature gas is much higher than that on the inner surface in contact with the cooling airflow. As a result, the outer surface is subjected to compressive thermal stress, and the inner surface to tensile thermal stress. This means that there is approximately a 180° phase difference between temperature and stress/strain on the outer surface of the blade, known as an out-of-phase (OP) cycle, while there is almost no phase difference between the two on the inner surface of the blade cooling channel, known as an in-phase (IP) cycle [3].

Previous studies [4–7] on TMF tests of various single crystal nickel-based superalloys have revealed that the temperature cycling leads to significant asymmetry in the stress–strain curves, and the value and direction of the mean stress/strain accumulated are closely related to testing conditions such as phase angle and load control mode. The mechanical behavior of single crystal superalloys could significantly alter the damage evolution, which ultimately affects the fatigue life of the materials under TMF conditions. Under the same mechanical loading, the TMF damage of a single crystal superalloy could be greater or less than the isothermal fatigue (IF) damage at the highest temperature of the cycles. Therefore, life assessment based on the IF testing may not always be the expected conservative in predicting TMF life.

However, under the same phase angle conditions, the trend of TMF life is similar to that of isothermal fatigue with changes in strain range, maximum temperature, and other factors. Therefore, to predict the TMF life under different loading conditions, several studies [8–11] have developed models using functions similar to the existing IF life models, such as strain-life models, strain/strain energy range partitionings, and Ostergren model, expressing the material constants as functions of phase angle and cyclic peak temperature. These models are relatively simple in form but have limited ranges constrained by temperature, strain, and their combinations.

The Neu-Sehitoglu damage model [12,13] assumes that the total damage of each TMF cycle is accumulated from fatigue, oxidation, and creep damage. By introducing phase factors in the oxidation damage term and the creep damage term, the competition between different damage types is characterized under different loads. This model can analyze the dominant failure factors according to the relative magnitudes of each damage term, thus achieving TMF life prediction under complex loads. However, the model requires a complex acquisition process due to numerous material constants, especially requiring special environmental tests to decouple the oxidation damage from the other two damage terms, which is still a challenge. In addition, the model has insufficient accuracy for cyclic asymmetry situations, as the fatigue damage is related only to the macroscopic mechanical strain range.

This paper focuses on investigating the influence of temperature cycling, phase angle, load control mode, and other factors on the mechanical behavior and fatigue life of single crystal nickel-based superalloy. TMF tests are conducted on thin-walled tubular specimens with mechanical strain control based on the typical service loads and structural characteristics of air-cooled turbine blades. Utilizing the viscoplastic constitutive model based on the slip system, the relationship between TMF damage and mesoscopic parameters of the slip system is established. Finally, a TMF life model based on mesoscopic parameters is established for the single crystal nickel-based superalloy, and a preliminary validation is carried out by combining existing test data.

## 2. Materials and Tests

### 2.1. Materials

Table 1 displays the nominal composition of DD6, a second-generation single crystal nickel-based superalloy material in the Chinese series that is comparable to CMSX-4 and PWA1484. The heat treatment process of DD6 is as follows: solution treatment at 1290 °C for 1 h, at 1300 °C for 2 h, and 1315 °C for 4 h, respectively, followed by a first-stage aging treatment at 1120 °C for 4 h and a second-stage aging treatment at 870 °C for 32 h. To align with the structural characteristics of air-cooled turbine blades [14], the TMF tests were carried out on tubular specimens with a wall thickness of 1 mm, as shown in Figure 1. The axial direction of the specimens represents the [001] orientation of the single crystal superalloy. The outer surface of the specimens was polished longitudinally, and the inner surface was honed to prevent any machining defects.

**Table 1.** Chemical composition of single crystal nickel-based superalloy DD6 (wt.%).

| Ni | Al | Ta | W | Co | Re | Hf | Cr | Mo |
|---|---|---|---|---|---|---|---|---|
| Balanced | 5.2~6.2 | 6.0~8.5 | 7.0~9.0 | 8.5~9.5 | 1.6~2.4 | 0.05~0.15 | 3.8~4.8 | 1.5~2.5 |

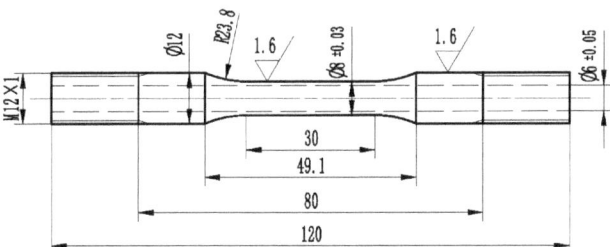

**Figure 1.** The schematic of thin-walled tubular specimens for the strain-controlled fatigue test (unit: mm).

## 2.2. Test Procedure

Previous Studies [3–7] have demonstrated a significant difference in the life trend of single crystal nickel-based materials between the strain-controlled and stress-controlled TMF tests. However, the actual loading conditions experienced by air-cooled turbine blades are somewhere between the two. In the "hot spot" or stress concentration area of the blade, the local deformation is constrained by the surrounding material, resulting in loads that are closer to those under the mechanical strain-controlled TMF test.

TMF tests were carried out using an MTS 810 servo-hydraulic test system in accordance with the ASTM E2368-10 test standard [15], including both out-of-phase and in-phase tests with a temperature range of 400~980 °C. In the out-of-phase TMF test, the phase angle between the temperature and mechanical loading is approximately 180 °C, to simulate the outer surface of the air-cooling blades when the inner cooling channels are strongly cooled. In the in-phase TMF test, however, it means that there is almost no phase difference between the temperature and mechanical loading, simulating the inner surfaces of the cooling channels. Meanwhile, the IF tests were performed with the maximum temperature of 980 °C in the TMF cycles to investigate the influence of temperature cycling.

The testing schemes of the TMF and IF tests are presented in Table 2. All the tests were tension-compression fatigue tests with a strain ratio of −1. The TMF frequency was approximately 0.01 Hz with a triangular waveform. The cyclic stress–strain data were recorded for each test so that variations in maximum and minimum strains could be examined. An electromagnetic induction heating system and a compressed air-cooling system were employed for the TMF tests. The surface temperature of the gauge section of the specimen was monitored using an S-type thermocouple combined with a thermal imager, while the axial strain was measured using a high-temperature extensometer.

**Table 2.** Testing matrix in strain-controlled TMF and IF test.

| Type | Temperature | Strain Amplitude | Strain Ratio | Cycling Time |
|---|---|---|---|---|
| - | °C | % | - | s |
| IP TMF | 400~980 | ±0.55 | −1 | 120 |
| IP TMF | 400~980 | ±0.50 | −1 | 120 |
| IP TMF | 400~980 | ±0.45 | −1 | 120 |
| IP TMF | 400~980 | ±0.40 | −1 | 120 |
| OP TMF | 400~980 | ±0.55 | −1 | 120 |
| OP TMF | 400~980 | ±0.50 | −1 | 120 |
| OP TMF | 400~980 | ±0.45 | −1 | 120 |
| OP TMF | 400~980 | ±0.40 | −1 | 120 |
| IF | 980 | ±1.0 | −1 | 8 |
| IF | 980 | ±0.80 | −1 | 6.4 |
| IF | 980 | ±0.60 | −1 | 4.81 |
| IF | 980 | ±0.50 | −1 | 4 |

## 3. Results

### 3.1. Stress–Strain Relationship

The hysteresis loops of single crystal specimens under different loading conditions are shown in Figure 2, and the evolutions of the maximum stress, the minimum stress, and the mean stress with the number of cycles are shown in Figure 3. The results indicated that the mechanical stress range, hysteresis loop width and cyclic stress range increase with the increase in the strain. For the IF cycle, the hysteresis loop is approximately symmetric (Figure 2a), and the mean stress tends to evolve over most of the life zone before fracture (Figure 3a) because the elastic modulus of the material is constant as a result of the stabilization of temperature over the whole test. However, in the TMF cycle, due to the temperature dependence of the mechanical behaviors of the single crystal superalloy, the asymmetric stress–strain response is significant under the symmetric mechanical loading.

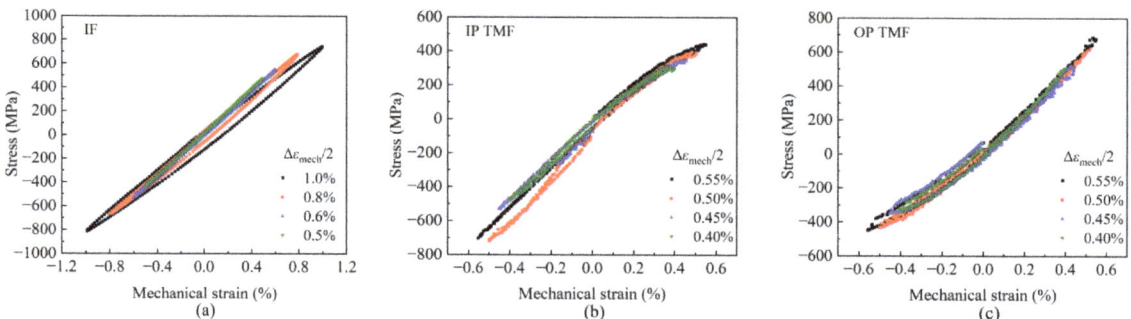

**Figure 2.** Hysteresis loops of the single crystal superalloy under different mechanical strain conditions: (**a**) IF, (**b**) IP TMF and (**c**) OP TMF.

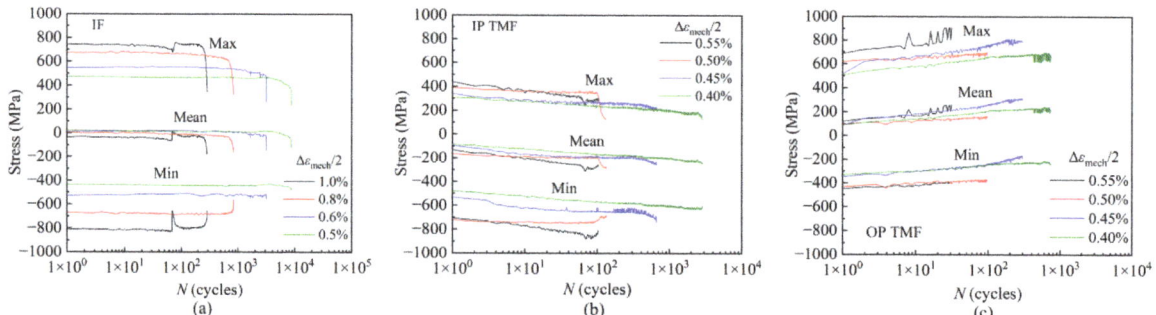

**Figure 3.** Cyclic maximum, minimum and mean stress versus cycle numbers for the single crystal superalloy under different mechanical strain conditions: (**a**) IF, (**b**) IP TMF and (**c**) OP TMF.

For the IP TMF cycle (Figure 2b), the material is subjected to tensile stress in the higher temperature stage and compressive stress in the lower temperature stage. With the increase in temperature, the ability of the single crystal superalloy DD6 to resist elasticity and inelasticity decreases, i.e., the elastic modulus decreases and the inelastic strain increases, resulting in a decrease in stress corresponding to the same mechanical strain, often called the tensile stress relaxation. This is manifested as a decrease in the stress slope of the high-temperature tensile section, and the hysteresis loop has a downward bending tendency in the tensile stress zone. These changes cause the stress peaks and valleys to shift downward, resulting in the compressive mean stress that accumulates in the compressive direction as the number of cycles increases (Figure 3b).

For the OP TMF cycle (Figure 2c), the material experiences tensile stress during the lower temperature stage and compressive stress during the higher temperature stage. Due to the relaxation of the compressive stress generated in the high-temperature section, the hysteresis loop tends to bend upwards in the compressive stress zone. These changes cause the peak and valley stress to shift upward, resulting in a mean tensile stress that evolves towards the tensile stress as the number of cycles increases (Figure 3c). Especially, it is worth noting that in the larger mechanical strain range (±0.55%), the maximum tensile stress of OP TMF gradually increases with the number of cycles, which can reach the value of the plastic flow point of the material, resulting in premature structural failure. This is deemed to be the main reason why the OP TMF cycle is the most damaging [16].

It should be noted that the TMF tests presented in this study were performed by the mechanical strain control, therefore the mean strain is maintained at 0 over the cycle. However, the thermal asymmetry of the DD6 material causes the evolution of mean stress related to the phase angle. In the stress-controlled TMF test of the same material with a stress ratio of −1, where the mean stress is kept at 0. Due to this the deformation resistance of the material decreases in the higher temperature section, the material appears softer, and its elastic modulus is lower at high-temperature loading. Therefore, for the OP TMF conditions, the hysteresis loop curves also showed upward bending in the compression loading section as the temperature increases, resulting in a mean compressive strain and accumulation in the compressive direction. For the IP TMF conditions, however, the hysteresis loops depicted a downward bending in the tensile stress zone due to a similar reason, leading to a mean tensile strain and accumulation in the tensile direction [3].

## 3.2. Fatigue Behavior

Figure 4 depicts the relationship between the mechanical strain amplitude and life under different test conditions. The results show that, at the same mechanical strain amplitude, the TMF life over the temperature range of 400~980 °C is an order of magnitude lower than the IF life at the maximum cycle temperature of 980 °C, indicating that the temperature cycling led to the additional material damage. Specifically, when the cyclic mechanical strain amplitude is in the range of ±0.50%, the lifetime for IF is approximately 100 times that in the OP TMF and IP TMF. Furthermore, in the test load range, the OP TMF life is generally lower than the IP TMF life under the same load conditions. This could be due to the fact that the OP cycle generates a gradually increasing mean tensile stress, which promotes microcrack propagation and aggravates material damage to some extent, while the IP cycle generates a mean compressive stress, which inhibits crack propagation to some extent. In other words, the mechanical damage in OP TMF is greater than that in IP TMF at the same strain amplitude level and temperature range under strain-controlled conditions.

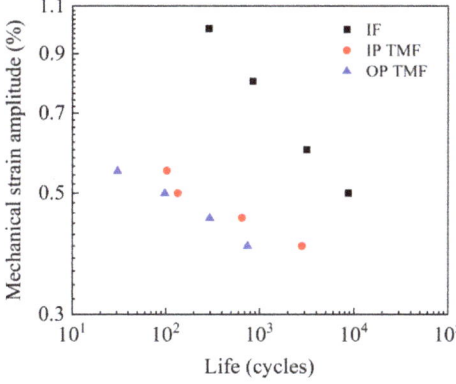

**Figure 4.** Mechanical strain amplitude versus life.

It is worth noting that the above life trend differs obviously from that observed in the stress-controlled test. Specifically, under strain-controlled conditions, the lifetime in TMF is consistently shorter than that in the IF, regardless of the phase angle between the mechanical and temperature. In contrast, under stress-controlled conditions, the life of the IP TMF is the shortest, while the life of the OP TMF is the longest. The life of the IF at the highest temperature falls in between. The results suggest that whether based on the IP TMF or OP TMF or IF tests data, life assessment may not always lead to the expected conservative design in predicting TMF life.

The loading control modes have different effects on the stress–strain response of the material, which in turn affects the evolution process of the material microstructure and its damage mechanism. Therefore, it may be more feasible to build a life prediction model that is independent of the load form, rather than considering all factors that affect life.

## 4. Discussion

### 4.1. Viscoplastic Constitutive Model Based on Slip System

The Walker viscoplastic constitutive model is based on the slip system and does not involve material yielding surfaces. The model is capable of accurately describing the effects of ratcheting and stress relaxation, which has been validated on various single crystal nickel-based superalloys such as DD6 and Hastelloy-X [17,18]. DD6 is a face-centered cube (FCC) structure material with 12 octahedral slip systems and 6 cubic slip systems. In the Walker model, the parameters for the octahedral slip systems and cubic slip systems are denoted by the superscript 'o' and 'c', respectively. The stress components of the stress tensor $\sigma$ can be written in the slip system coordinates using equations as follows:

$$\pi_{mn}^r = (m_r^o)^T \sigma n_r^o; \pi_{mm}^r = (m_r^o)^T \sigma m_r^o; \pi_{nn}^r = (n_r^o)^T \sigma n_r^o; \tag{1}$$

$$\pi_{zz}^r = (z_r^o)^T \sigma z_r^o; \pi_{mz}^r = \pi_{zm}^r = (m_r^o)^T \sigma z_r^o; \pi_{nz}^r = \pi_{zn}^r = (n_r^o)^T \sigma z_r^o; \tag{2}$$

where $m$, and $n$ are the normal direction and the slip direction of the slip system, and $z$ is the cross-product vector of $m$ and $n$. The unit vectors in the coordinate axes are denoted by $m_r^a$, $n_r^a$ and $z_r^a = m_r^a \times n_r^a$, $a$ denotes the type of slip system when $a = o$ for the octahedral slip systems and $a = c$ for the cube slip systems, $r$ is the number of slip systems, 12 or 6. $\pi_{mn}^r$ is the Schmid stress, i.e., the resolved shear stress along the slip direction in the slip plane of the stress tensor $\sigma$.

For the octahedral slip systems, the flow rule can be expressed as a power law as follows:

$$\dot{\gamma}_r^o = \left| \frac{\pi_r^o - \omega_r^o}{K_r^o} \right|^{n-1} \left( \frac{\pi_r^o - \omega_r^o}{K_r^o} \right) \tag{3}$$

where $\dot{\gamma}_r^o$ is the shear strain rate in the $r$-th octahedral slip system, $\pi_r^o = \pi_{mn}^o$ is the Schmid stress, $n$ is the material constant on the sensitivity of rate, $\omega_r^o$ and $K_r^o$ are the back stress and the drag stress, and can describe the kinematic hardening and the isotropic hardening to the deformation of the material, respectively.

The back stress and the drag stress can be written in equations as follows:

$$\dot{\omega}_r^o = \rho_1^o \dot{\gamma}_r^o - \rho_2^o \left| \dot{\gamma}_r^o \right| \omega_r^o - \rho_3^o |\omega_r^o|^{p-1} \omega_r^o \tag{4}$$

$$K_r^o = K_0^o + \rho_4^o \pi_{nz}^o + \rho_5^o |\Psi_r| \tag{5}$$

where $\rho_1^o, \rho_2^o, \rho_3^o, \rho_4^o, \rho_5^o$, $p$, and $K_0^o$ are the material constants depending on the temperature, $\pi_{nz}^o$ and $\Psi_r$ are the non-Schmid stress and Takeuchi-Kuramoto stress, respectively. The Takeuchi-Kuramoto stress, $\Psi_r$, can be calculated according to the equations in Table 3.

**Table 3.** Takeuchi-Kuramoto stress in Walker viscoplastic constitutive model.

| | | | |
|---|---|---|---|
| $\Psi_1 = (m_1^o)^T \sigma j$ | $\Psi_2 = (m_2^o)^T \sigma k$ | $\Psi_3 = (m_3^o)^T \sigma i$ | $\Psi_4 = (m_4^o)^T \sigma i$ |
| $\Psi_5 = (m_6^o)^T \sigma k$ | $\Psi_6 = (m_6^o)^T \sigma j$ | $\Psi_7 = (m_7^o)^T \sigma j$ | $\Psi_8 = (m_8^o)^T \sigma k$ |
| $\Psi_9 = (m_9^o)^T \sigma i$ | $\Psi_{10} = (m_{10}^o)^T \sigma i$ | $\Psi_{11} = (m_{11}^o)^T \sigma k$ | $\Psi_{12} = (m_{12}^o)^T \sigma j$ |

Similarly, the flow rule, the back stress, and the drag stress in the $r$-th cube slip systems can be expressed as the following equations:

$$\dot{\gamma}_r^c = \left| \frac{\pi_r^c - \omega_r^c}{K_r^c} \right|^{m-1} \left( \frac{\pi_r^c - \omega_r^c}{K_r^c} \right) \quad (6)$$

$$\dot{\omega}_r^c = \rho_1^c \dot{\gamma}_r^c - \rho_2^c |\dot{\gamma}_r^c| \omega_r^c - \rho_3^c |\omega_r^c|^{q-1} \omega_r^c \quad (7)$$

$$K_r^c = K_0^c \quad (8)$$

where $\dot{\gamma}_r^c$ is the shear strain rate in the $r$-th cube slip system, $\pi_r^c = \pi_{mn}^c$ is the Schmid stress, $\rho_1^c, \rho_2^c, \rho_3^c, q$, and $K_0^c$ are the material constants depending on the temperature.

The material constants in the Walker model at 780 °C and 960 °C are given in Table 4 [19]. Using the Walker viscoplastic constitutive model, the macroscopic and mesoscopic stress–strain responses of the single crystal superalloy DD6 were calculated. As shown in Figure 5, the macroscopic stress predicted by the Walker constitutive model for the DD6 specimen is well coincidental with that in the experiment.

**Table 4.** Material constants in Walker viscoplastic constitutive model (760 °C and 980 °C).

| | Octahedral Slip System | | | Cube Slip System | |
|---|---|---|---|---|---|
| $\rho_1^o$ | $5.0 \times 10^6$ | $0.5 \times 10^6$ | $\rho_1^c$ | $5.0 \times 10^6$ | $0.5 \times 10^6$ |
| $\rho_2^o$ | $2.083 \times 10^4$ | $0.25 \times 10^4$ | $\rho_2^c$ | $1.852 \times 10^4$ | $0.442 \times 10^4$ |
| $\rho_3^o$ | 0.0 | 0.0 | $\rho_3^c$ | 0.0 | 0.0 |
| $\rho_4^o$ | 0.15 | 0.0 | - | - | - |
| $\rho_5^o$ | 0.1 | -3.5 | - | - | - |
| $n$ | 9.033 | 4.362 | $m$ | 10.52 | 6.293 |
| $p$ | 3.0 | 3.0 | $q$ | 3.0 | 3.0 |
| $K_0^o$ | 428.5 | $1.018 \times 10^3$ | $K_0^c$ | 441.3 | $6.243 \times 10^2$ |

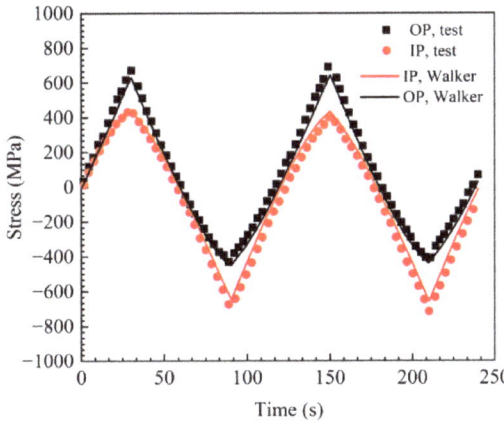

**Figure 5.** Calculated stress versus testing results under TMF.

## 4.2. Mesomechanical Parameters in TMF Testing

The evolutions of Schmid stress and slip shear strain over time for both the mechanical strain-controlled TMF (denoted by $\Delta\varepsilon_{mech}$) and the stress-controlled TMF (denoted by $\Delta\sigma$) are shown in Figure 6. The stress-controlled TMF data are derived from the literature [3]. Due to the difference in cycle periods between the two load control modes, the horizontal axis in Figure 6 uses a normalized time, which is the ratio of the current time to the cycle period. This paper only presents the calculation results at the typical loads, as the evolutions of mesoscopic parameters under other conditions are similar.

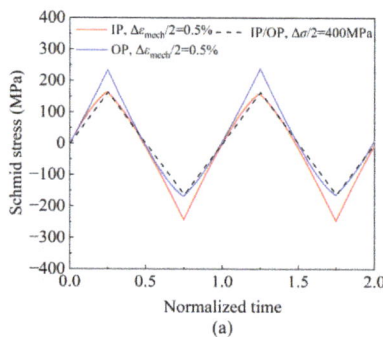 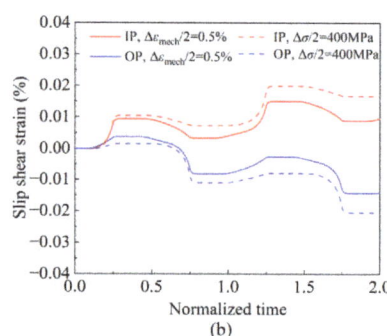

**Figure 6.** Mesoscopic parameters on slip systems versus time under mechanical strain-controlled and stress-controlled TMF: (**a**) Schmid stress versus normalized time and (**b**) slip shear strain versus normalized time.

The Schmid stress is obtained by decomposing the macroscopic stress tensors. In the stress-controlled TMF tests, the Schmid stress remains symmetrical both in the IP and in the OP, as shown in Figure 6a, because the macroscopic mechanical stress is symmetric under the stress ratio of −1. In contrast, in the mechanical strain-controlled TMF tests with the strain ratio of −1, the Schmid stress exhibits a noticeable asymmetry due to the asymmetry of the macroscopic mechanical stress induced by stress relaxation. Specifically, the mean Schmid stress of the IP cycle is negative and shifts in the compressive direction with the number of cycles, while the average value of the Schmid stress of the OP cycle was positive and shifted in the tensile direction with the number of cycles. This behavior is consistent with the evolution of measured macroscopic stress shown in Figure 3b,c. Therefore, the results depict that the Schmid stress and its related quantities could be regarded as the representative mesoscopic parameters of the material mechanical behavior.

Regardless of the control mode, the slip shear strain accumulated significantly shown in Figure 6b, indicating the irreversibility of slip. For the IP cycle, the slip shear strain remained positive, and the cycle maximum and average values of the slip shear strain gradually increased with the number of cycles. In contrast, for OP cycles, the slip shear strain is initially positive, transitions to negative values in the subsequent compression section, and accumulates in negative directions during subsequent cycles. Its trend of change reflects the influence of temperature cycles of TMF on the mechanical behavior of materials. Therefore, in the same way, the results depict that the slip shear strain and its related quantities also could be regarded as the representative mesoscopic parameters of the material mechanical behavior.

## 4.3. TMF Life Model Based on Mesoscopic Parameters

The mechanical response of the material and its damage evolution are dependent on the test conditions, such as the temperature-strain phase angle and the load control mode. Therefore, it is essential to select appropriate parameters for accurate TMF life modeling based on the mesoscopic stress–strain characteristics of the single crystal superalloy under different loading conditions. Previous studies [20–22] have used some parameters in the damage rate model, including the maximum Schmid stress, the maximum slip shear strain rate, and the slip shear strain range directly related to plastic deformation. By performing calculations based on the second cycle, typical mesoscopic parameters for each slip system were determined, and then the relationship curves between the mesoscopic parameters and the TMF life were obtained, as shown in Figure 7. It is evident from Figure 7 that, in the logarithmic coordinate system, the maximum Schmid stress, the maximum slip shear strain rate, and the slip shear strain range for each test condition exhibit an approximately linear relationship with TMF life.

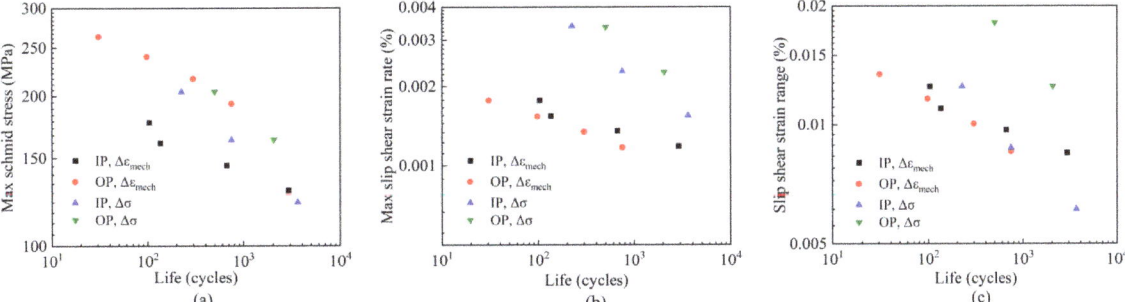

**Figure 7.** Microscopic parameters on slip systems versus life under TMF: (**a**) Max Schmid stress versus life, (**b**) Max slip shear strain rate versus life and (**c**) Slip shear strain range versus life.

Considering the difference in TMF life caused by the load control modes, based on mesoscopic parameters such as the maximum Schmid stress, maximum slip shear strain rate, slip shear strain range and Schmid stress ratio, a TMF life model in a similar form to the CDA model [22,23] is established by taking them as damage parameters based on the critical plane methods. The model is expressed as follows:

$$\log N_{\text{TMF}} = A_{\text{TMF}} + m_{\text{TMF}}\log|\pi_{max}^{\alpha}| + n_{\text{TMF}}\log\left(\left|\dot{\gamma}^{\alpha}\right|_{max}\right) + z_{\text{TMF}}\log|\Delta\gamma^{\alpha}| + a_{\text{TMF}}\left(\pi_{min}^{\alpha}/\pi_{max}^{\alpha}\right) \tag{9}$$

where $N_{\text{TMF}}$ is the life of TMF, $\alpha$ denotes the type of slip system, $\pi_{min}^{\alpha}$ and $\pi_{max}^{\alpha}$ represent the corresponding Schmid stress values for the minimum and the maximum macroscopic stress, respectively, $\dot{\gamma}^{\alpha}$ is the slip shear strain rate and $\Delta\gamma^{\alpha}$ is the range of the slip shear strain, $A_{\text{TMF}}$, $m_{\text{TMF}}$, $n_{\text{TMF}}$, $z_{\text{TMF}}$, and $a_{\text{TMF}}$ are the material constants, which are related to the maximum temperature of the TMF cycle and the crystal orientation and are independent of the phase angle between the temperature and the mechanical loading.

Based on the results of the strain-controlled TMF test in this paper and that of the stress-controlled TMF test previously conducted [3], combined with the calculations according to the viscoplastic constitutive model, multiple linear regression is performed on the above Equation (9), and the material constants of the model were then determined and are presented in Table 5. The results predicted by the TMF life model for different phase angles and load control modes were compared and found to fall within the twofold scatter band, as shown in Figure 8. Therefore, this model can be used to accurately evaluate the TMF life of air-cooling blades in the single crystal nickel-based material.

**Table 5.** Material constants in TMF life model (400~980 °C).

| $A_{\text{TMF}}$ | $m_{\text{TMF}}$ | $n_{\text{TMF}}$ | $z_{\text{TMF}}$ | $a_{\text{TMF}}$ |
|---|---|---|---|---|
| 35.0782 | −11.3985 | −0.6103 | 3.3247 | 1.7894 |

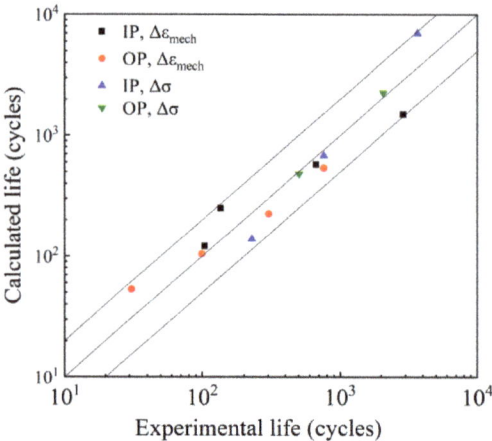

**Figure 8.** TMF life prediction under different phase shifts and load-controlled modes.

## 5. Conclusions

(1) Under strain-controlled conditions, for the same mechanical strain amplitude, the TMF life is significantly lower than the isothermal fatigue life at the maximum temperature of the TMF cycle. Moreover, the out-of-phase TMF life is generally lower than the in-phase TMF life under the same load over the range of loads investigated.

(2) Under strain-controlled conditions, temperature cycling can induce significant stress asymmetry. For out-of-phase cycles, the hysteresis loop bends upwards during the higher-temperature compression half-cycle, producing a mean tensile stress that gradually increases, which can promote micro-crack propagation and exacerbate material damage. In contrast, for in-phase cycles, the mean compressive stress generated moves gradually towards the compression direction as the number of cycles increases, thereby inhibiting crack propagation.

(3) The stress–strain response of the material can vary significantly under different load control modes, leading to different effects on the microstructure evolution and changing the damage mechanism and the life of the material.

(4) Based on the primary damage parameters identified from the stress and strain on the slip planes using the Walker viscoplasticity constitutive model, a TMF life model was developed. The model accurately predicts the TMF life for different phases and load-controlled modes within a twofold dispersion band.

**Author Contributions:** Conceptualization, methodology, J.C., F.J. and J.Y.; data curation, writing—original draft preparation, J.C. and F.J.; writing—review and editing, F.J. and J.Y.; project administration, J.C.; funding acquisition, J.C. and F.J. All authors have read and agreed to the published version of the manuscript.

**Funding:** This research was funded by the National Natural Science Foundation of China (Grant No. 51375031), and the National Science and Technology Major Project, grant number J2019-IV-0003-0070 and J2019-V-0006-0099.

**Data Availability Statement:** Not applicable.

**Acknowledgments:** The authors gratefully acknowledge the financial support from the funding.

**Conflicts of Interest:** The authors declare no conflict of interest.

## References

1. Weijun, Z. Thermal mechanical fatigue of single crystal superalloys: Achievements and challenges. *Mater. Sci. Eng. A* **2016**, *650*, 389–395.
2. Jones, J.; Whilttaker, M.; Lancaster, R.; Williams, S. The influence of phase angle, strain range and peak cycle temperature on the TMF crack initiation behaviour and damage mechanisms of the nickel based superalloy, RR1000. *Int. J. Fatigue* **2017**, *98*, 279–285. [CrossRef]
3. Fulei, J.; Kanghe, J.; Bin, Z.; Dianyin, H.; Rongqiao, W. Experimental research on thermomechanical fatigue in nickel based single crystal superalloy DD6. *J. Aerosp. Power* **2018**, *33*, 2965–2971.
4. Amaro, R.; Antolovich, S.; Neu, R.; Staroselsky, A. On thermo-mechanical fatigue in single crystal Ni-base superalloys. *Procedia Eng.* **2010**, *2*, 815–824. [CrossRef]
5. Hynunuk, H.; Jeong-gu, K.; Baiggyu, C.; Insoo, K.; Youngsoo, Y.; Changyong, Y. A comparative study on thermomechanical and low cycle fatigue failures of a single crystal nickel-based superalloy. *Int. J. Fatigue* **2011**, *33*, 1592–1599.
6. Jinjiang, Y.; Guoming, H.; Zhaokuang, C.; Xiaofeng, S.; Tao, J.; Zhuangqi, H. High temperature thermo-mechanical and low cycle fatigue behaviors of DD32 single crystal superalloy. *Mater. Sci. Eng. A* **2014**, *592*, 164–172.
7. Junjie, Y.; Fulei, J.; Zhengmao, Y.; Kanghe, J.; Dianyin, H.; Bin, Z. Thermomechanical fatigue damage mechanism and life assessment of a single crystal Ni-based superalloy. *J. Alloys Compd.* **2021**, *872*, 159578.
8. Cunha, F.; Dahmer, M.; Chyu, M. Thermal-mechanical life prediction system for anisotropic turbine components. *J. Turbomach.* **2006**, *128*, 240–250. [CrossRef]
9. Gomez, T.; Awarke, A.; Pischinger, S. A new low cycle fatigue criterion for isothermal and out-of-phase thermomechanical loading. *Int. J. Fatigue* **2010**, *32*, 769–779. [CrossRef]
10. Amaro, R.; Antolovich, S.; Neu, R.; Fernandez-Zelaia, P. Thermomechanical fatigue and bithermal–thermomechanical fatigue of a nickel-base single crystal superalloy. *Int. J. Fatigue* **2012**, *42*, 165–171. [CrossRef]
11. Vacchieri, E.; Holdsworth, S.; Poggio, E.; Villari, P. Service-like TMF tests for the validation and assessment of a creep-fatigue life procedure developed for GT blades and vanes. *Int. J. Fatigue* **2017**, *99*, 216–224. [CrossRef]
12. Vose, F.; Becker, M.; Fischersworring, B.; Hackenberg, H. An approach to life prediction for a nickel-base superalloy under isothermal and thermo-mechanical loading conditions. *Int. J. Fatigue* **2013**, *53*, 49–57. [CrossRef]
13. Abdullahi, A.; Samir, E.; Panagiotis, L.; Riti, S. Aero-engine turbine blade life assessment using the Neu/Sehitoglu damage model. *Int. J. Fatigue* **2014**, *61*, 160–169.
14. Rongqiao, W.; Kanghe, J.; Fulei, J.; Dianyin, H. Thermomechanical fatigue failure investigation on a single crystal nickel superalloy turbine blade. *Eng. Fail. Anal.* **2016**, *66*, 284–295.
15. *ASTM E2368-10*; Standard Practice for Strain Controlled Thermomechanical Fatigue Testing. ASTM: West Conshohocken, PA, USA, 2010; pp. 1–10.
16. Kersey, R.; Staroselsky, A.; Dudzinski, D.; Genest, M. Thermomechanical fatigue crack growth from laser drilled holes in single crystal material. *Int. J. Fatigue* **2013**, *55*, 183–193. [CrossRef]
17. Rongqiao, W.; Fulei, J.; Dianyin, H. Fatigue life prediction model based on critical plane of nickel-based single crystal superalloy. *J. Aerosp. Power* **2013**, *28*, 2587–2592.
18. Jordan, E.; Shixiang, S.; Walker, K. The viscoplastic behavior of Hastelloy-X single crystal. *Int. J. Plast.* **1993**, *9*, 119–139. [CrossRef]
19. Fulei, J.; Shibai, T.; Junjie, Y. Influence of small hole on thermal mechanical fatigue in single crystal superalloy. *J. Aerosp. Power* **2021**, *36*, 1669–1679.
20. Tinga, T.; Brekelmans, W.; Geers, M. Time-incremental creep-fatigue damage rule for single crystal Ni-base superalloys. *Mater. Sci. Eng. A* **2009**, *508*, 200–208. [CrossRef]
21. Rongqiao, W.; Bin, Z.; Dianyin, H.; Kanghe, J.; Jianxin, M.; Fulei, J. A critical-plane-based thermomechanical fatigue lifetime prediction model and its application in nickel-based single-crystal turbine blades. *Mater. High Temp.* **2019**, *36*, 325–334.
22. Fulei, J.; Rongqiao, W.; Dianyin, H.; Kanghe, J. Damage parameter determination and life modeling for high temperature fatigue of nickel-based single crystal superalloys. *Acta Aeronaut. Astronaut. Sin.* **2016**, *37*, 2749–2756.
23. Rongqiao, W.; Kanghe, J.; Dianyin, H.; Fulei, J.; Xiuli, S. High temperature fatigue life model for single crystal nickel superalloy based on principal component analysis. *J. Aerosp. Power* **2016**, *31*, 1359–1367.

**Disclaimer/Publisher's Note:** The statements, opinions and data contained in all publications are solely those of the individual author(s) and contributor(s) and not of MDPI and/or the editor(s). MDPI and/or the editor(s) disclaim responsibility for any injury to people or property resulting from any ideas, methods, instructions or products referred to in the content.

Article

# Investigation on Structural, Tensile Properties and Electronic of Mg–X (X = Zn, Ag) Alloys by the First-Principles Method

Yan Gao [1], Wenjiang Feng [2], Chuang Wu [1], Lu Feng [2] and Xiuyan Chen [2,*]

[1] Experimental Teaching Center, Shenyang Normal University, Shenyang 110034, China; gaoy@synu.edu.cn (Y.G.)
[2] College of Physics Science and Technology, Shenyang Normal University, Shenyang 110034, China
* Correspondence: chenxy@synu.edu.cn

**Abstract:** In order to study the strengthening effect of Mg–X (X = Zn, Ag) alloys, solid solution structures of $Mg_{54}$, $Mg_{53}X_1$ and $Mg_{52}X_2$ (X = Zn, Ag) with atomic contents of 1.8 at.% and 3.7 at.% were established, respectively. The structural stability, tensile properties and electronic properties were investigated by first-principles simulation. The calculated results of cohesive energies show that all solid solution structures were stable under different tensile strains, and $Mg_{52}Ag_2$ had the best stability. The results of tensile tests show that Zn and Ag atoms promoted the Mg-based alloy's yield strength and tensile strength. In addition, through comparative analyses, we have demonstrated that the tensile property of Mg-based alloys was also affected by solid solubility. Finally, the electronic density of states (DOS) and electron density difference of several solid solution structures were analyzed.

**Keywords:** first-principles calculation; solid solubility; stability; tensile; electronic

**Citation:** Gao, Y.; Feng, W.; Wu, C.; Feng, L.; Chen, X. Investigation on Structural, Tensile Properties and Electronic of Mg–X (X = Zn, Ag) Alloys by the First-Principles Method. Crystals 2023, 13, 820. https://doi.org/10.3390/cryst13050820

Academic Editor: Duc Nguyen-Manh

Received: 29 March 2023
Revised: 11 May 2023
Accepted: 11 May 2023
Published: 16 May 2023

**Copyright:** © 2023 by the authors. Licensee MDPI, Basel, Switzerland. This article is an open access article distributed under the terms and conditions of the Creative Commons Attribution (CC BY) license (https://creativecommons.org/licenses/by/4.0/).

## 1. Introduction

As light metal materials used in industrial production, magnesium alloys are known as a green engineering material with excellent performance characteristics [1–3]. They have the characteristics of low density, good stiffness, high strength-to-weight ratio, high heat dissipation, outstanding casting performance, and high recycling rate. The application of magnesium alloys is always a research hot spot. Magnesium alloys have a wide range of use in the automotive manufacturing, aerospace, 3C electronics, rail transportation and biomedical fields [4–6]. On the one hand, magnesium alloys can meet the performance requirements in practical applications; on the other hand, they are a lightweight material [7–9]. Therefore, magnesium alloys have advantages that other alloy materials cannot compete with. However, the poor ductility and heat resistance of magnesium alloys limit their further development. Researchers have been committed to improving the performance of magnesium alloys, so as to develop new light alloy materials with high performance capacities.

Solid solution strengthening is an effective method used to enhance the mechanical properties of magnesium alloys by adding solid alloying elements dissolved in Mg-based alloys [10,11]. In the solid solutions' structures, the solid atoms replace the magnesium atoms in some areas of the lattice dot matrix, which causes atomic misalignment and lattice distortion [12]. This lattice distortion will prevent dislocation movement and slip, thus concentrating the stress on Mg-based alloys, as a result of which the strength of the magnesium alloys will be improved [13–15]. Zn and Ag are two important alloying elements that are used to improve the properties of magnesium alloys. The atomic structures of Zn and Ag are similar to that of Mg, and have high solid solubility in magnesium alloys; they can also be used as an effective solid solution in Mg-based alloy. The solid solubility of Zn and Ag in magnesium alloys can reach 8.4% and 15.5% at eutectic temperatures. However, it is worth noting that little research has been undertaken on the tensile properties of Mg–Zn and Mg–Ag alloys at the atomic scale. In particular, comparative studies of their tensile properties at different levels of solid solubility are scarce.

The tensile strength of materials has always been a significant indicator of the solid solution strength [16,17]. Tensile strength can be predicted using the first principle tensile simulation method, after which the solution strengthening effect of alloy elements can be analyzed. The average stress on the structure can be calculated using the first principle tensile simulation method, after which the stresses and strains in the structures can be analyzed. The tensile strength values of structures can be predicted using their stress values at yield. The first principle tensile simulation method was first applied to the calculation of simple crystal structures such as Si, Cu, Al, Mo, Ge, SiC and diamond [18–20], and good prediction results were obtained. These data have helped in guiding experimental operations, and have also helped in saving a lot of money and time.

In the past several years, with the continuous development of computer equipment, it has become possible to assess more complex crystal structures using the first principle tensile simulation method. Zhang et al. [21] found that the maximum stress value of a pure aluminum structure is reached at the tensile strain of 16%, with a value of $9.5 \times 10^3$ MPa. Pei et al. investigated the tensile properties of the Al(111)/Al$_3$Ti(112) interface using first principle tensile simulation. The results indicate that the maximum stress value was $14.38 \times 10^3$ MPa, and the Al side was able to absorb most of the deformation energy. Liu et al. [22] studied the solid solution strengthening of Al and Er when used in magnesium alloys through first principles tensile simulations. The study found that Er has a better reinforcement effect. Luo et al. [23] studied the solid solution strengthening effects of Al, Zn and Y on a magnesium matrix using first principles tensile simulations. The strengthening effect of Y is superior to that of other atoms. Wang and Han et al. [24] studied the tensile properties of $Mg_{53}Al$ and $Mg_{51}Al_3$ using first principles simulation. Collectively, these data indicate that the tensile strength of the Mg-based alloy could be enhanced by covalent bonding between two types of atoms, and the maximum stress value of $Mg_{51}Al_3$ was increased by 9.4% compared to that of $Mg_{54}$ using this approach. Wang et al. [25] also investigated the effects of the distribution of Zn and Al atoms on the tensile strength of the magnesium matrix, and found that a uniformly distributed structure had greater tensile strength than a separated one. Wang et al. [25] also studied the relationship between the tensile strength of the magnesium matrix and atomic distribution, and their results have revealed that the tensile strength of the structure was better when the atoms were uniform.

Within the maximum atomic solid solubility range of Zn and Ag atoms in magnesium, the structures of $Mg_{53}X_1$ (X = Zn, Ag) and $Mg_{52}X_2$ (X = Zn, Ag) with solid solubilities of 1.8% at.% and 3.7 at.% have been established and studied. First principle tensile simulations were used to test the stability, as well as the stress properties and electronic characteristics, of solid solution structures under 0–20% strains. The effects and mechanisms of the solid solution strengthening of elements Zn and Ag when used in a Mg-based alloy were predicted.

## 2. Computational Methods

Based on the first principle method with density functional theory (DFT) and generalized gradient approximation (GGA), tensile tests of solid solution structures ($Mg_{53}X_1$ and $Mg_{52}X_2$ (X = Zn, Ag)) were carried out. The CASTEP software was used for the calculations, and the exchange correlation functional was PW91. The ultrasoft pseudopotential was used to assess the interaction between the ion nucleus and electron. For the optimization of the solid solution structure, the settings were as follows: the total energy threshold was $1.0 \times 10^{-5}$ eV/atom, the interatomic force threshold was $3 \times 10^{-2}$ eV/nm, the maximum internal stress threshold was $0.05 \times 10^3$ MPa, and the tolerance deviation threshold was 0.001 Å. Regarding the electronic settings, the cutoff energy and k-point mesh were 340 eV and $3 \times 3 \times 3$, respectively. The tolerance value and maximum number of convergence steps of the SCF self-consistent iteration were $2.0 \times 10^{-6}$ eV/atom and 150.

Tensile strain was applied along the c-axis of the crystal structures, which was achieved by changing the lattice constant c value in the simulation. During this test, the upper limit of strain was set to 20%. The strain interval between 6 and 10% strain was 1%, and for the remaining range the strain interval was 2%. After each incremental increase in strain, the solid solution structures need to be geometrically re-optimized, but only the atomic occupation coordinates were optimized, and the lattice constants were not optimized.

In tensile simulations, according to Nielsen-Martin method [26], the average stress acting on the whole cell can be expressed as:

$$\sigma_{\alpha\beta} = \frac{1}{\Omega} \frac{\partial E_{tot}}{\partial \varepsilon_{\alpha\beta}} \tag{1}$$

In the above formula, $E_{tot}$ represents the energy of the entire cell, $\varepsilon_{\alpha\beta}$ represents the strain tensor, and $\Omega$ represents the volume. The tensile strain $\varepsilon$ was calculated as follows:

$$\varepsilon = \frac{(l_\varepsilon - l_0)}{l_0} \times 100\% \tag{2}$$

In the above formula, $l_0$ and $l_\varepsilon$ are the initial length and tensile length in the c-axis direction of the cell. In addition, in order to reduce the time of calculation, the effect of c-axis stretching on lattice constants in the other two directions was not considered in the stretching simulation process; that is, the influence of the Poisson effect was ignored.

## 3. Results and Discussion

### 3.1. Structural Properties

The crystal structure of pure magnesium is shown in Figure 1a. Pure magnesium has an Hcp structure, and its space group is $P6_3/mmc$. The calculated lattice constants are a = b = 0.3209 nm and c = 0.5211 nm, which are very similar to the results of other simulations [27] wherein a = b = 0.3195 nm and c = 0.5178 nm, and experimental data [28] wherein a = b = 0.3210 nm and c = 0.5211 nm. A 3 × 3 × 3 supercell was established through the software's Supercell function [22,29], based on pure magnesium's crystal structure. The 3 × 3 × 3 supercell is shown in Figure 1b. Geometric optimization was performed on the supercell, and the optimized lattice constants are a = b = 0.9628 nm and c = 1.5632 nm. The supercell contains 54 atoms, hereinafter referred to as $Mg_{54}$. The structure of the solid solution with an X content of 1.8 at.% is shown in Figure 1c, where the X atom coordinate is (x = 0.5552, y = 0.4448, z = 0.5802), hereinafter referred to as $Mg_{53}X_1$ (X = Zn, Ag). The structure of the solid solution with an X content of 3.7 at.% is shown in Figure 1d, where the X atoms coordinates are (x = 0.5552, y = 0.4448, z = 0.5802) and (x = 0.5552, y = 0.4447, z = 0.2505), respectively, hereinafter referred to as $Mg_{52}X_2$ (X = Zn, Ag). The atomic positions of $Mg_{52}X_2$ are given in Table 1. During the model establishment process, the energy values of the crystal structures at different positions of the X atoms are calculated and compared, and the most reasonable positions for the placement of the atoms are ultimately determined, and are used as the positions of the X atoms here.

**Table 1.** The atomic positions of $Mg_{54}X_2$.

| Element | Atom Number | Fractional Coordinates of Atoms | | |
|---|---|---|---|---|
| | | x | y | z |
| Mg | 1 | 0.1109 | 0.2220 | 0.0829 |
| Mg | 2 | 0.2273 | 0.1170 | 0.2505 |
| Mg | 3 | 0.4464 | 0.2264 | 0.0865 |
| Mg | 4 | 0.5563 | 0.1169 | 0.2505 |
| Mg | 5 | 0.7780 | 0.2220 | 0.0828 |
| Mg | 6 | 0.8889 | 0.1111 | 0.2497 |
| Mg | 7 | 0.1109 | 0.5551 | 0.0842 |
| Mg | 8 | 0.2279 | 0.4434 | 0.2505 |

**Table 1.** *Cont.*

| Element | Atom Number | Fractional Coordinates of Atoms | | |
|---|---|---|---|---|
| | | x | y | z |
| Mg | 9 | 0.4465 | 0.5535 | 0.0865 |
| Mg | 10 | 0.7736 | 0.5536 | 0.0865 |
| Mg | 11 | 0.8831 | 0.4437 | 0.2505 |
| Mg | 12 | 0.1108 | 0.8892 | 0.0842 |
| Mg | 13 | 0.2222 | 0.7778 | 0.2501 |
| Mg | 14 | 0.4449 | 0.8891 | 0.0842 |
| Mg | 15 | 0.5566 | 0.7728 | 0.2505 |
| Mg | 16 | 0.7780 | 0.8891 | 0.0829 |
| Mg | 17 | 0.8830 | 0.7727 | 0.2505 |
| Mg | 18 | 0.1103 | 0.2217 | 0.4169 |
| Mg | 19 | 0.2272 | 0.1169 | 0.5823 |
| Mg | 20 | 0.4479 | 0.2291 | 0.4172 |
| Mg | 21 | 0.5563 | 0.1170 | 0.5823 |
| Mg | 22 | 0.7783 | 0.2217 | 0.4170 |
| Mg | 23 | 0.8890 | 0.1110 | 0.5833 |
| Mg | 24 | 0.1108 | 0.5549 | 0.4171 |
| Mg | 25 | 0.2271 | 0.4434 | 0.5823 |
| Mg | 26 | 0.4479 | 0.5521 | 0.4173 |
| Mg | 27 | 0.7709 | 0.5521 | 0.4172 |
| Mg | 28 | 0.8830 | 0.4437 | 0.5823 |
| Mg | 29 | 0.1107 | 0.8893 | 0.4170 |
| Mg | 30 | 0.2222 | 0.7778 | 0.5828 |
| Mg | 31 | 0.4451 | 0.8892 | 0.4171 |
| Mg | 32 | 0.5566 | 0.7729 | 0.5823 |
| Mg | 33 | 0.7783 | 0.8897 | 0.4169 |
| Mg | 34 | 0.8831 | 0.7728 | 0.5823 |
| Mg | 35 | 0.1109 | 0.2221 | 0.7501 |
| Mg | 36 | 0.2236 | 0.1115 | 0.9170 |
| Mg | 37 | 0.4465 | 0.2265 | 0.7474 |
| Mg | 38 | 0.5545 | 0.1116 | 0.9171 |
| Mg | 39 | 0.7779 | 0.2221 | 0.7501 |
| Mg | 40 | 0.8890 | 0.1110 | 0.9170 |
| Mg | 41 | 0.1109 | 0.5551 | 0.7489 |
| Mg | 42 | 0.2237 | 0.4453 | 0.9171 |
| Mg | 43 | 0.4466 | 0.5534 | 0.7474 |
| Mg | 44 | 0.5556 | 0.4444 | 0.9170 |
| Mg | 45 | 0.7735 | 0.5535 | 0.7474 |
| Mg | 46 | 0.8884 | 0.4455 | 0.9171 |
| Mg | 47 | 0.1108 | 0.8892 | 0.7489 |
| Mg | 48 | 0.2221 | 0.7779 | 0.9170 |
| Mg | 49 | 0.4449 | 0.8891 | 0.7489 |
| Mg | 50 | 0.5547 | 0.7763 | 0.9171 |
| Mg | 51 | 0.7779 | 0.8890 | 0.7501 |
| Mg | 52 | 0.8885 | 0.7764 | 0.9170 |
| X | 1 | 0.5552 | 0.4447 | 0.2505 |
| X | 2 | 0.5552 | 0.4448 | 0.5802 |

To compare the stability of the two kinds of solid solution structures, the cohesive energy has been calculated. The whole crystal structure's stability depends on the cohesive energy. When the cohesive energy is negative, the structure is stable. In addition, the stability values of different structures can be compared with each other according to the absolute value of cohesive energy. The cohesive energy can be calculated by means of the following formula [30–32]:

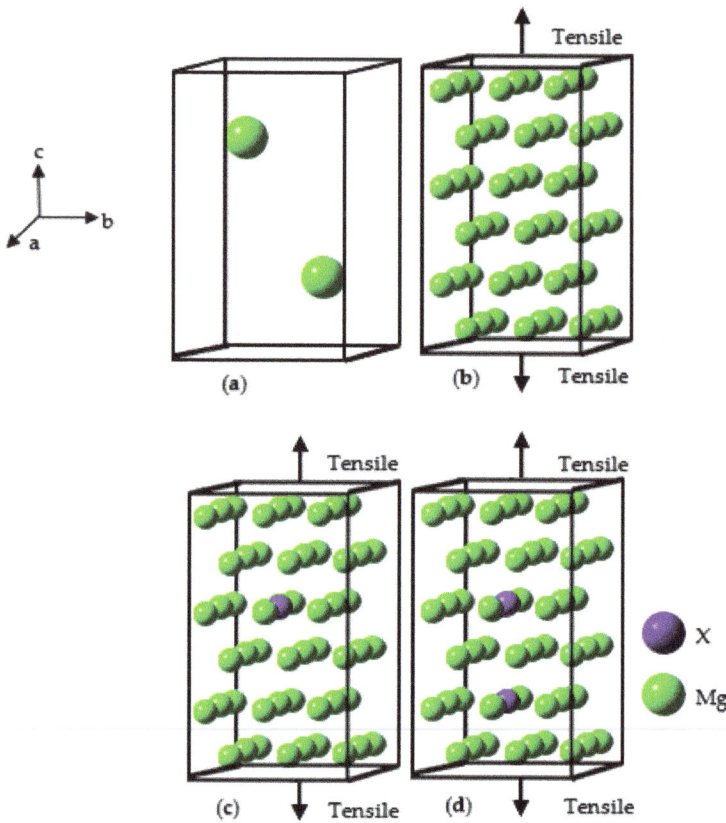

**Figure 1.** Crystal structure diagram: (**a**) Mg, (**b**) Mg$_{54}$, (**c**) Mg$_{53}$X$_1$, (**d**) Mg$_{52}$X$_2$ (X = Zn, Ag).

$$E_{coh} = \frac{E_{tot} - N_A E_{atom}^A - N_B E_{atom}^B}{N_A + N_B} \quad (3)$$

In the above formula, $E_{coh}$ and $E_{tot}$ are the cohesive energy and total energy of the structures. $E_{atom}^A$ and $E_{atom}^B$ are the free state energy of Mg and X atoms. The free state energy values of Mg, Zn and Ag atoms are −972.5823 eV/atom, −1708.9884 eV/atom and −1024.9934 eV/atom, respectively. $N^A$ and $N^B$ are the numbers of Mg and X atoms in the structures.

The calculation results of Mg$_{54}$, Mg$_{53}$X$_1$ and Mg$_{52}$X$_2$ (X = Zn, Ag) are listed in Table 2. The calculated cohesive energies representing the strain are shown in Figure 2. It can be seen that the cohesive energy values for all the structures are negative in the range of 0–20% strain, which indicates that several structures remain stable during the tensile process. When there is no strain, the absolute order of cohesive energy for the five structures is Mg$_{54}$ < Mg$_{52}$Zn$_2$ < Mg$_{53}$Ag$_1$ < Mg$_{52}$Ag$_2$, which indicates that the inclusion of Zn and Ag can promote the stability of Mg-based alloys. At the same time, the stability of a Ag-containing solid solution structure is greater than that of on containing Zn. In addition, in the range of 0–20% strain, the cohesive energy values of Mg$_{53}$X$_1$ and Mg$_{52}$X$_2$ are always negative, but the absolute values are decreased. It can be concluded that with an increase in strain, Mg$_{53}$X$_1$ and Mg$_{52}$X$_2$ remain stable, but their stability shows a downward trend.

Table 2. Table 2. The cohesive energy of $Mg_{54}$, $Mg_{53}X_1$ and $Mg_{52}X_2$ (X = Zn, Ag) under different strains.

| Strain (%) | $E_{coh}$, kJ/mol | | | | |
|---|---|---|---|---|---|
| | $Mg_{54}$ | $Mg_{53}Zn_1$ | $Mg_{52}Zn_2$ | $Mg_{53}Ag_1$ | $Mg_{52}Ag_2$ |
| 0 | −193.46 | −193.77 | −194.01 | −196.93 | −200.27 |
| 2 | −193.01 | −193.25 | −193.44 | −196.29 | −199.58 |
| 4 | −192.44 | −192.60 | −192.73 | −195.64 | −198.84 |
| 6 | −191.70 | −191.82 | −191.92 | −194.84 | −197.99 |
| 7 | −191.41 | −191.46 | −191.50 | −194.33 | −197.24 |
| 8 | −191.37 | −191.22 | −191.23 | −194.13 | −197.05 |
| 9 | −190.90 | −190.91 | −190.89 | −193.84 | −196.77 |
| 10 | −190.47 | −190.47 | −190.42 | −193.42 | −196.36 |
| 12 | −189.77 | −189.70 | −189.59 | −192.72 | −195.61 |
| 14 | −188.75 | −188.63 | −188.49 | −191.74 | −194.64 |
| 16 | −187.46 | −187.31 | −187.13 | −190.48 | −193.44 |
| 18 | −186.00 | −185.86 | −185.66 | −189.33 | −192.01 |
| 20 | −184.60 | −184.44 | −184.26 | −187.57 | −190.99 |

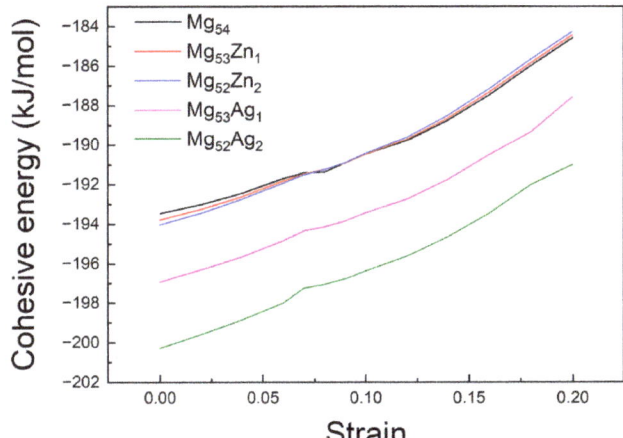

Figure 2. The calculated cohesive energies as a function of the strain.

*3.2. Tensile Properties*

The stress values of $Mg_{54}$, $Mg_{53}X_1$ and $Mg_{52}X_2$ (X = Zn, Ag) under different strains have been calculated using first principle tensile simulations, and are listed in Table 3. The strain should not exceed 20%. For a more intuitive analysis, the corresponding stress–strain curves are shown in Figures 3 and 4. The abscissa represents crystal strain and the ordinate represents stress. It is clear that the types of deformation experienced by the different solid solution structures are the same, as elastic deformation, uneven plastic deformation and uniform plastic deformation successively occurred. In the elastic deformation zone, the solution structure will undergo elastic deformation when stretching, and the structure will return to its original length when stretching ceases. In the non-uniform plastic deformation zone, elastic deformation and plastic deformation will occur simultaneously with the application of stress. Finally, only plastic deformation occurs in the uniform plastic deformation zone.

**Table 3.** The stress values of $Mg_{54}$, $Mg_{53}X_1$ and $Mg_{52}X_2$ (X = Zn, Ag) under different strains.

| Strain (%) | Stress, $10^3$ MPa | | | | |
|---|---|---|---|---|---|
| | $Mg_{54}$ | $Mg_{53}Zn_1$ | $Mg_{52}Zn_2$ | $Mg_{53}Ag_1$ | $Mg_{52}Ag_2$ |
| 0 | 0.00 | 0.00 | 0.00 | 0.00 | 0.00 |
| 2 | 1.74 | 2.00 | 2.21 | 2.05 | 2.34 |
| 4 | 2.35 | 2.51 | 2.67 | 2.55 | 2.77 |
| 6 | 2.69 | 2.86 | 3.03 | 2.83 | 2.95 |
| 7 | 1.21 | 1.46 | 1.70 | 1.25 | 1.37 |
| 8 | 1.85 | 1.99 | 2.20 | 1.82 | 1.81 |
| 9 | 2.22 | 2.33 | 2.52 | 2.23 | 2.22 |
| 10 | 2.47 | 2.63 | 2.84 | 2.52 | 2.58 |
| 12 | 3.11 | 3.29 | 3.45 | 3.05 | 3.10 |
| 14 | 4.15 | 4.29 | 4.41 | 4.04 | 3.96 |
| 16 | 4.97 | 5.01 | 5.10 | 4.89 | 4.78 |
| 18 | 5.15 | 5.16 | 5.17 | 5.03 | 5.34 |
| 20 | 5.02 | 5.06 | 5.04 | 5.27 | 5.28 |

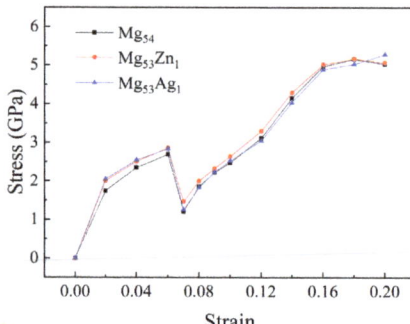

**Figure 3.** Stress–strain curve of materials with Zn and Ag content of 1.8 at.%.

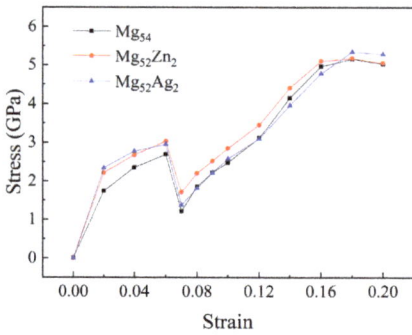

**Figure 4.** Stress–strain curve of materials with Zn and Ag contents of 3.7 at.%.

In the elastic deformation area, the stress and strain change in direct proportion to one another, and the image change trend shows a straight line. Figures 3 and 4 show that the elastic deformation areas of several structures are very small, and the strain range is 4–8%. With the increase in strains, non-uniform plastic deformation will rapidly develop. Thereafter, elastic deformation will be accompanied by elastic plastic deformation, after which it will be difficult for this material to completely return to its original length. In industrial production, yield strength is the strength index of a material that can yield, and materials are typically selected for their yield strength. Figure 3 shows that, with the increase in strain, several solid solution structures undergo obvious yielding, accompanied by the appearance of an upper yield point and a lower yield point. Because lower yield is a stable form of the yield process, the lower yield point is usually selected as the yield strength if the structure shows both upper and lower yield points [33]. It can be seen from Table 3 and Figure 3 that when the solid's solubility is 1.8 at.%, the solid solution structure will have a lower yield point at the strain of 7%, and the yield strengths of $Mg_{54}$, $Mg_{53}Zn_1$ and $Mg_{53}Ag_1$ will be $1.21 \times 10^3$ MPa, $1.46 \times 10^3$ MPa and $1.25 \times 10^3$ MPa, respectively. When the solid solubility is 1.8 at.%, the inclusion of both Zn and Ag will increase the yield strength of the Mg-based alloy. The yield strengths of the Mg-based alloys with Zn and Ag are increased by 20.67% and 3.31%, respectively, and the strengthening effect of Zn is obviously higher than that of Ag. Additionally, it can be seen from Table 3 and Figure 4 that the yield strengths of $Mg_{52}Zn_2$ and $Mg_{52}Ag_2$ are $1.70 \times 10^3$ MPa and $1.37 \times 10^3$ MPa, respectively, when the solid solubility is 3.7 at.%, which values are 40.50% and 13.22% higher than those of the Mg-based alloys. It can be seen from the comparison that the inclusion of both Zn and Ag can promote the Mg-based alloy materials' yield strength, and the effect of Zn is more obvious. In addition, the strengthening effect is also affected by the value of solid solubility; in particular, when the solid solubility is 3.7 at.%, the strengthening effect is more obvious.

Tensile strength is another important index that is used to measure the tensile properties of materials. It is the maximum stress that materials can withstand under tension, which can be used to reflect the material's ability to resist damages. When the structure exhibits yielding, a plastic deforming area will appear as the strain continues to increase. In the region of plastic deformation, the stress increases with the increase in strain until the tensile strength is reached. If the strain continues to increase, the stress will drop, which indicates that the structure has been destroyed. The figures show that several solid solution structures of $Mg_{54}$, $Mg_{53}X_1$ and $Mg_{52}X_2$ (X = Zn, Ag) reach their tensile strength value in the plastic deformation zone. When the solid solubility is 1.8 at.%, the tensile strengths of $Mg_{54}$, $Mg_{53}Zn_1$ and $Mg_{53}Ag_1$ are $5.15 \times 10^3$ MPa, $5.16 \times 10^3$ MPa and $5.27 \times 10^3$ MPa, respectively. The results show that Zn hardly improves the tensile strength of Mg-based alloy materials, while Ag can improve the tensile strength of Mg-based alloys. When the solid solubility is 3.7 at.%, the tensile strengths of $Mg_{52}Zn_2$ and $Mg_{52}Ag_2$ are $5.17 \times 10^3$ MPa and $5.34 \times 10^3$ MPa, respectively. Further analysis shows that Zn has little effect on the tensile strength of Mg-based alloy materials. Conversely, the strengthening effect of Ag on Mg-based alloy materials is obvious, and will be made more obvious with the increase in solid solubility.

Through the above research, we found that a suitable Ag content can enhance the yield strength and tensile strength of Mg-based alloys, and this strengthening effect is better than that of Zn. The results of this study validate other experimental results. Feng et al. [34] found via microhardness tests that the microhardness of magnesium alloy could be enhanced by adding Ag, and the alloy exhibited good mechanical properties. Ben Hamu et al. [35] found that adding Ag could improve the mechanical properties of Mg–Zn alloys. Zhao et al. [36] found through experiments that Mg–Zn–Ag alloys display their best mechanical properties as the concentration of Ag is changed, with the results being particularly striking for yield strength, ultimate tensile strength, and elongation. From these results, it can be inferred that the method used in this study is correct and feasible for application. Ag belongs to the group of precious metals, which greatly limits the smooth progression of this experiment. This study employed the method of simulated stretching, which not only saves time and costs, but can also be used to obtain reliable data, providing a new path for the study of precious and rare metals.

*3.3. Electronic Properties*

3.3.1. Density of States

To further understand the stability and strengthening mechanisms of the structures, the densities of different states (total and partial) of $Mg_{52}Zn_2$ and $Mg_{52}Ag_2$ structures have been calculated without the application of strain. The bonding characteristics of the structures can be inferred through the total density of states and the partial density of states. The total density of states (TDOS) and the partial density of states (PDOS) of $Mg_{52}Zn_2$ and $Mg_{52}Ag_2$ are depicted in Figures 5 and 6, respectively.

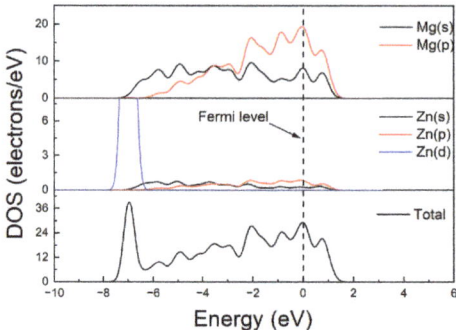

**Figure 5.** The total and partial density of states of $Mg_{52}Zn_2$.

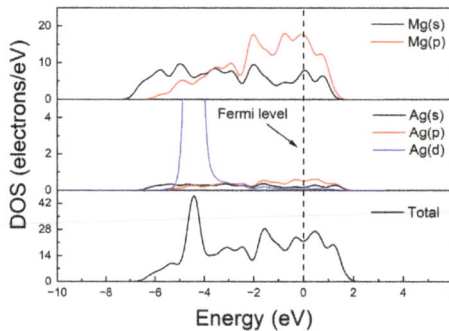

**Figure 6.** The total and partial density of states of $Mg_{52}Ag_2$.

The figures show that at no point does the energy range approach the Fermi level, whereat there are no electron states, i.e., there is no band gap. The valence and conduction bands overlap, and electrons are delocalized, so the structures exhibit a metallic character. Figure 5 shows that the bonding peaks of $Mg_{52}Zn_2$ are mainly distributed in the range from −7.5 to 1.5 eV. Here, the valence band is associated with the valence electrons of Mg s, Mg p, Zn s, Zn p and Zn d orbitals, and the conduction band is associated with the contribution of valence electrons of Mg s, Mg p, Zn s and Zn p orbitals. It is worth noting that there is a peak at −7 eV, which is mainly provided by the valence electrons of the Zn d orbital. In addition, there is sp hybridization between Mg s−Mg p and Zn s−Zn p orbitals in the valence band. Figure 6 shows that the bonding peaks in $Mg_{52}Ag_2$ are mainly distributed between −6.5 and 2 eV. Here, the valence band is associated with the valence electrons of the Mg s, Mg p, Ag s, Ag p and Ag d orbitals, and the conduction band is mostly associated with the valence electrons of Mg s and Mg p orbitals. Overall, the contribution of the Ag s and Ag p orbital is limited. There is a peak at −4.5 eV, which is mainly contributed by the valence electrons of the Ag d orbital. At the same time, the density of states of the Ag d orbital overlaps with those of other orbitals in the entire valence band range, indicating the presence of spd hybridization amongst Mg s, Mg p, and Ag d orbitals. In general, the density of state of $Mg_{52}Ag_2$ at the Fermi level is higher than that of $Mg_{52}Zn_2$, which shows more active metallicity. In addition, the hybridization of $Mg_{52}Ag_2$ is also more pronounced than that of $Mg_{52}Zn_2$.

### 3.3.2. Electron Density Difference

In order to reveal the mechanism of the strengthening effect of Zn and Ag on a Mg-based alloy, the electron density differences of $Mg_{54}$, $Mg_{52}Zn_2$ and $Mg_{52}Ag_2$ have been calculated and analyzed. Usually, the electron density difference can be used to analyze the electron transfer between atoms in the crystal structures. At the same time, the state of bonding in the crystal structures can also be assessed using an electron density difference diagram. The electron density difference diagrams of $Mg_{54}$, $Mg_{52}Zn_2$ and $Mg_{52}Ag_2$ are shown in Figure 7. It can be seen from Figure 7 that the electron clouds around each nucleus in $Mg_{54}$, $Mg_{52}Zn_2$ and $Mg_{52}Ag_2$ are uniformly and clearly distributed, and there are no obvious overlaps between electron clouds. This indicates that in the structures of $Mg_{54}$, $Mg_{52}Zn_2$ and $Mg_{52}Ag_2$, the atoms bond with each other through metallic bonds. From Figure 7, charge accumulation can be seen around the nuclei of Mg, Zn, and Ag atoms, but the scales of the color cards in the three color maps are different. In the scaling of the color card, the reddest area (at the bottom) indicates the maximum charge density. It can be found that the maximum charge density of $Mg_{54}$ is $2.425 \times 10^{-2}$ e/Å$^3$, the maximum charge density of $Mg_{52}Zn_2$ is $5.247 \times 10^{-2}$ e/Å$^3$, and the maximum charge density of $Mg_{52}Ag_2$ is $1.047 \times 10^{-1}$ e/Å$^3$. It is clear that $Mg_{52}Zn_2$ and $Mg_{52}Ag_2$ aggregate more electrons in their bonding regions than $Mg_{54}$, indicating that the addition of Zn and Ag atoms can enhance the stability of the matrix. In addition, $Mg_{52}Ag_2$ aggregates more electrons in the bonding region than $Mg_{52}Zn_2$, indicating that Ag atoms have a stronger capacity to promote the stability of the matrix than Zn atoms. This also corresponds to previous calculations and the results of the analysis of cohesive energies and tensile properties from an electronic perspective, indicating that the previous analysis results are correct.

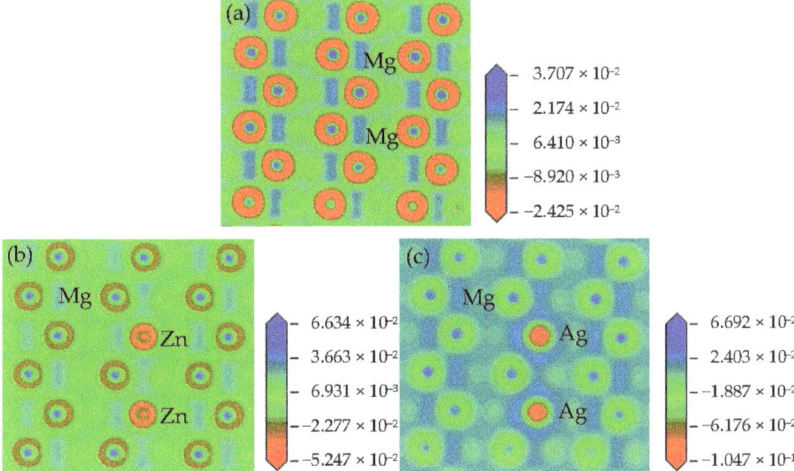

**Figure 7.** Electron density differences of $Mg_{54}$ (**a**), $Mg_{52}Zn_2$ (**b**) and $Mg_{52}Ag_2$ (**c**).

### 4. Conclusions

In this work, the structural, electronic and tensile properties of the solid solution structures of $Mg_{54}$, $Mg_{53}X_1$ and $Mg_{52}X_2$ (X = Zn, Ag) subjected to 0–20% strains are investigated using the first principle calculation method. The conclusion is that the solid solution structures of $Mg_{53}X_1$ and $Mg_{52}X_2$ (X = Zn, Ag) are stable in the 0–20% strain range. In addition, the stability of the Ag structures is greater than that of the Zn structures when two levels of solid solubility are employed. With the increase in strains, several solid solution structures at these two levels of solid solubility show obvious yielding. The results show that the application of both Zn and Ag can increase the yield strength of Mg-based alloys. The strengthening effect of Zn is more obvious. In addition, the strengthening effect is also affected by the solution's concentration. When the solubility is 3.7 at.%, the strengthening effect is greater. The stress–strain results of tensile strength show that Zn does not promote the tensile strength of Mg-based alloys, while Ag can improve the tensile strength. The strengthening effect is strong at a solubility level of 3.7 at.%, compared to 1.8 at.%. Finally, the analyses of the density of states and the electron density difference verify the mechanisms of the strengthening effects of elements Zn and Ag on Mg-based alloys from the electronic perspective.

**Author Contributions:** Writing—original draft preparation, Y.G.; revising W.F.; writing—review and editing, L.F. and C.W.; project administration, Y.G. and X.C. All authors have read and agreed to the published version of the manuscript.

**Funding:** This study was funded by the Natural Science Foundation of Liaoning Province (Nos. 2021-BS-156 and 2021-BS-154), the Ministry of Education industry-school cooperative education project (Nos. 220606517061413, 220502116230206 and 220606517061317), the Graduate Education and Teaching Reform Project of Shenyang Normal University (No. YJSJG320210102), the ninth batch of teaching reform project of Shenyang Normal University (No. JG2021-YB063), and Innovative Talent Program of Shenyang Normal University, grant number 222-51300417.

**Institutional Review Board Statement:** Not applicable.

**Informed Consent Statement:** Not applicable.

**Data Availability Statement:** Not applicable.

**Conflicts of Interest:** The authors declare no conflict of interest.

## References

1. Xie, J.; Zhang, J.; You, Z.; Liu, S.; Guan, K.; Wu, R.; Wang, J.; Feng, J. Towards developing Mg alloys with simultaneously improved strength and corrosion resistance via RE alloying. *J. Magnes. Alloys* **2021**, *9*, 41–56. [CrossRef]
2. Sharma, S.K.; Saxena, K.K.; Malik, V.; Mohammed, K.A.; Prakash, C.; Buddhi, D.; Dixit, S. Significance of Alloying Elements on the Mechanical Characteristics of Mg-Based Materials for Biomedical Applications. *Crystals* **2022**, *12*, 1138. [CrossRef]
3. Ji, H.; Liu, W.; Wu, G.; Ouyang, S.; Gao, Z.; Peng, X.; Ding, W. Influence of Er addition on microstructure and mechanical properties of as-cast Mg-10Li-5Zn alloy. *Mater. Sci. Eng. A* **2019**, *739*, 395–403. [CrossRef]
4. Tian, Y.; Hu, H.J.; Zhao, H.; Zhang, W.; Liang, P.C.; Jiang, B.; Zhang, D.F. An extrusion−shear−expanding process for manufacturing AZ31 magnesium alloy tube. *Trans. Nonferrous Met. Soc. China* **2022**, *32*, 2569–2577. [CrossRef]
5. Wang, G.G.; Weiler, J.P. Recent developments in high-pressure die-cast magnesium alloys for automotive and future applications. *J. Magnes. Alloys* **2023**, *11*, 78–87. [CrossRef]
6. Liu, J.; Sun, J.; Chen, Q.; Lu, L. Intergranular Cracking in Mg-Gd-Y Alloy during Tension Test. *Crystals* **2022**, *12*, 1040. [CrossRef]
7. Wang, M.; Xu, X.Y.; Wang, H.Y.; He, L.H.; Huang, M.X. Evolution of dislocation and twin densities in a Mg alloy at quasi-static and high strain rates. *Acta Mater.* **2020**, *201*, 102–113. [CrossRef]
8. Du, H.Q.; Li, F.; Huo, P.D.; Wang, Y. Microstructure evolution and ductility improvement mechanisms of magnesium alloy in interactive alternating forward extrusion. *Trans. Nonferrous Met. Soc. China* **2022**, *32*, 2557–2568. [CrossRef]
9. Guo, P.C.; Li, L.C.; Liu, X.; Ye, T.; Cao, S.F.; Xu, C.C.; Li, S.K. Compressive deformation behavior and microstructure evolution of AM80 magnesium alloy under quasi-static and dynamic loading. *Int. J. Impact. Eng.* **2017**, *109*, 112–120. [CrossRef]
10. Pan, F.S.; Mao, J.J.; Zhang, G.; Tang, A.; She, J. Development of high-strength, low-cost wrought Mg–2.0 mass% Zn alloy with high Mn content. *Prog. Nat. Sci.-Mater.* **2016**, *26*, 630–635. [CrossRef]
11. Jiang, B.; Dong, Z.H.; Zhang, A.; Song, J.F.; Pan, F.S. Recent advances in micro-alloyed wrought magnesium alloys: Theory and design. *Trans. Nonferrous Met. Soc. China* **2022**, *32*, 1741–1780. [CrossRef]
12. Huang, W.S.; Chen, J.H.; Yan, H.G.; Li, Q.; Xia, W.J.; Su, B.; Zhu, W.J. Solid solution strengthening and damping capacity of Mg−Ga binary alloys. *Trans. Nonferrous Met. Soc. China* **2022**, *32*, 2852–2865. [CrossRef]
13. Miao, J.S.; Sun, W.H.; Klarner, A.D.; Luo, A.A. Interphase boundary segregation of silver and enhanced precipitation of $Mg_{17}Al_{12}$ Phase in a Mg-Al-Sn-Ag alloy. *Scr. Mater.* **2018**, *154*, 192–196. [CrossRef]
14. Cáceres, C.H.; Blake, A. The Strength of Concentrated Mg-Zn Solid Solutions. *Phys. Status Solidi A* **2002**, *194*, 147–158. [CrossRef]
15. Wang, J.F.; Wang, K.; Hou, F.; Liu, S.J.; Peng, X.; Wang, J.X.; Pan, F.S. Enhanced strength and ductility of Mg-RE-Zn alloy simultaneously by trace Ag addition. *Mater. Sci. Eng. A* **2018**, *728*, 10–19. [CrossRef]
16. Tong, L.B.; Li, X.H.; Zhang, H.J. Effect of long period stacking ordered phase on the microstructure, texture and mechanical properties of extruded Mg-Y-Zn alloy. *Mater. Sci. Eng. A* **2013**, *563*, 177–183. [CrossRef]
17. Giusepponi, S.; Celino, M. The ideal tensile strength of tungsten and tungsten alloys by first-principles calculations. *J. Nucl. Mater.* **2013**, *435*, 52–55. [CrossRef]
18. Roundy, D.; Cohen, M.L. Ideal strength of diamond, Si, and Ge. *Phys. Rev. B* **2001**, *64*, 212103. [CrossRef]

19. Luo, W.D.; Roundy, D.; Cohen, M.L.; Morris, J.W. Ideal strength of bcc molybdenum and niobium. *Phys. Rev. B* **2002**, *66*, 094110. [CrossRef]
20. Lu, G.H.; Deng, S.H.; Wang, T.M.; Kohyama, M.; Yamamoto, R. Theoretical tensile strength of an Al grain boundary. *Phys. Rev. B* **2004**, *69*, 134106. [CrossRef]
21. Zhang, Y.; Lü, G.H.; Deng, S.H.; Wang, T.M. First-principles computational tensile test on an Al grain boundary. *Acta Phys. Sin-Ch. Ed.* **2006**, *55*, 2901–2907. (In Chinese) [CrossRef]
22. Liu, X.; Liu, Z.; Liu, G.; Wang, W.; Li, J. First-principles study of solid solution strengthening in Mg–X (X=Al, Er) alloys. *Bull. Mater. Sci.* **2019**, *42*, 16. [CrossRef]
23. Luo, S.Q.; Tang, A.T.; Pan, F.S.; Song, K.; Wang, W.Q. Effect of mole ratio of Y to Zn on phase constituent of Mg-Zn-Zr-Y alloys. *Trans. Nonferrous Met. Soc. China* **2011**, *21*, 795–800. [CrossRef]
24. Wang, C.; Han, P.D.; Zhang, L.; Zhang, C.L.; Yan, X.; Xu, B.S. The strengthening effect of Al atoms into Mg-Al alloy: A first-principles study. *J. Alloys Compd.* **2009**, *482*, 540–543. [CrossRef]
25. Wang, C.; Huang, T.L.; Wang, H.Y.; Xu, X.N.; Jiang, Q.C. Effects of distributions of Al, Zn and Al+Zn atoms on the strengthening potency of Mg alloys: A first-principles calculations. *Comput. Mater. Sci.* **2015**, *104*, 23–28. [CrossRef]
26. Gironcoli, S.D.; Baroni, S.; Resta, R. Piezoelectric properties of III-V semiconductors from first-principles linear-response theory. *Phys. Rev. Lett.* **1989**, *62*, 2853–2856. [CrossRef]
27. Ganeshan, S.; Shang, S.L.; Wang, Y.; Liu, Z.K. Effect of alloying elements on the elastic properties of Mg from first-principles calculations. *Acta Mater.* **2009**, *57*, 3876–3884. [CrossRef]
28. Batchelder, F.W.; Raeuchle, R.F. Lattice Constants and Brillouin Zone Overlap in Dilute Magnesium Alloys. *Phys. Rev. B* **1957**, *105*, 59–61. [CrossRef]
29. Zhang, C.; Han, P.; Li, J.; Chi, M.; Yan, L.; Liu, Y.; Liu, X.; Xu, B. First-principles study of the mechanical properties of NiAl microalloyed by M (Y, Zr, Nb, Mo, Tc, Ru, Rh, Pd, Ag, Cd). *J. Phys. D Appl. Phys.* **2008**, *41*, 095410. [CrossRef]
30. Mao, P.L.; Yu, B.; Liu, Z.; Wang, F.; Ju, Y. Mechanical, electronic and thermodynamic properties of $Mg_2Ca$ Laves phase under high pressure: A first-principles calculation. *Comput. Mater. Sci.* **2014**, *88*, 61–70. [CrossRef]
31. Wang, F.; Sun, S.J.; Wang, Z.; Yu, B.; Mao, P.L.; Liu, Z. Microstructure, mechanical properties and first-principle analysis of vacuum die-cast Mg-7Al alloy with Sn addition. *Rare Metals* **2015**, *41*, 1961–1967. [CrossRef]
32. Mao, P.L.; Yu, B.; Liu, Z.; Wang, F.; Ju, Y. First-principles calculations of structural, elastic and electronic properties of AB2 type intermetallics in Mg–Zn–Ca–Cu alloy. *J. Magnes. Alloys* **2013**, *1*, 256–262. [CrossRef]
33. Li, C.; Zhang, K.; Ru, J.G. Pressure dependence of structural, elastic and electronic of $Mg_2Y$: A first principles study. *J. Alloys Compd.* **2015**, *647*, 573–577. [CrossRef]
34. Feng, Y.; Zhu, S.; Wang, L.; Chang, L.; Hou, Y.; Guan, S. Fabrication and characterization of biodegradable Mg-Zn-Y-Nd-Ag alloy: Microstructure, mechanical properties, corrosion behavior and antibacterial activities. *Bioact. Mater.* **2018**, *3*, 225–235. [CrossRef] [PubMed]
35. Ben-Hamu, G.; Eliezer, D.; Kaya, A.; Na, Y.G.; Shin, K.S. Microstructure and corrosion behavior of Mg–Zn–Ag alloys. *Mater. Sci. Eng. A* **2006**, *5*, 579–587. [CrossRef]
36. Zhao, H.; Wang, L.Q.; Ren, Y.P.; Yang, B.; Li, S.; Qin, G.W. Microstructure, Mechanical Properties and Corrosion Behavior of Extruded Mg–Zn–Ag Alloys with Single-Phase Structure. *Acta Metall. Sin. (Engl. Lett.)* **2018**, *31*, 575–583. [CrossRef]

**Disclaimer/Publisher's Note:** The statements, opinions and data contained in all publications are solely those of the individual author(s) and contributor(s) and not of MDPI and/or the editor(s). MDPI and/or the editor(s) disclaim responsibility for any injury to people or property resulting from any ideas, methods, instructions or products referred to in the content.

Article

# Effect of Aging Temperature on Precipitates Evolution and Mechanical Properties of GH4169 Superalloy

Anqi Liu [1,2], Fei Zhao [1,2,*], Wensen Huang [1,2], Yuanbiao Tan [1,2], Yonghai Ren [3], Longxiang Wang [3] and Fahong Xu [4]

1. College of Materials and Metallurgy, Guizhou University, Guiyang 550025, China; liuanqi9797@163.com (A.L.); huangws89@163.com (W.H.); ybtan1@gzu.edu.cn (Y.T.)
2. Key Laboratory for Materials Structure and Strength of Guizhou Province, Guiyang 550025, China
3. Guiyang Anda Aerospace Material Engineering Co., Ltd., Guiyang 550009, China; yonghairen@163.com (Y.R.); sonilgtcl@163.com (L.W.)
4. Guizhou Special Equipment Inspection and Testing Institute, Guiyang 550016, China; xfh598@163.com
* Correspondence: fzhao@gzu.edu.cn

**Abstract:** GH4169 is primarily strengthened through precipitation, with heat treatment serving as a crucial method for regulating the precipitates of the alloy. However, the impact of aging temperature on the microstructure and properties of GH4169 has not been thoroughly studied, hindering effective regulation of its microstructure and properties. This study systematically investigated the effects of aging temperature on the evolution of precipitates and mechanical properties of GH4169 alloy using various techniques such as OM, SEM, XRD and TEM. The results indicate that raising the aging temperature leads to an increase in the sizes of both the $\gamma''$ and $\gamma'$ phases in the alloy, as well as promoting the precipitation of δ phase at grain boundaries. Notably, the increase in $\gamma''$ phase size enhances the strength of the alloy, while the presence of δ phase is detrimental to its strength but greatly enhances its elongation. The yield strength of the alloy aged at 750 °C exhibits the highest yield strength, with values of 1135 MPa and 1050 MPa at room temperature and elevated temperature, respectively. As the aging temperature increases, the Portevin-Le Châtelier (PLC) effect during elevated temperature tensile tests at 650 °C gradually weakens. The PLC effect disappears almost completely when the aging temperature reaches 780 °C.

**Keywords:** GH4169 alloy; aging temperature; $\gamma''$ phase; δ phase; mechanical properties

Citation: Liu, A.; Zhao, F.; Huang, W.; Tan, Y.; Ren, Y.; Wang, L.; Xu, F. Effect of Aging Temperature on Precipitates Evolution and Mechanical Properties of GH4169 Superalloy. *Crystals* **2023**, *13*, 964. https://doi.org/10.3390/cryst13060964

Academic Editors: Daniel Medyński, Grzegorz Lesiuk and Anna Burduk

Received: 24 May 2023
Revised: 12 June 2023
Accepted: 15 June 2023
Published: 17 June 2023

**Copyright:** © 2023 by the authors. Licensee MDPI, Basel, Switzerland. This article is an open access article distributed under the terms and conditions of the Creative Commons Attribution (CC BY) license (https://creativecommons.org/licenses/by/4.0/).

## 1. Introduction

GH4169 (Inconel718) is a highly utilized superalloy in various industries such as aerospace and petroleum due to its exceptional mechanical properties [1,2]. The key contributors to its excellent mechanical properties are the precipitates within the matrix, which include the $\gamma''$ (Ni$_3$Nb) phase, $\gamma'$ (Ni$_3$(Al,Ti)) phase and δ (Ni$_3$Nb) phase. The $\gamma''$ phase is the main strengthening phase with an ordered body-centered tetragonal DO$_{22}$ crystal structure, and the precipitation temperature range is 595–870 °C [3]. The strengthening effect of $\gamma'$ phase is less than $\gamma''$ phase, which is a secondary strengthening phase with a face-centered cubic LI$_2$ crystal structure. The precipitation temperature range is 593–816 °C [4]. The formation of the $\gamma''$ phase and $\gamma'$ phase occurs during the two-stage aging process. In the conventional double aging process (720 °C/8 h + 620 °C/8 h), the purpose of aging at 720 °C is to mainly precipitate the $\gamma''$ phase, while aging at 620 °C is mainly to precipitate the $\gamma'$ phase [5,6]. The δ phase is typically incoherently precipitated with the matrix, with a precipitation temperature range of 750–1020 °C [7]. It is worth noting that the $\gamma''$ phase is a metastable phase in GH4169 alloy. The $\gamma''$ phase transforms into the δ phase when the alloy is subjected to long-term aging or high temperatures, thereby reducing the strengthening effect of the alloy. This limits the maximum temperature at which the alloy can be used to less than 650 °C [8,9].

The precipitation behavior of the $\gamma''$ and $\gamma'$ phases in GH4169 alloy is affected by temperature, with their nucleation and growth closely tied to aging temperatures. Altering the aging temperature can impact precipitate precipitation, which in turn affects the mechanical properties of the alloy. Rafiei et al. [10] discovered that the precipitation kinetics of the $\gamma''$ phase is highly dependent on aging temperature, with the maximum precipitation rate occurring at 780 °C. But Drexler et al. [11] discovered that the temperature at which precipitation occurs most rapidly is 750 °C. In a study by Fisk et al. [12], it was found that the aging temperature has a significant impact on the hardness of Inconel718. By analyzing the change in hardness, they determined that the sample aged at 760 °C for 8 h exhibited the highest level of hardness. The $\gamma''$ and $\gamma'$ phases are coherent with the matrix and can be cut by dislocations, resulting in a strengthening effect. The increase in strength caused by dislocation cutting the second phase is related to the size of the second phase. Qin et al. [13] found that when the volume fraction was similar, the strength displayed a rising trend with the increase of $\gamma''$ phase size. A similar phenomenon was reported in the study by Ran et al. [14]. They quantitatively calculated the contribution of the precipitated phase to the strength and found that the strength change is sensitive to the size of the $\gamma''$ phase. They also observed that the strengthening effect of large size $\gamma''$ phase is more significant. The precipitation of the $\delta$ phase in GH4169 alloy can enhance its plasticity [15,16]. However, since the $\delta$ phase has the same chemical composition as the $\gamma''$ phase, excessive precipitation of the former can result in a decrease in the formation element Nb of the latter, ultimately leading to a reduction in strength [17].

In summary, the precipitation behavior of the $\gamma''$ and $\gamma'$ phases in the alloy is significantly influenced by the aging temperature. Additionally, a competitive relationship exists between the $\gamma''$ and $\delta$ phases. The $\gamma''$ phase positively impacts the strength of the alloy, while the acquisition of the $\delta$ phase can improve its plasticity. Heat treatment is a crucial method for regulating the precipitates in GH4169 superalloy. However, current research on the precipitation behavior of precipitates through heat treatment remains insufficient. This study involved the preparation of three distinct GH4169 samples, each with a different heat treatment scheme. The impact of aging temperature on the second phase of GH4169 alloy was investigated using various techniques including OM, XRD, SEM and TEM. The study aimed to establish the relationship between the microstructure characteristics and mechanical properties of the alloy by examining the effects of the second phase characteristics on the tensile properties of the alloy at both room and elevated temperatures.

## 2. Materials and Methods

The material used in this study was commercial GH4169 superalloy, whose chemical compositions is shown in Table 1. Before undergoing the aging treatment, the samples underwent a solution treatment at a temperature of 1030 °C for a duration of 1 h to dissolve any precipitates present in the matrix. The microstructure after solution treatment is shown in Figure 1, which is composed of equiaxed grains and a small amount of carbides, and the average grain size is 81 μm. The samples after solid solution were treated with different heat treatments: (1) isothermal at 720 °C for 8 h, followed by furnace cooling to 620 °C at the rate of 50 °C/h, isothermal at 620 °C for 8 h, and then air cooling, recorded as A720; (2) isothermal at 750 °C for 8 h, followed by furnace cooling to 620 °C at the rate of 50 °C/h, isothermal at 620 °C for 8 h, and then air cooling, recorded as A750; (3) isothermal at 780 °C for 8 h, followed by furnace cooling to 620 °C at the rate of 50 °C/h, isothermal at 620 °C for 8 h, and then air cooling, recorded as A780. Aging treatments were carried out in a box-type heat treatment furnace with air as the medium.

Table 1. Chemical composition of GH4169 alloy.

| Element | Ni | Si | Mn | C | Cr | Mo | Al | Ti | Nb | Fe |
|---|---|---|---|---|---|---|---|---|---|---|
| Wt.% | 51.72 | 0.048 | 0.02 | 0.028 | 18.86 | 3.007 | 0.519 | 0.947 | 5.212 | Bal |

**Figure 1.** Microstructure of GH4169 alloy after solution treatment.

The aged samples were cut into tensile samples of the size indicated in Figure 2. Among them, the thickness of the tensile sample at room temperature was 2 mm. Prior to testing, the tensile samples were smoothed with sandpaper to eliminate cutting marks and oxide layers. The MTS 810 electro-hydraulic servo testing machine was used to conduct the tensile tests at a rate of 1 mm/min. To eliminate the influence of temperature gradient during elevated temperature tensile test (650 °C), the test was carried out after 20 min of heat preservation. The samples were cooled to room temperature in the air after fracture. In order to observe the fracture morphology, samples were cut along the direction perpendicular to the tensile direction near the fracture, and SEM observation was performed after ultrasonic cleaning with alcohol.

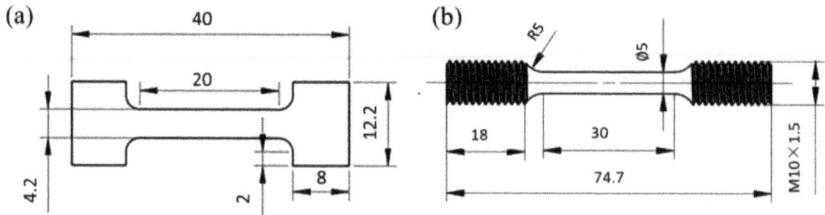

**Figure 2.** Schematic diagram with dimensions (in mm) of tensile samples: (**a**) room temperature tensile specimen; (**b**) elevated temperature tensile specimen.

OM, SEM and TEM were used to characterize the microstructure of the samples. Before OM and SEM observation, the samples need to be polished first and then corroded with 2.5 g $CuCl_2$ + 50 mL HCl + 50 mL $CH_3CH_2OH$ solution at room temperature. TEM samples were sampled on the samples after aging treatment. The samples were mechanically ground to about 50 μm, and the samples were punched into small round pieces with a diameter of 3 mm by a sample puncher. Then, electrolytic polishing was carried out. The precipitates of the samples were characterized by FEI F200X TEM, and the test voltage was 200 kV. The phase characteristics were determined by XRD using Cu Kα radiation in a range for 2θ of 20–100° at a scanning speed of 2°/min. The size and content of precipitates in SEM and TEM images were analyzed by Image-Pro Plus 6.0 software. Five pictures of each sample were selected for statistics.

## 3. Results and Discussions

*3.1. Effect of Aging Temperature on Mechanical Properties*

3.1.1. Microhardness

Table 2 shows the microhardness of the alloys that were aged at three different temperatures. The results indicate that the hardness of A720, A750 and A780 alloys are 418.9 HV, 471.8 HV and 453.5 HV, respectively. As the aging temperature increases, the hardness of the alloy initially increases and then decreases. Additionally, the hardness of the A750 and A780 samples is greater than that of the A720 samples.

**Table 2.** Microhardness of samples at different aging temperatures.

| Samples | A720 | A750 | A780 |
|---|---|---|---|
| Hardness/Hv | 418.9 | 471.8 | 453.5 |

3.1.2. Tensile Properties at Room Temperature

Figure 3 shows the engineering stress–strain curves and specific mechanical properties data histogram for the three heat-treated alloys at room temperature. The graph in Figure 3b indicates that as the aging temperature increases, the yield strength (YS) and ultimate tensile strength (UTS) of the alloy experience a significant increase initially, followed by a slight decrease. The yield strength and ultimate tensile strength of the A750 alloy are the highest, 1135 MPa and 1345 MPa, respectively, which are 13.5% and 8.9% higher than that of the A720 alloy. Furthermore, it can be found that the elongation (EL) of A750 and A780 alloys is not much different, but it is significantly lower than that of the A720 alloy. The A750 alloy has the highest yield strength and ultimate tensile strength, measuring at 1135 MPa and 1345 MPa, respectively. These values are 13.5% and 8.9% higher than those of the A720 alloy.

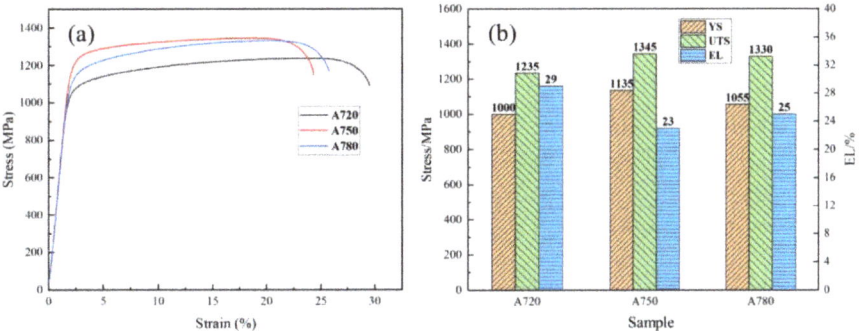

**Figure 3.** Tensile test at room temperature of specimens at different aging temperatures: (**a**) stress–strain curves; (**b**) strength and ductility values.

3.1.3. Tensile Properties at Elevated Temperature

Figure 4a shows the hot tensile stress–strain curves of the three aged samples at 650 °C. The specific tensile strength data are presented in Figure 4b, which shows that the yield strength of the alloy increases initially and then decreases with the increase in aging temperature. The yield strength of A750 and A780 alloys is considerably higher than that of the A720 alloy. The yield strength of the A750 alloy is 1050 MPa, which represents an 11% increase compared to that of the A720 alloy. Both the A720 and A780 alloys have good plasticity of 26%, which is obviously better than A750 alloy. Additionally, the ultimate tensile strength of the alloys increases with the aging temperature, as shown in Figure 4b. The ultimate tensile strengths of A720, A750 and A780 alloys are 1105 MPa, 1185 MPa and 1210 MPa, respectively. During high temperature tensile tests, both A720 and A750 alloys exhibit typical dynamic strain aging phenomenon, except for the A780 alloy, as illustrated in Figure 4a.

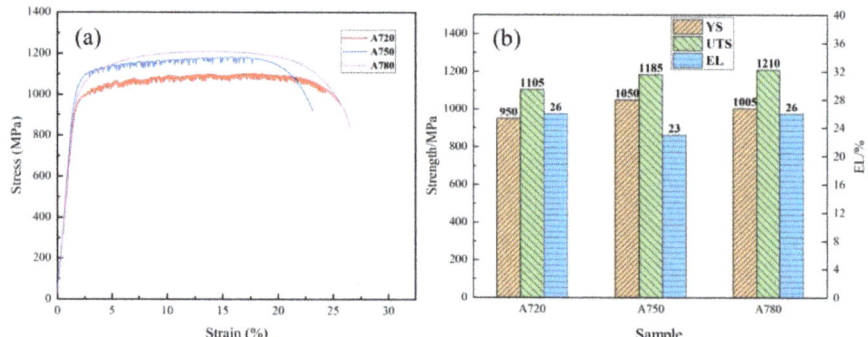

**Figure 4.** Tensile test at elevated temperature of specimens at different aging temperatures: (**a**) stress–strain curves; (**b**) strength and ductility values.

*3.2. Relationship between Microstructure Evolution and Mechanical Properties*

Figure 5 shows the optical microscope images of GH4169 alloy following three aging processes. The images reveal that the grains remain equiaxed, with a small presence of twins and bulk precipitates. The bulk precipitate was analyzed using energy dispersive spectroscopy (EDS), with the result presented in Figure 5d. The EDS analysis indicates that these bulk precipitates are carbides such as NbC or TiC. The grain size of the aged alloy was measured using the transversal method in Image-Pro Plus 6.0 software. The average grain sizes of A720, A750 and A780 alloys are 84 μm, 86 μm and 87 μm, respectively. Upon comparing the metallographic image of the solid solution alloy shown in Figure 1, it was observed that the aging temperature had no effect on the grain size of the alloy.

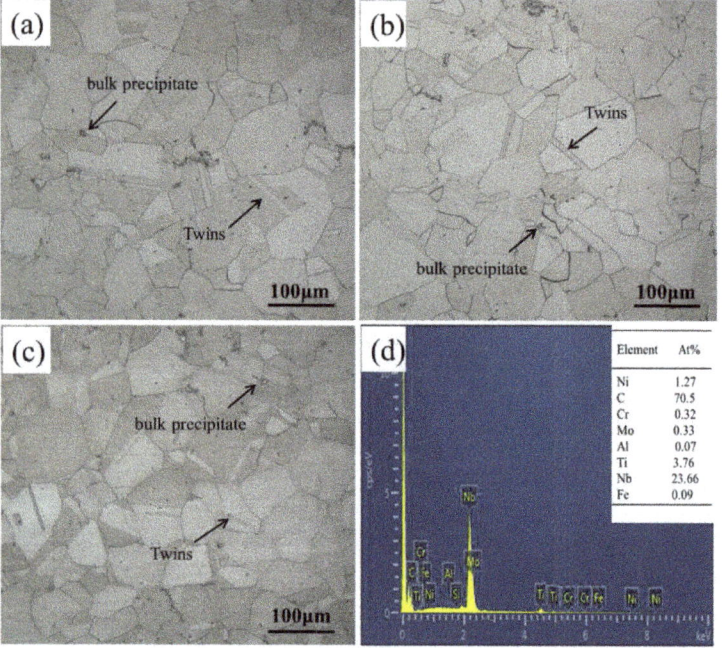

**Figure 5.** Microstructure of GH4169 treated at different aging temperatures: (**a**) A720; (**b**) A750; (**c**) A780; (**d**) EDS analysis of bulk precipitates.

Figure 6 shows the results of XRD phase analysis of the three heat-treated alloys. From the diagram, there is no obvious difference between the diffraction patterns of the three alloys. The diffraction peaks with higher diffraction intensity correspond to γ (111), γ (200) and γ (220), respectively. No diffraction peak of δ phase was observed in Figure 6, which may be due to the absence of δ phase precipitation or the low content of δ phase in GH4169 alloy after heat treatment. In addition, the absence of diffraction peaks of γ″ and γ′ phases in the diagram can be attributed to their coherence with the γ matrix. As a result, the peaks of γ′ (111) and γ (111) coincide; γ′ (200) and γ″ (200) peaks coincide with the γ (200) peak; and γ′ (220) and γ″ (220) peaks coincide with the γ (220) peak.

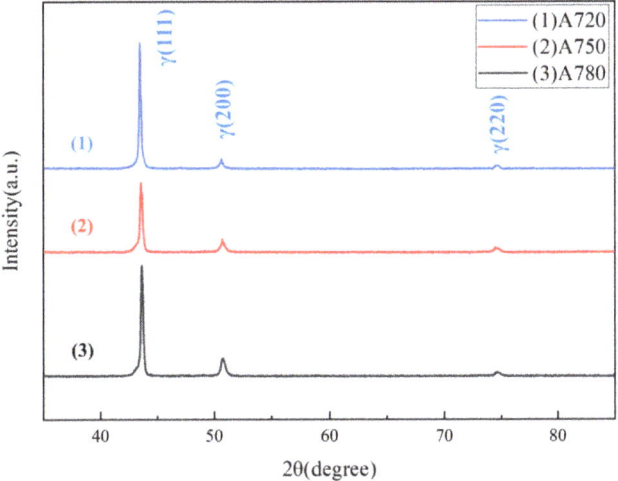

**Figure 6.** XRD patterns of three heat-treated samples.

The results from Figures 3 and 4 indicate that the yield strength and ultimate tensile strength of the A750 alloy and the A780 alloy are significantly improved compared with the A720 alloy. In general, the main strengthening mechanisms for nickel-based superalloys include solution strengthening, precipitation strengthening, dislocation strengthening and grain boundary strengthening [18,19]. The contribution of solution strengthening to strength was consistent across the three aged samples due to the presence of the same alloying elements in the matrix. The grain size of the alloy was not affected by the aging temperature, as evidenced by Figure 5, leading to similar grain boundary strengthening effects. Additionally, the XRD pattern in Figure 6 showed similar peak widths across all three aging treatment samples. This indicates that the lattice strain ($\varepsilon$) is similar. According to the calculation formula of dislocation density $\rho = 16.1\varepsilon^2/b^2$ ($\rho$ is dislocation density, and $b$ is the Burgers vector of GH4169 alloy) [20], there is no significant difference in dislocation density among the three aged samples, which leads to the similar contribution of dislocation strengthening to strength. Therefore, the strength difference of the three aged samples mainly comes from precipitation strengthening.

The TEM images of the alloy after three aging processes are presented in Figure 7. The bright field images of the alloy are shown in Figure 7a,d,g, accompanied by their corresponding selected area electron diffraction patterns in Figure 7b,e,h. The diffraction patterns of the aged alloys exhibit similar characteristics, consisting of distinct matrix diffraction spots and faint precipitates diffraction spots. The identified weak diffraction spots are characteristic of γ″ and γ′ superlattice diffraction, indicating that the fine precipitates dispersed in the matrix are γ″ and γ′ phases. The γ″ and γ′ phases cannot be distinguished and counted in the bright field image due to the lack of contrast between the precipitates and the matrix. To determine the size and volume fraction of the γ″ and γ′ phases, dark field images analysis with (002) γ″ was performed, as illustrated in Figure 7c,f,i. It can be seen from the dark field images that the size of the precipitates increases significantly with the increase in the aging temperature, which is consistent with the phenomenon observed by SEM.

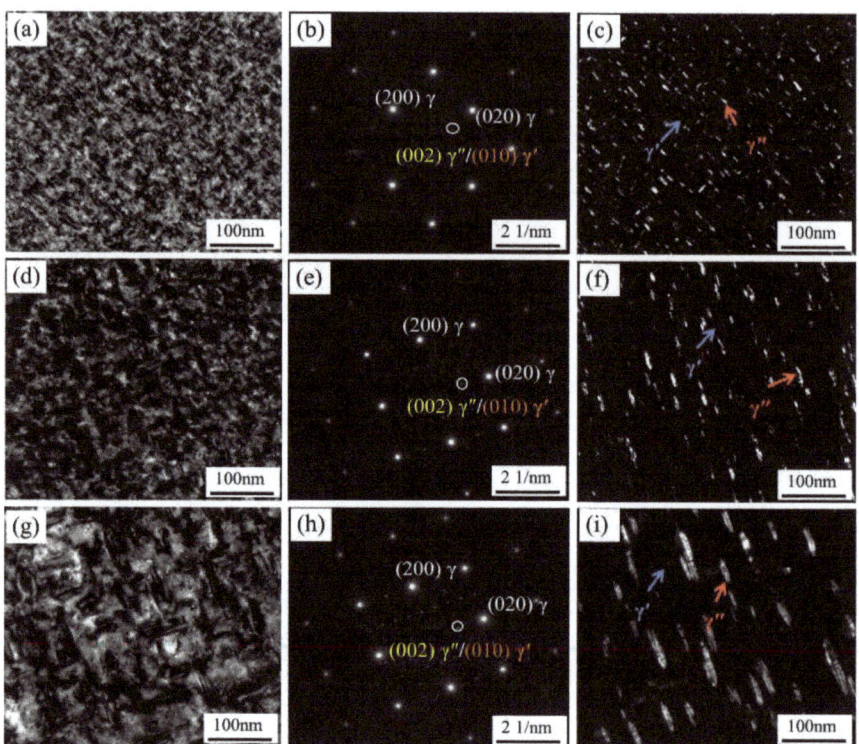

**Figure 7.** TEM image of $\gamma''$ and $\gamma'$ phases: (**a**), (**d**) and (**g**) are the bright field images of samples A720, A750 and A780, respectively; (**b**), (**e**) and (**h**) are the selected electron diffraction patterns of $\gamma''$ and $\gamma'$ in A720, A750 and A780 samples, respectively; (**c,f,i**) are the dark field images of A720, A750 and A780 samples made by (002) $\gamma''$ diffraction spots.

The size of about 400 precipitates in the dark field image of each aged sample was measured, and the volume fraction was counted. For the disk-like $\gamma''$ phase, the dimensions of its long axis and short axis are measured respectively, and for the spherical $\gamma'$ phase, the diameter is measured. The results are shown in Table 3. As the aging temperature increases, the growth of $\gamma''$ and $\gamma'$ phases becomes more apparent, with $\gamma''$ growing particularly along the long axis. The size of the $\gamma''$ phase along the long axis increases from 13.6 nm in the A720 alloy to 47 nm in the A780 alloy, while the diameter of the $\gamma'$ phase increases from 7.1 nm in the A720 alloy to 17.5 nm in the A780 alloy. The volume fraction of the second phases does not change significantly.

**Table 3.** Size and volume fraction statistics of $\gamma''$ and $\gamma'$ phases.

| Specimen | $\gamma''$ | | | | $\gamma'$ | |
|---|---|---|---|---|---|---|
| | Mean Long Axis (R)/nm | Mean Short Axis (H)/nm | H/R | Volume Fraction/% | Diameter /nm | Volume Fraction/% |
| A720 | 13.6 | 4 | 0.294 | 13.4 | 7.1 | 3.4 |
| A750 | 30 | 8.5 | 0.283 | 12.4 | 13 | 2.7 |
| A780 | 47 | 9.6 | 0.204 | 12.3 | 17.5 | 2.4 |

In order to understand the interface relationship between the $\gamma''$ phase and the matrix in different aged alloys, high-resolution transmission electron microscopy (HRTEM) and inverse fast Fourier transform (IFFT) observations were carried out. The results are shown in Figure 8. Based on the IFFT diagrams presented in Figure 8d–f, it is evident that the $\gamma''$ phases in the A720, A750 and A780 samples exhibit a strong coherence with the $\gamma$ matrix interface, indicating a well-maintained coherent relationship between $\gamma''$ and the matrix. Slama et al. [21] believed that $\gamma''$ will lose the coherent relationship when the long axis size of $\gamma''$ is larger than 120 nm, while Devaux et al. [22] believed that the critical size is 95 nm. Combined with the data in Table 2, it can be seen that the size of $\gamma''$ obtained in this study is significantly smaller than the critical size reported in the literatures.

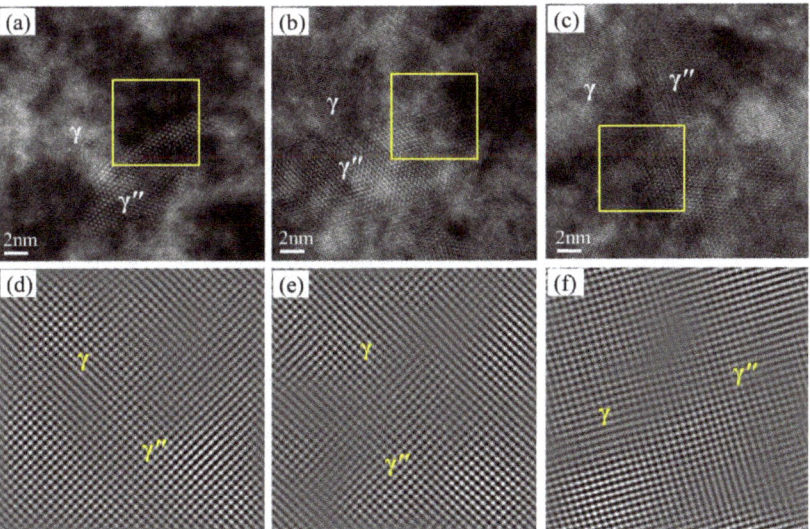

**Figure 8.** Interface relationship between $\gamma''$ phase and the matrix: (**a**–**c**) are HRTEM images of $\gamma''$ in A720, A750 and A780 samples, respectively; (**d**–**f**) are the IFFT images in the yellow box.

The strength of GH4169 alloy is primarily derived from the $\gamma''$ phase, which is caused by the coherent strain resulting from the lattice mismatch between the $\gamma''$ phase and the matrix. In the case of ellipsoidal precipitates with tetragonal distortion perpendicular to the habit plane, the relationship between coherent strain ($\varepsilon^c$) and stress-free strain ($\varepsilon^T$) can be expressed as [23]:

$$\varepsilon^c = \left[1 - \frac{(1-2\nu)}{2(1-\nu)}\frac{\alpha}{R}\right]\varepsilon^T \tag{1}$$

where $\alpha = 4h/3$ is the thickness of ellipsoidal particles with long axis $R$ and short axis $h$; $\nu$ is Poisson's ratio ($\nu = 1/3$), $\varepsilon^T = 0.0286$. It can be seen that the ratio of the short axis to the long axis of the $\gamma''$ phase can result in a greater coherent strain, which is consistent with previous research by Lu et al. [24]. The $\gamma''$ phases in the three aged alloys in this study maintain a good coherent relationship with the matrix. As the aging temperature increased, the increase of $\gamma''$ phase size was accompanied by the decrease of short axis and long axis ratio, which led to the increase in coherent strain. When the dislocation passed through the coherent strain zone, the ability to hinder the dislocation was enhanced, resulting in a greater strengthening effect.

In precipitation-strengthened nickel-based superalloys, the strengthening effect is primarily achieved by impeding dislocations through precipitates. The main mechanism for strengthening is through dislocation shearing or bypassing. Qin et al. [25] discovered that dislocation cut-through is the main mechanism for strengthening the GH4169 alloy when the long axis size of the $\gamma''$ phase is less than 90 nm. The long axis sizes of $\gamma''$

phases of the A720, A750 and A780 alloys in this study are 13.6 nm, 30 nm and 47 nm, respectively, which are all less than 90 nm. Therefore, the main strengthening mechanism can be considered as dislocation shearing. The increase of strength caused by the resistance of coherent precipitation to dislocation motion can be calculated by $\Delta \tau = \beta G \varepsilon^{\frac{3}{2}} (\frac{r}{b})^{\frac{1}{2}} f^{\frac{1}{2}}$ ($\beta$ is a constant related to the dislocation type; $G$ is the shear modulus; $r$ and $f$ are the size and content of the precipitated phases, respectively; and $b$ is the Burgers vector of GH4169 alloy) [26]. It can be seen from the formula that the contribution of dislocation shearing to strength is closely related to the size of precipitates, and the larger size precipitation has a more prominent contribution to the strength improvement. In this paper, the volume fraction of $\gamma''$ phase obtained by the three aging processes is similar. The sizes of $\gamma''$ phases in the A750 and A780 alloys are significantly larger than that of the A720 alloy, which has a greater strengthening effect on the alloy. This corresponds to the increase in yield strength and ultimate tensile strength of the alloy in Figures 3 and 4 and the increase in hardness in Table 2.

The SEM images of the aged alloys are displayed in Figure 9. In the A720 alloy, the grain boundary appears to be relatively clean, as depicted in Figure 9a. The yellow arrow points out the absence of any noticeable precipitates after aging at 720 °C. On the other hand, in the A750 alloy, a few fine granular precipitates, measuring about 216 nm, emerge at the grain boundaries. EDS analysis indicates that the phase is a Nb-rich δ phase, as illustrated in Figure 9e. At an aging temperature of 780 °C, δ phases with a granular and short rod-like structure, measuring approximately 538 nm, precipitate at the grain boundary. These δ phases are larger than those found in A750 alloy. Additionally, needle-like δ phases with a length of about 1.5 μm appear within grains. It is worth noting that δ phase is not observed in the A720 alloy due to its precipitation temperature range of 750–1020 °C [27].

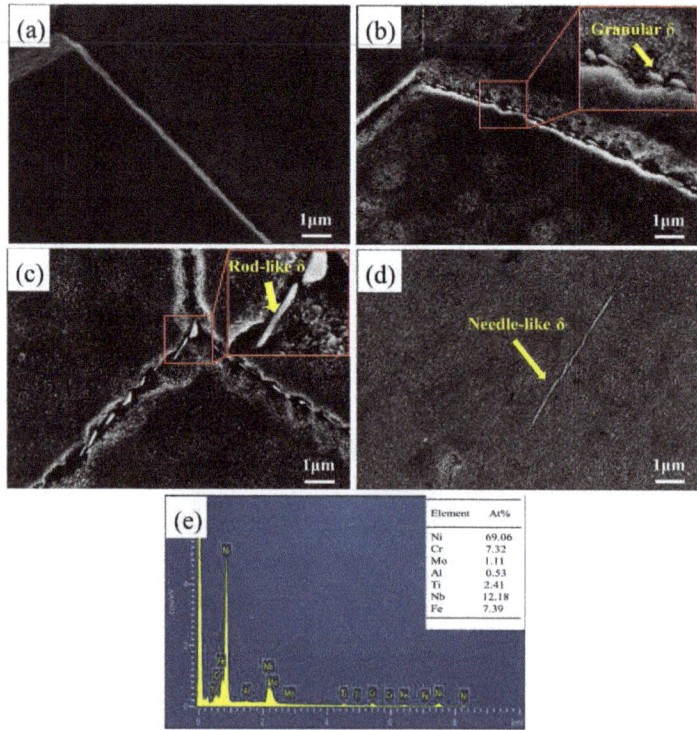

**Figure 9.** SEM images of δ phase of samples at different aging temperatures: (**a**) A720; (**b**) A750; (**c**) A780; (**d**) Intragranular δ phase of A780; (**e**) EDS analysis of grain boundary precipitates.

The strength of the A750 and A780 alloys is significantly improved compared with the A720 alloy due to the growth of the $\gamma''$ phase and the increase in dislocation hindrance. At the same time, the slip of dislocation is hindered, resulting in a decrease in plasticity. Upon analysis of the elevated temperature and room temperature tensile curves, it is evident that the sample aged at 750 °C has a higher yield strength than the sample aged at 780 °C, but the plasticity of the latter sample is better. This is due to the increase of the volume fraction of δ phase at the grain boundary, which is consistent with the results of Andersong et al. [28], who discovered that increasing the volume fraction of δ phases at grain boundaries can improve plasticity by reducing yield strength. This is because the formation of the δ phase will produce a $\gamma''$ phase missing zone around it, which has high plasticity and can alleviate stress concentration, thereby improving the plasticity of the alloy [29]. The increase in δ phase content also leads to a lower hardness of A780 alloy compared to A750 alloy, as shown in Table 2, which is consistent with the results of Rafiei et al. [10].

The fracture surfaces after room temperature tensile tests of the aged alloys are shown in Figure 10. There are a small amount of micro holes and a large number of dimples in the fracture surfaces of the alloys. In addition, a small number of microcracks were found in the fracture of the alloys, indicating that a large plastic deformation was experienced during the tensile tests. Figure 10b,c show the fracture surfaces of the A750 and A780 alloys at room temperature. The existence of dimples and holes can be observed, but the dimples are smaller and shallower than that of A720 alloy. In addition, cleavage faces were also found in the fracture surfaces of the A750 and A780 alloys, and the number of cleavage faces of A750 is more than that of the A780 alloy. According to the characteristics of fracture surfaces, the alloy aged at 720 °C has better plasticity, which just explains the law in Figure 3b.

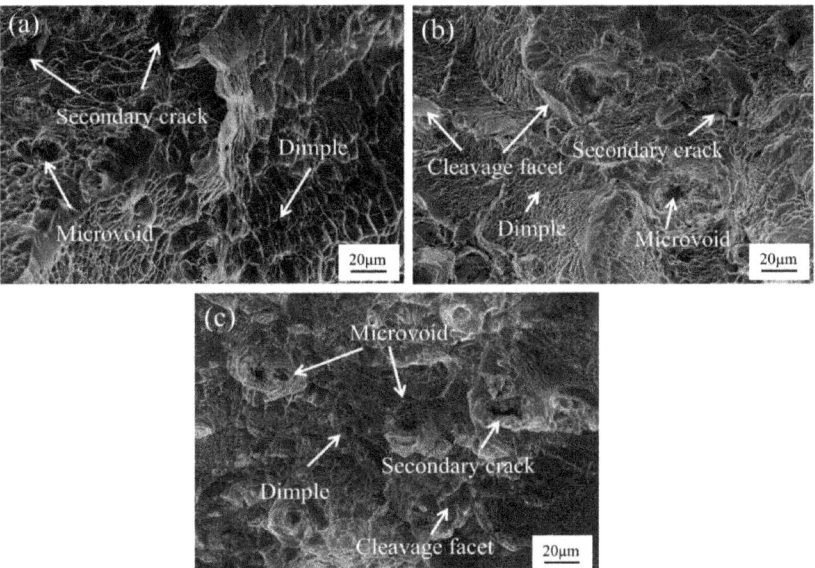

**Figure 10.** Fracture surfaces of tensile samples at room temperature: (**a**) A720; (**b**) A750; (**c**) A780.

The fractures of the tensile samples tested at 650 °C were analyzed by SEM, and the results are shown in Figure 11. Figure 11a,c are the morphologies of the fracture center of the A720 and A780 alloys after hot tensile, respectively, with a large number of dimples and a small amount of micro-pores. Figure 11b is the morphology of the central part of the fracture of the A750 alloy, which is significantly different from that of the A720 and A780 alloys. The dimple size in the fracture of the A750 alloy is significantly reduced, which corresponds to the deterioration of plasticity. In addition, there are inclusions at the bottom of the pores in the fracture of the alloys. It can be judged by EDS analysis that these inclusions are NbC and TiC, as shown in Figure 11d. The plastic law presented by the fracture surfaces is consistent with the data in Figure 4b.

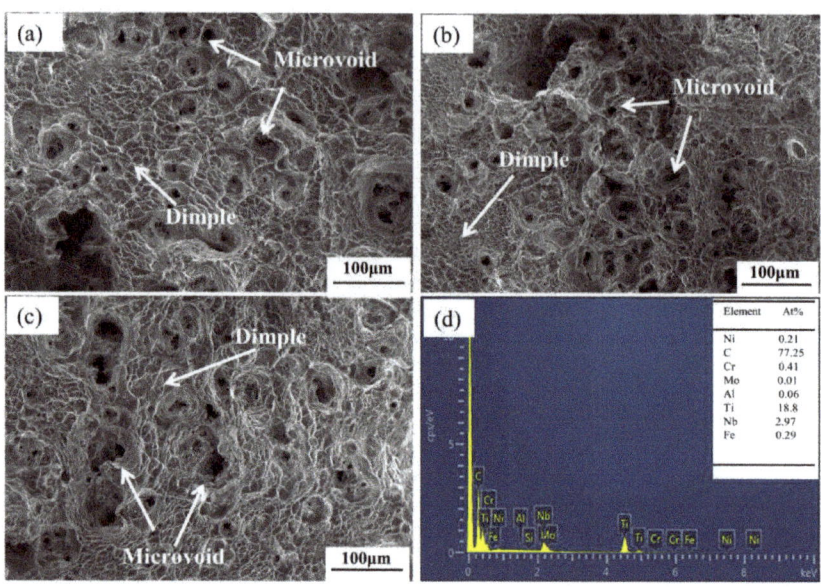

**Figure 11.** Fracture surfaces of tensile specimen at elevated temperature: (**a**) A720; (**b**) A750; (**c**) A780; (**d**) EDS analysis of inclusions at hole bottom.

The stress–strain curves in Figure 4a demonstrate the presence of PLC effects in alloys aged at 720 °C and 750 °C during elevated temperature tensile processes. As the aging temperature increases to 780 °C, the PLC effect becomes almost negligible. The PLC effect is generated by the interaction between moving dislocations and solute atoms, which is shown as a zigzag curve. The space of serration on the curve represents the time interval between the dislocation fixed by the atom and the dislocation separated from the atom. The space of serration on the curve gradually increases with the aging temperature. The size of the $\gamma''$ phase and the precipitation of $\delta$ phase significantly increase with the rise in aging temperature, as evidenced by Figures 7 and 9. The enlarged $\gamma''$ phase enhances the hindrance to dislocations, leading to a longer time for dislocations to pass through it. This results in an increase in the spacing of serration [30]. In addition, as the aging temperature increases, the formation and growth of $\gamma''$ phase and the precipitation of $\delta$ phase utilize a significant amount of solute atoms, leading to a reduction in solute atoms in the matrix. This decrease in solute atoms results in a longer time needed for dynamic strain aging, causing the serration space to increase [31]. In summary, the enhancement of the ability of $\gamma''$ phase to hinder dislocations and the reduction of solute atoms in the matrix lead to the disappearance of the PLC effect during the tensile process of the A780 alloy. The PLC effect will cause the surface roughness of the material and affect the mechanical properties of the alloy, which is not conducive to the processing and use of the alloy [32]. Increasing the aging temperature can effectively inhibit the PLC effect.

## 4. Conclusions

In this paper, the effects of aging temperatures (720 °C, 750 °C and 780 °C) on the precipitation behavior of $\gamma''$, $\gamma'$ and $\delta$ phases in GH4169 alloy were systematically studied, and the correlation between the characteristics of precipitates and the mechanical properties of the alloy at room temperature and elevated temperature was clarified. The following conclusions were drawn.

(1) The size of the $\gamma''$ and $\gamma'$ phases is significantly affected by the aging temperature, while the volume fraction remains relatively unchanged. As the aging temperature increases, the size of the $\gamma''$ and $\gamma'$ phases increases greatly. Specifically, the size of the $\gamma''$ phase in the long axis direction increases from 13.6 nm in the A720 alloy to 47 nm in the A780 alloy. It is worth noting that the $\gamma''$ phase maintains a good coherent relationship with the matrix. Additionally, the diameter of the $\gamma'$ phase increases from 7.1 nm in the A720 alloy to 17.5 nm in the A780 alloy.

(2) The precipitation of the $\delta$ phase is promoted with an increase in aging temperature. When aged at 750 °C, granular $\delta$ phase precipitates at the grain boundary of the alloy. However, when aged at 780 °C, not only does granular $\delta$ phase form at the grain boundary, but needle-like $\delta$ phase is also formed within the grains. The size of $\delta$ phase formed at 780 °C is larger than that of the alloy aged at 750 °C.

(3) The yield strength and ultimate tensile strength of alloys at room and elevated temperatures are increased due to the larger sizes of the $\gamma''$ phases, which create a greater coherent strain and strengthen the hindrance to dislocations. The alloy aged at 750 °C has the highest yield strength, measuring 1135 MPa at room temperature and 1050 MPa at elevated temperature. Compared to the alloy aged at 720 °C, the yield strength of the alloy increased by 13.5% and 10.5%, respectively. The $\delta$ phase is detrimental to the strength of the alloy, but it significantly increases the elongation of the alloy. When the aging temperature is 780 °C, the increase in the size of the $\gamma''$ phases enhances the hindrance of dislocations, and the precipitation of the $\gamma''$ and $\delta$ phases reduces the solute atoms in the matrix, both of which inhibit the PLC effect.

**Author Contributions:** Conceptualization, A.L., Y.R., L.W. and F.X.; investigation, A.L., Y.R., L.W. and F.X.; resources, A.L., L.W. and F.Z.; writing—original draft, A.L., W.H. and F.Z.; writing—review and editing, A.L., F.Z., W.H. and Y.T.; funding acquisition, F.Z., Y.R. and L.W.; supervision, A.L., W.H., F.Z. and Y.T. All authors have read and agreed to the published version of the manuscript.

**Funding:** This study was supported by Science and Technology granted by Guiyang city with Grant No. [2021]1-7 and the Guizhou science and technology project with Grant No. ZK [2022]023.

**Data Availability Statement:** Not applicable.

**Conflicts of Interest:** The authors declare no conflict of interest.

# References

1. Tian, S.G.; Li, Z.R.; Zhao, Z.G.; Chen, L.Q.; Sun, W.R.; Liu, X.H. Influence of deformation level on microstructure and creep behavior of GH4169 alloy. *Mater. Sci. Eng. A* **2012**, *550*, 235–242.
2. Ran, R.; Wang, Y.; Zhang, Y.X.; Fang, F.; Wang, G.D. Microstructure, precipitates and mechanical properties of Inconel 718 alloy produced by two-stage cold rolling method. *Mater. Sci. Eng. A* **2020**, *793*, 139860. [CrossRef]
3. Zhu, J.; Yuan, W. Effect of pretreatment process on microstructure and mechanical properties in Inconel 718 alloy. *J. Alloys Compd.* **2023**, *939*, 168707. [CrossRef]
4. Deschamps, A.; Hutchinson, C.R. Precipitation kinetics in metallic alloys: Experiments and modeling. *Acta Mater.* **2021**, *220*, 117338. [CrossRef]
5. Franco-Correa, J.C.; Martínez-Franco, E.; Alvarado-Orozco, J.M.; Cáceres-Díaz, L.A.; Espinosa-Arbelaez, D.G.; Villada, J.A. Effect of Conventional Heat Treatments on the Microstructure and Microhardness of IN718 Obtained by Wrought and Additive Manufacturing. *J. Mater. Eng. Perform.* **2021**, *30*, 7035. [CrossRef]
6. Chamanfar, A.; Sarrat, L.; Jahazi, M.; Asadi, M.; Weck, A.; Koul, A.K. Microstructural characteristics of forged and heat treated Inconel-718 disks. *Mater. Design* **2013**, *52*, 791–800. [CrossRef]
7. He, D.; Lin, Y.C.; Tang, Y.; Li, L.; Chen, J.; Chen, M.; Chen, X. Influences of solution cooling on microstructures, mechanical properties and hot corrosion resistance of a nickel-based superalloy. *Mater. Sci. Eng. A* **2019**, *746*, 372–383. [CrossRef]
8. Firoz, R.; Basantia, S.K.; Khutia, N.; Bar, H.N.; Sivaprasad, S.; Murthy, G.V.S. Effect of microstructural constituents on mechanical properties and fracture toughness of Inconel 718 with anomalous deformation behavior at 650 °C. *J. Alloys Compd.* **2020**, *845*, 156276. [CrossRef]
9. Chen, Y.; Yeh, A.; Li, M.; Kuo, S. Effects of processing routes on room temperature tensile strength and elongation for Inconel 718. *Mater. Design* **2017**, *119*, 235–243. [CrossRef]
10. Rafiei, M.; Mirzadeh, H.; Malekan, M. Precipitation kinetics of $\gamma''$ phase and its mechanism in a Nb-bearing nickel-based superalloy during aging. *Vacuum* **2020**, *178*, 109456. [CrossRef]

11. Drexler, A.; Oberwinkler, B.; Primig, S.; Turk, C.; Povoden-Karadeniz, E.; Heinemann, A.; Ecker, W.; Stockinger, M. Experimental and numerical investigations of the $\gamma''$ and $\gamma'$ precipitation kinetics in Alloy 718. *Mater. Sci. Eng. A* **2018**, *723*, 314–323. [CrossRef]
12. Fisk, M.; Andersson, J.; du Rietz, R.; Haas, S.; Hall, S. Precipitate evolution in the early stages of ageing in Inconel 718 investigated using small-angle x-ray scattering. *Mater. Sci. Eng. A* **2014**, *612*, 202–207. [CrossRef]
13. Qin, H.; Bi, Z.; Yu, H.; Feng, G.; Zhang, R.; Guo, X.; Chi, H.; Du, J.; Zhang, J. Assessment of the stress-oriented precipitation hardening designed by interior residual stress during ageing in IN718 superalloy. *Mater. Sci. Eng. A* **2018**, *728*, 183–195. [CrossRef]
14. Ran, Q.; Xiang, S.; Tan, Y. Improving Mechanical Properties of GH4169 Alloys by Reversing the Deformation and Aging Sequence. *Adv. Eng. Mater.* **2021**, *23*, 2100386. [CrossRef]
15. Ye, N.; Cheng, M.; Zhang, S.; Song, H.; Zhou, H.; Wang, P. Effect of $\delta$ Phase on Mechanical Properties of GH4169 Alloy at Room Temperature. *J. Iron Steel Res. Int.* **2015**, *22*, 752–756. [CrossRef]
16. Deng, H.; Wang, L.; Liu, Y.; Song, X.; Meng, F.; Yu, T. Microstructure and tensile properties of IN718 superalloy aged with temperature/stress coupled field. *J. Mater. Res. Technol.* **2023**, *23*, 4747–4756. [CrossRef]
17. Ramalho Medeiros, M.A.; de Melo, C.H.; Pinto, A.L.; de Almeida, L.H.; Araújo, L.S. The $\delta$ phase precipitation during processing and the influence on grain boundary character distribution and mechanical properties of superalloy 718. *Mater. Sci. Eng. A* **2018**, *726*, 187–193. [CrossRef]
18. Roth, H.A.; Davis, C.L.; Thomson, R.C. Modeling solid solution strengthening in nickel alloys. *Metall. Mater. Trans. A* **1997**, *28*, 1329. [CrossRef]
19. Sui, S.; Tan, H.; Chen, J.; Zhong, C.; Li, Z.; Fan, W.; Gasser, A.; Huang, W. The influence of Laves phases on the room temperature tensile properties of Inconel 718 fabricated by powder feeding laser additive manufacturing. *Acta Mater.* **2019**, *164*, 413–427. [CrossRef]
20. Zhang, Y.; Lan, L.; Zhao, Y. Effect of precipitated phases on the mechanical properties and fracture mechanisms of Inconel 718 alloy. *Mater. Sci. Eng. A* **2023**, *864*, 144598. [CrossRef]
21. Slama, C.; Servant, C.; Cizeron, G. Aging of the Inconel 718 alloy between 500 and 750 °C. *J. Mater. Res.* **1997**, *12*, 2298–2316. [CrossRef]
22. Devaux, A.; Nazé, L.; Molins, R.; Pineau, A.; Organista, A.; Guédou, J.Y.; Uginet, J.F.; Héritier, P. Gamma double prime precipitation kinetic in Alloy 718. *Mater. Sci. Eng. A* **2008**, *486*, 117–122. [CrossRef]
23. Sundararaman, M.; Mukhopadhyay, P.; Banerjee, S. Some aspects of the precipitation of metastable intermetallic phases in INCONEL 718. *Metall. Trans. A* **1992**, *23*, 2015–2028. [CrossRef]
24. Lu, X.D.; Du, J.H.; Deng, Q. High temperature structure stability of GH4169 superalloy. *Mater. Sci. Eng. A* **2013**, *559*, 623–628. [CrossRef]
25. Qin, H.; Bi, Z.; Yu, H.; Feng, G.; Du, J.; Zhang, J. Influence of stress on $\gamma''$ precipitation behavior in Inconel 718 during aging. *J. Alloys Compd.* **2018**, *740*, 997–1006. [CrossRef]
26. Oblak, J.M.; Duvall, D.S.; Paulonis, D.F. An estimate of the strengthening arising from coherent, tetragonally-distorted particle. *Mater. Sci. Eng.* **1974**, *13*, 51–56. [CrossRef]
27. He, D.G.; Lin, Y.C.; Jiang, X.Y.; Yin, L.X.; Wang, L.H.; Wu, Q. Dissolution mechanisms and kinetics of $\delta$ phase in an aged Ni-based superalloy in hot deformation process. *Mater. Design* **2018**, *156*, 262–271. [CrossRef]
28. Anderson, M.; Thielin, A.L.; Bridier, F.; Bocher, P.; Savoie, J. $\delta$ Phase precipitation in Inconel 718 and associated mechanical properties. *Mater. Sci. Eng. A* **2017**, *679*, 48–55. [CrossRef]
29. Cai, D.; Zhang, W.; Nie, P.; Liu, W.; Yao, M. Dissolution kinetics of $\delta$ phase and its influence on the notch sensitivity of Inconel 718. *Mater. Charact.* **2007**, *58*, 220–225. [CrossRef]
30. Lian, X.T.; An, J.L.; Wang, L.; Dong, H. A New Strategy for Restraining Dynamic Strain Aging in GH4169 Alloy During Tensile Deformation at High Temperature. *Acta Metall. Sin.* **2022**, *35*, 1895–1902. [CrossRef]
31. Zhao, J.; Hung, F.; Lui, T. Microstructure and tensile fracture behavior of three-stage heat treated inconel 718 alloy produced via laser powder bed fusion process. *J. Mater. Res. Technol.* **2020**, *9*, 3357–3367. [CrossRef]
32. Cui, C.; Zhang, R.; Zhou, Y.; Sun, X. Portevin-Le Châtelier effect in wrought Ni-based superalloys: Experiments and mechanisms. *J. Mater. Sci. Technol.* **2020**, *51*, 16–31. [CrossRef]

**Disclaimer/Publisher's Note:** The statements, opinions and data contained in all publications are solely those of the individual author(s) and contributor(s) and not of MDPI and/or the editor(s). MDPI and/or the editor(s) disclaim responsibility for any injury to people or property resulting from any ideas, methods, instructions or products referred to in the content.

*Article*

# Detection of Porosity in Impregnated Die-Cast Aluminum Alloy Piece by Metallography and Computer Tomography

Mihály Réger [1], József Gáti [1], Ferenc Oláh [1,2], Richárd Horváth [1,*], Enikő Réka Fábián [1] and Tamás Bubonyi [3]

[1] Bánki Donát Faculty of Mechanical and Safety Engineering, University of Óbuda, H-1081 Budapest, Hungary; reger.mihaly@uni-obuda.hu (M.R.); gati@uni-obuda.hu (J.G.); olah.ferenc@bgk.uni-obuda.hu (F.O.); fabian.reka@bgk.uni-obuda.hu (E.R.F.)
[2] Doctoral School on Materials Sciences and Technologies, University of Óbuda, H-1081 Budapest, Hungary
[3] Institute of Metal Formation and Nanotechnology, University of Miskolc, H-3515 Miskolc, Hungary; fembubo@uni-miskolc.hu
* Correspondence: horvath.richard@bgk.uni-obuda.hu

**Abstract:** The porosity of die-cast aluminum alloys is a determining factor for the quality of the product. In this paper, we studied the porosity of a selected part of a die-cast AlSi9Cu3(Fe) compressor part by computer tomography and metallography. In the case of this part, the achievable resolution by CT, a non-destructive testing method, was 30 µm—this method could not detect smaller cavities. Based on metallographic analysis, the percentage of defects larger than 30 µm ranges from 10 to 30% of the total number of defects, which represents 75–95% of the defective area (area ratio). Impregnation with methacrylate resin (used to seal cavities to prevent leakage) can be detected with UV-illuminated optical microscopic examination on metallographically prepared specimens. As confirmed by scanning electron microscopy, partial filling and partial impregnation can occur in a system of shrinkage cavities.

**Keywords:** aluminum alloy; die casting; gas tightness requirement; porosity; impregnation; computed tomography; metallography; UV illumination; scanning electron microscope

## 1. Introduction, Background

The solidification of aluminum alloys is accompanied by a shrinkage of about 5–7% [1,2]. The change in crystallization volume and the degree of gas porosity and air entrapment can be compensated for with a variety of die-casting processes (e.g., melt treatment, post-pressure, post-compression). With these methods, porosity can be reduced to 0.2–0.5% in quality casting [3–5]. This amount of porosity is unacceptable where gas permeability is not allowed. Porosity cavities can be sealed, and leakage can be prevented by impregnation with methacrylate resin. Cast products are usually impregnated when they are fully finished. The main steps of the process are as follows: vacuuming, immersion in resin under vacuum; application of atmospheric pressure on the liquid surface of the impregnation resin, lifting out; washing with cold water; and washing with hot water, during which the resin is cured between 90 and 95 °C [6]. The technology is generally efficient and reliable, but leaking is nevertheless common. In order to detect leaking castings, a helium leakage test is carried out on each casting in the assembly line.

Leakage is primarily associated with the machined surfaces, as the external surface of a casting usually seals the internal shrinkage cavity system in a gas-tight manner (the bi-film oxide layer effect on leakage is not investigated in this study). The inherently closed cavity network inside a cast piece can become an open system during machining. The internal surface of holes and threads can cut spatial porosity cavity systems, and, as a result, leakage often occurs through the internal holes.

The reason for leakage in impregnated castings is assumed to be the limited impregnation efficiency of the complex-shaped shrinkage cavity system with different cross-sections. Both vacuuming and saturation with resin may be impeded by small cross-sections in a cavity system with complex geometry, particularly if the cavity system includes capillary sections.

In aluminum alloy die-cast pieces, porosity can develop, mainly in the larger cross-sections. Even if porosity is adequate for the casting as a whole, the degree of local porosity may be critical from the point of view of gas tightness in certain casting sections. Technological solutions to reduce the porosity of the casting will result in the compaction of the porous material area. These solutions reduce the volume of porosity to a significant degree but do not necessarily eliminate the shrinkage cavity system. Compression may also result in a reduction of cavity and channel dimensions in the cavity system in the typical cross-sections.

X-ray computed tomography (CT) is among the most widespread non-destructive examination methods used to analyze the porosity and shrinkage cavity system of a sample [7–9]. CT is an imaging method based on the absorption of X-rays in materials and makes non-destructive three-dimensional analysis of samples and their internal structures possible.

As the thickness of the sections increases, the resolution of the X-ray image decreases (larger focal spot used), and less detailed information can be extracted from the images. Different types of CT equipment work with different sample sizes; thus, their resolution varies. The resolution of industrial CT is 5–150 µm, that of micro-CT is 1–100 µm, and that of nano-CT is around 0.5 µm. With X-ray microscopy, resolutions as low as 100 nm can be achieved [10]. Recently, Zhuang et al. [11] developed a method to analyse porosities below the resolution of CT. The accuracy and analysis of the results are influenced by a number of additional factors, such as transparency and segmentation [12–14].

Limodin [15] studied the relationship between the porosity and mechanical properties of an Al–Si–Mg alloy. The micro-CT study was performed with a voxel size of 1.7 µm. The author notes that a compromise between resolution and sample size had to be found, as a high resolution provides less representative results. In the end, the decision was made to analyse the volume fractions of 3.8 and 18 mm$^3$. The images were processed and binarized with the ImageJ software. In the 18 mm$^3$ sample, 1104 pores were identified (61 pores/mm$^3$). The size distribution of the complex-shaped pores was given as a function of the Feret diameter, with the most common value being 16 pores/mm$^3$ at Feret diameters between 30 and 40 µm.

The porosity of the aluminum alloys produced by die casting is of crucial importance, not only because of gas tightness but also because of the fatigue properties. To this end, Garb and co-workers [16] conducted an extensive series of micro-CT studies to statistically characterize the microporosity of aluminum castings. They used two different resolutions (3 and 8 µm) to find out how resolution affects the accuracy of determining the porosity distribution. Resolution limits the size of the volume, which can be investigated. A cylindrical sample of 6 mm in diameter with a length of 5.4 mm was used at a resolution of 3 µm, and a 15.5 mm long sample was used at an 8 µm resolution (the sample volumes were 0.155 and 0.490 cm$^3$). In the 0.155 cm$^3$ volume investigated at a 3 µm resolution, approximately 7000 pores were detected. Different statistical distribution functions were fitted to the equivalent diameters of the identified pores; the maximum for the distribution curve was between 8 and 10 µm. The authors found that the shrinkage pores often contain large-volume patches connected by thin cavities with dimensions below the resolution. Investigations under nearly identical conditions and similar results are reported in a study by Weidt et al. [17], who reported investigations under nearly identical conditions and similar results in a study on Al–Si–Cu alloys.

The relationship between fatigue characteristics and pores, revealed by micro-CT and metallographic examination, is discussed in the study by Nicoletto et al. [18]. The tested volume of the AlSi7Mg sample was 12.5 mm$^3$, and the voxel size was 1.7 µm. The authors presented some pore shapes revealed by CT and metallography but no statistical data on the dimensions and distribution. They concluded that the identified pores could be considered a complex system of larger hollow spaces and narrow channels connecting them. Recently, additive manufacturing of aluminum alloys by selective laser melting (SLM) is also of research interest. Zhang et al. [19] investigated the SLM-built AlSi10Mg thin-walled parts and their macro-mechanical behavior in correlation with the relative densities.

There is very little reliable test information available in the literature on the impregnation of hollow systems with complex geometries in aluminum alloy castings. Publications of impregnant producers provide some informative data. According to laboratory impregnation experiments on sintered metal alloys with a controlled pore structure [20], cavities with a size below 100 μm are totally filled in every case. The impregnation problem may be mainly the sealing of larger pores between 100 and 500 μm, about one-third of which were not sealed in the experiments. Above a pore size of 500 μm, impregnation has been shown to be essentially ineffective in ensuring the gas tightness of sintered samples. The pore structure of Al–Si–X die-cast components with gas tightness requirements contains cavities and channels typically below 100 μm; however, impregnation often fails to seal the leak paths.

Soga et al. [21] detected and studied the impregnating resin in a cavity system by computed tomography. The technique is based on the detection of contrast differences caused by the impregnating resin, the aluminum matrix and the gases in the cavities. They reported that, in simple-geometry castings, the filling of larger (several mm) cavities with resin could be assessed (voxel size 100 μm). The authors do not discuss the possibilities of investigating smaller cavities (10–100 μm) or more complex spatial geometries.

Many details of porosity problems and the formation of leakage paths are still unclear. In this study, an impregnated cast part—produced with a gas tightness requirement and nevertheless showing leakage, despite impregnation—was examined in detail. The aim of the investigations was to determine the location and geometric characteristics of potential leaking paths in the casting. The shape and size of the cavities and porosities in the investigated castings were determined using metallographic and micro-CT methods. The parallel metallographic and micro-CT analyses of the continuity defects identified in a given cross-section of the cast part provide an opportunity to compare the advantages and disadvantages, as well as the limitations of both methods.

## 2. Materials and Testing Methods

We determined the porosity map of a selected part of a mass-produced EN AC 46000 AlSi9Cu3(Fe) die-casting (compressor) part. The impregnated casting, finished after pressure die-casting, showed leakage between the inner and outer space in the helium inspection test prior to assembly and was therefore taken out of production. The leakage was identified in the greater wall thickness of the thin-walled cylindrical casting. The front face of the shell is machined, and the sides have cast surfaces. This part contains holes for connecting the associated elements, one of which has a threaded internal surface. Leakage to the external space is through the inner part of the threaded hole.

A detailed structural analysis was conducted on the cast piece (Figure 1). The machined specimen contains a volumetric size of about 4.5 cm$^3$, including the threaded bore. Therefore, it can be assumed that there are discontinuities associated with leakage in this region. The defect distribution characteristics were determined by computed tomography (micro-CT) and metallographic analysis, for which the sample was embedded in metallographic resin. The tests were carried out within the planes presented in Figure 1. The examined planes were parallel to the base plane. The sample was studied by light optical microscopy (LOM), without etching, after metallographic preparation (grinding and polishing). Lower magnification metallographic images, showing the complete cross-section of the sample, were produced with an OLYMPUS DSX1000 (Olympus Scientific Solutions Americas Corp., Waltham, MA, USA) opto-digital microscope with a resolution of 3.4 μm. Higher-resolution images of the elements of the revealed structure were produced with the opto-digital microscope (LOM) and a JEOL JSM 5310 LA scanning electron microscope (SEM).

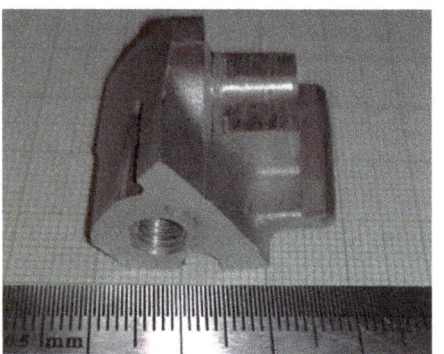 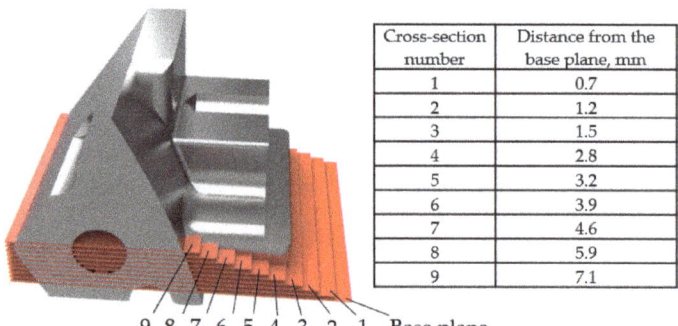

| Cross-section number | Distance from the base plane, mm |
|---|---|
| 1 | 0.7 |
| 2 | 1.2 |
| 3 | 1.5 |
| 4 | 2.8 |
| 5 | 3.2 |
| 6 | 3.9 |
| 7 | 4.6 |
| 8 | 5.9 |
| 9 | 7.1 |

**Figure 1.** The sample from the leakage region and positions of the cross-sections examined by metallography.

The micro-CT scans were performed on the YXLON FF35 (YXLON International GmbH, Hamburg, Germany) machine at the 3D Laboratory of the University of Miskolc. Porosity was studied by porosity/inclusion analyses using the VGStudio MAX 3.0 software at the 3D laboratory of Óbuda University. The relatively large sample volume allowed a resolution of 30 μm, which meant the detection of $2 \cdot \times 10^8$ spatial pixels (the size of the resulting file was 196 GByte). Limodin [14], Garb [15], and Weidt [16] measured approximately the same number of voxels but used a significantly smaller sample to achieve a better resolution. The results of the porosity analysis in a 3D reconstructed image are shown in Figure 2.

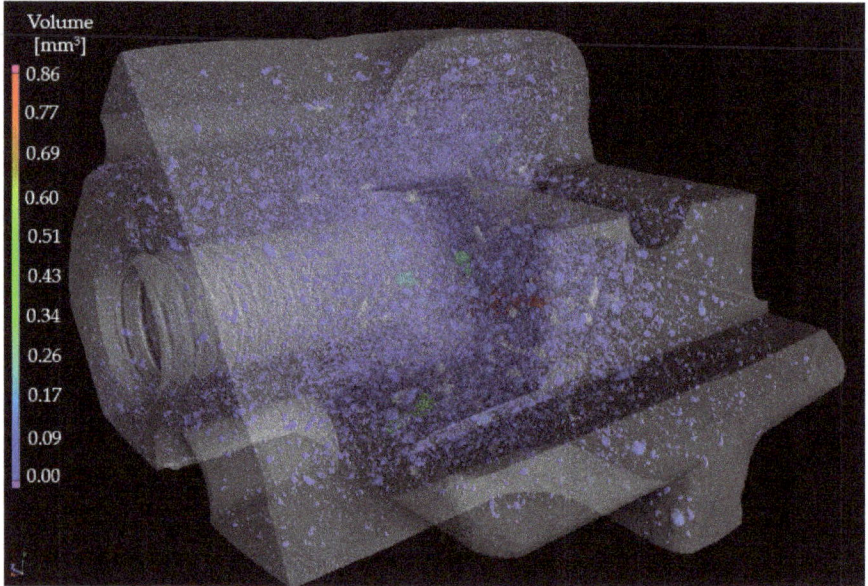

**Figure 2.** Volumetric distribution of porosities in a 3D reconstructed CT image at 60% transparency.

The position of the plane in the optical image can also be reconstructed from the spatial micro-CT results and thus, the metallographic and CT scan results can be combined for a given scan section. The ImageJ 1.52a image processing software was used to analyse the images acquired by both methods.

The impregnation resin fluoresces in ultraviolet light, and thus, the filling of cavities in metallographic surfaces with impregnation resin can be checked with the use of UV illumination. The UV-illuminated micrographs were produced with an OLYMPUS DSX1000 opto-digital microscope with a Labino 135 UV reflector.

We could not detect the impregnation resin in the cavities by micro-CT scanning.

## 3. Results and Discussion

Shrinkage cavities, gas porosities and other discontinuity defects in the polished surface section of the test specimen, revealed by metallography, were identified by optical microscopy. We observed gas porosities and microcavities on the metallographically prepared surfaces. Figure 3 shows some characteristic defects of the studied part at a distance of 0.7 mm from the base plane (level 1). Spherical gas pores and discrete shrinkage cavities are clearly visible at low magnifications; however, the characterization of microcavities requires higher resolutions. Figure 3a,b show the typical appearance of gas pores, shrinkage cavities and microcavities.

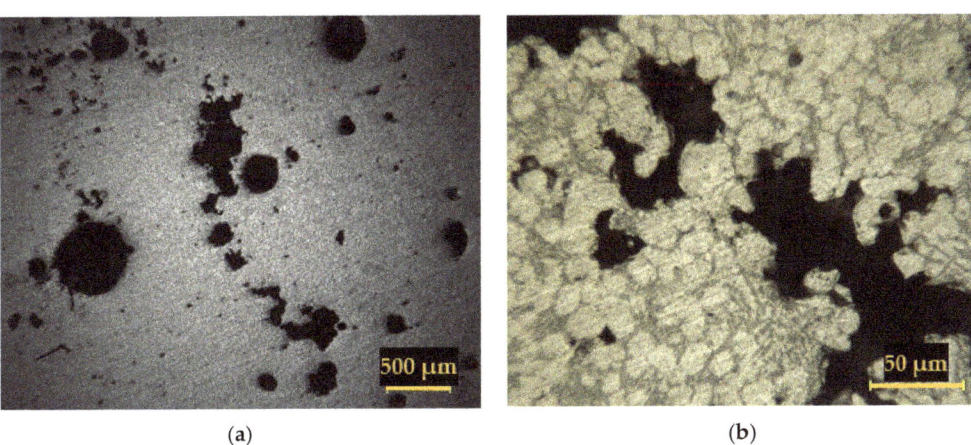

**Figure 3.** Porosity, as shown by optical microscopy; (**a**) gas porosities and shrinkage; (**b**) microporosity.

A high-resolution view of several cm$^2$ sample areas was produced from a series of panoramic images. Depending on the actual size of the sample area, each macro image was produced by automatically stitching 48–72 photographs at a magnification of 100×.

From the 3D CT-scan data set, the position of the metallographic plane can be determined from the geometric data of the specimen; thus, a two-dimensional CT image of the investigated plane can be reconstructed. The comparison of the image produced by the optical microscope and the CT image showing the same plane allows a joint analysis of the results of the destructive and non-destructive tests.

We performed a systematic comparison of the optical and CT examination results in nine examined planes. The sections were marked in a position parallel to the base plane (see Figure 1) at a distance of 0.7–7.1 mm from it. Figure 4 shows the cross-section explored in a plane that is 3.9 mm from the base plane (cross-section 6). The images in Figure 4a,b show the total area of the sample; in this section, the convex area of the sample is 6.3 cm$^2$. The position of the threaded hole and the location of the larger individual defects show that the metallographic section and the 2D CT image represent the same plane. The cross-section examined shows a large number of small to large continuity gaps, most of which can be matched with the two inspection methods. The enlarged images of the section marked in Figure 4a are shown in Figure 4c. Again, the larger individual continuity defects look the same with both methods; however, the identification of the smaller defects is not straightforward. The one order of magnitude lower resolution of the CT image and its inherently poorer image quality result in blurred contours of smaller defects and, thus, an uncertain identification of shapes. This difference is clearly visible with a higher magnification of the surface area, as shown in Figure 4b (Figure 4c). The contour of the large cavity is significantly blurred, and some small cavities are not even identifiable in the CT image. Although the resolution of the CT images is 30 µm, the lower limit of the sizes of objects that can be identified with confidence is around 50 µm due to image quality.

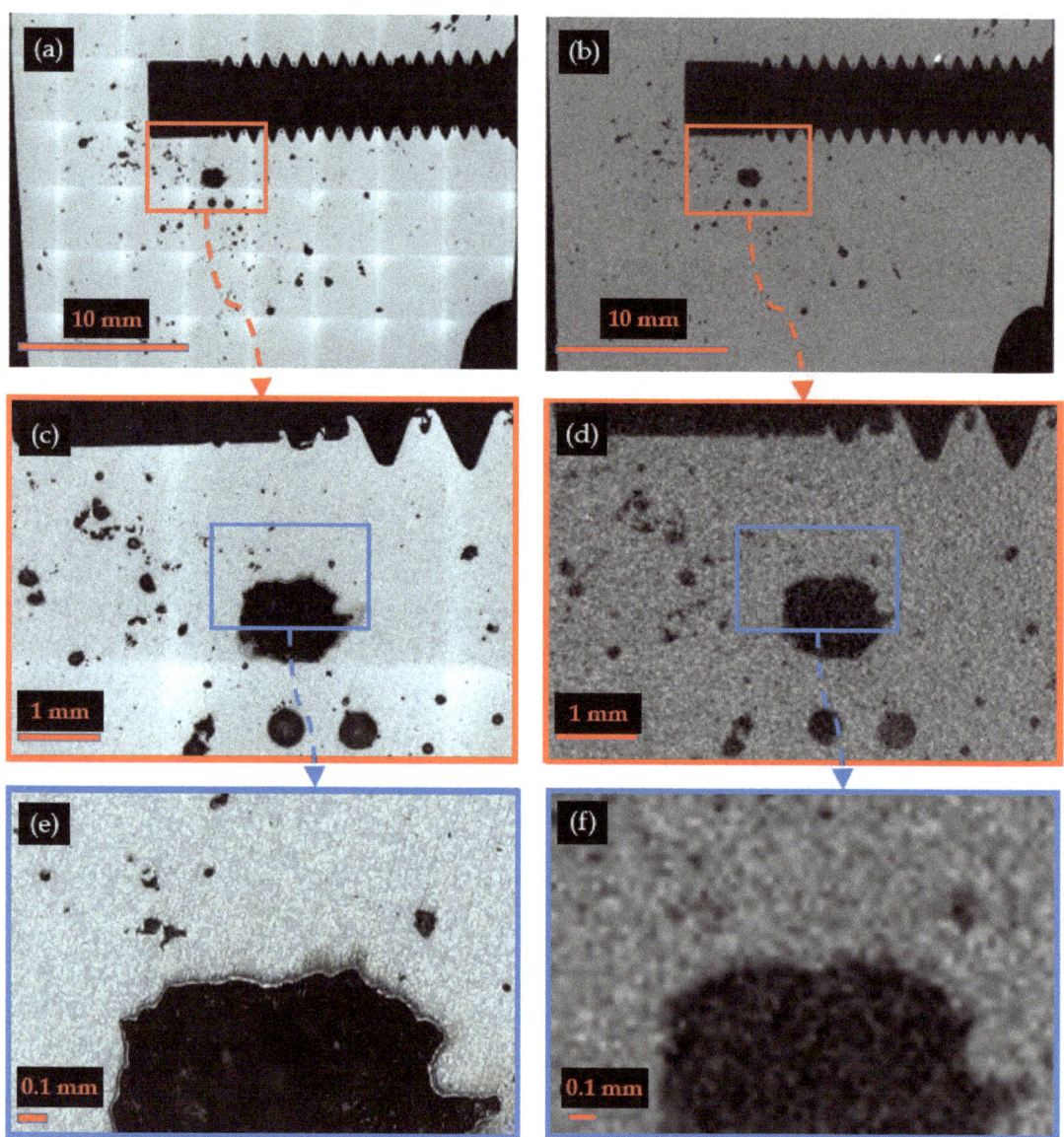

**Figure 4.** Optical microscopy (**left**) and micro-CT (**right**) images of cross-section 6. (**a**) Image of the complete section by optical microscopy (LOM); (**b**) image of the section by CT; (**c**) magnification of the marked area in (**a**) (LOM); (**d**) magnification of the marked area in (**b**); (**e**) magnification of the marked area in (**c**); (**f**) magnification of the marked area in (**d**).

The cross-sectional images were binarized with the use of the automatically offered threshold of the ImageJ software. No image correction was applied. In the case of the CT images, due to the poorer image quality, the number of false signals of a few pixels in size increased during binarization. We did not modify the binarized image in any way. Figure 5 shows the optical and CT images of Figure 4d after binarization. The binarized CT image on the right shows several continuity gaps of 1 pixel in size, which are not found in the optical image (left), which we consider the reference. In order to reliably evaluate the area of larger cavities, we did not apply erosion and dilation to the image. Optical microscopy images can have similar defects, but to a lesser extent, due to the significantly better image quality.

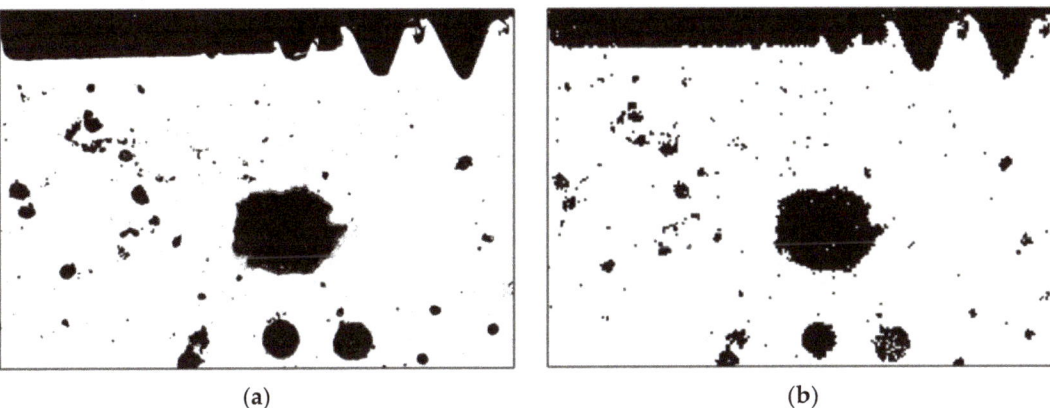

**Figure 5.** Images of cross-section 6 after binarization; (**a**) optical microscopy; (**b**) micro-CT.

In the full cross-section of the sample (Figure 5), 3901 defects can be detected in the optical microscopy image and 849 defects in the CT image.

The diagram in Figure 6a shows the size distribution of the identified defects. In this diagram, the diameter of a circle with the same area as the defect area, the so-called equivalent circle diameter, is used to characterize the size of the defect. The frequency of defects increases sharply with decreasing size, based on both the metallographic and CT examination results. The higher value identified in the CT image in the 35 μm equivalent defect size category is caused by the binarization of the gray image, as explained above. The 35 μm equivalent defect size category is representative of the presence of continuity defects of a few pixels in the CT image.

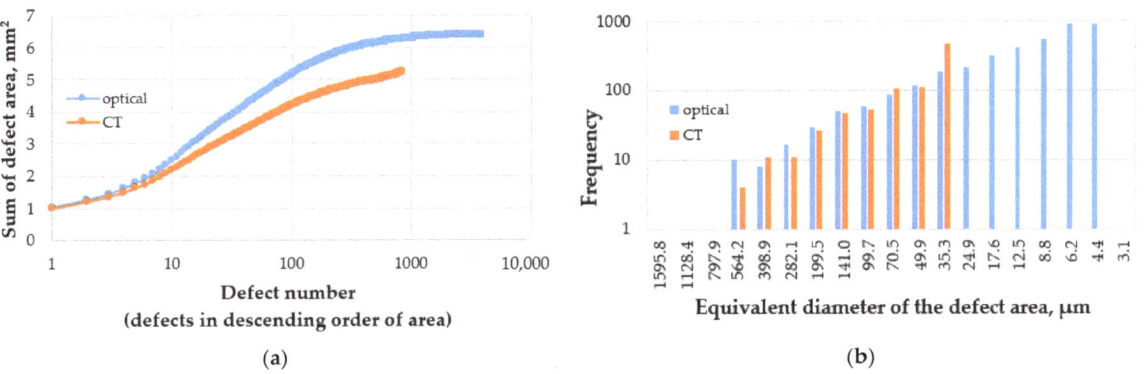

**Figure 6.** Assessment of the discontinuities identified in cross-section 6. (**a**) Frequency distribution of defect size (equivalent circular diameter); (**b**) sum of defect areas in descending order.

More than 3.000 defects, or about 78% of the total number of defects, cannot be detected by CT because their size falls below the resolution of CT, in the equivalent diameter range of 4–30 μm. However, this amount of defects account for about 19% of the area of defects, i.e., more than 80% of the defect area in the examined section was detectable by CT. We sorted the detected defects in decreasing order as a function of their area measured in the plane and then plotted the sum of the defect areas 1 to $n$ ($n \leq$ number of defects) (Figure 6b). The value of the points representing the sum function provides the sum of the area of the detected defects, with a maximum value of 6.4 mm for optical microscopy and 5.3 mm for CT.

The area sum function for CT is continuously below the area sum function determined by optical microscopy. The area of the first defect, i.e., the largest defect (the largest defect in the images in Figure 2), is 1.03 mm$^2$ in the optical microscopy image and 0.98 mm$^2$ in the CT image. The summed areas of the first ten largest defects show a similar difference, with 2.48 and 2.17 mm$^2$, respectively. The cumulative defect area up to the first 849 defects (all the defects detected by CT) is 6.259 mm$^2$, determined by optical microscopy, and 5.222 mm$^2$, determined by CT. The cumulative area of the 3901 defects detected by optical microscopy is 6.340 mm$^2$. The 3901 − 849 = 3052 defects identified on the metallographic section account for 6.340 − 6.259 = 0.081 mm$^2$ of the cumulative defect area, proportionally 1.3% of the total value.

As shown above, there is a systematic discrepancy between the defect area sum functions determined by optical microscopy and CT scanning. This is proved by the linear correlation coefficient ($R$ = 0.993) between the defect area data pairs determined with the two methods. The tangent of the line fitted to the data pairs is $m$ = 0.912, i.e., the defect area on the sample surface determined by CT is, on average, 91.2% of the defect area measured with the help of optical microscopy. The sections shown in Figure 5 have the same field of view at the same magnification; the area of the shapes extracted from the CT images is smaller than that of the shapes detected by optical microscopy. The difference is presumably due to the different imaging quality and different resolution.

We performed the above analysis on nine sections of the test sample. Table 1 shows the results of the measurement series. Based on the mean value in Table 1, 83% of the defect area detected by metallographic (optical microscopy) examination can be identified on CT scans, i.e., approximately 17% of the defect areas are not detected by CT scans. This 17% area ratio represents 80% of the total number of defects. We calculated the linear correlation of the defect area data pairs that were measured by optical microscopy and CT scans; the data were sorted in descending order of size. The average correlation coefficient for the nine sections is $R$ = 0.981. The proportion of defect areas identified by CT scanning and optical microscopy ranges from 0.644 to 0.942, with an average value of $m$ = 0.827.

Table 1. Results of a series of measurements in nine sections of the sample.

| Cross-Section Number | Distance from the Base, mm | Defective Area in the Sample Area, % | | | Specific Defect Number, Defect/mm$^2$ | | | Opt/CT Defect Area Correlation | |
|---|---|---|---|---|---|---|---|---|---|
| | | Optical Microscopy | CT | CT/Opt | Optical | CT | CT/Opt | $m$ | $R$ |
| 1 | 0.7 | 2.20 | 1.09 | 0.79 | 24.81 | 1.79 | 0.07 | 0.792 | 0.974 |
| 2 | 1.2 | 1.53 | 0.83 | 0.80 | 21.61 | 1.46 | 0.07 | 0.799 | 0.990 |
| 3 | 1.5 | 1.18 | 0.83 | 0.80 | 12.78 | 1.37 | 0.11 | 0.801 | 0.974 |
| 4 | 2.8 | 0.83 | 0.78 | 0.94 | 2.65 | 0.27 | 0.10 | 0.942 | 0.982 |
| 5 | 3.2 | 1.13 | 0.80 | 0.64 | 13.02 | 1.25 | 0.10 | 0.644 | 0.968 |
| 6 | 3.9 | 1.23 | 1.05 | 0.91 | 7.48 | 1.70 | 0.23 | 0.912 | 0.993 |
| 7 | 4.6 | 1.05 | 0.71 | 0.77 | 4.75 | 1.50 | 0.32 | 0.769 | 0.983 |
| 8 | 5.9 | 1.06 | 0.75 | 0.91 | 3.51 | 1.30 | 0.37 | 0.906 | 0.980 |
| 9 | 7.1 | 1.39 | 1.13 | 0.88 | 4.83 | 2.00 | 0.41 | 0.877 | 0.986 |
| | average | 1.29 | 0.88 | 0.83 | 10.60 | 1.41 | 0.20 | 0.827 | 0.981 |
| | deviation | 0.40 | 0.16 | 0.09 | 8.10 | 0.49 | 0.14 | 0.093 | 0.008 |

Based on metallography and optical analyses, the number of defects larger than 100 μm is 3–5%, and this represents 65–80% of the discontinuity area. The percentage of defects larger than 30 μm (i.e., the resolution of CT) ranges from 10 to 30% of the total number of defects in the nine sections. This 10–30% number of defects represents 75–95% of the total number of defects detected (Table 2).

**Table 2.** Defect size rates in nine cross-sections of the sample by equivalent diameter range.

| Equivalent Diameter, μm | Number of Defects, % | Percentage of Defective Area % |
| --- | --- | --- |
| >100 μm | 3–5 | 65–80 |
| >30 μm | 10–30 | 75–95 |
| >10 μm | 40–60 | 95–98 |

## 4. The Filling of the Cavities

The gas tightness of castings can be ensured by sealing the leakage paths and impregnating the cavity systems connecting the outer and inner spaces. Thousands of cavities, identified by optical microscopic metallographic analysis and CT scans, as described in Figure 6, may be part of an interconnected microcavity system, even running to the machined surface, or a localized isolated continuity gap. The latter has essentially no effect on gas tightness. Due to the complex spatial shape of a leaking cavity system, it is not possible to determine from a cavity identified in a planar section whether it plays a role in leakage or not. If the cavity contains impregnation resin, its continuation is certainly connected to an external surface; however, if the presence of resin cannot be verified and the casting is proven to leak, the cavity is part of a leaking pathway. The CT scan does not detect sections of the spatial cavity system below the detection limit. In this case, at 30 μm, the filling of the cavity with resin cannot be demonstrated with this scan. Conventional bright-field microscopy cannot be used to show resin in the cavities since the cavity appears as a dark-toned spot, as seen in Figures 4 and 5, whether the resin is present or not. The detection of resin in the cavities is only possible by scanning electron microscopy on a metallographically prepared section, fractured surface or by examination under UV light. Both methods are only suitable for the identification of the impregnating agent on the examined surface.

We detected impregnation resin in the cavities by scanning electron microscopy in sections 5, 6, and 9 and by UV-illuminated optical microscopy in section 9. Figure 7 shows some typical images of impregnation resin-saturated and non-saturated cavities.

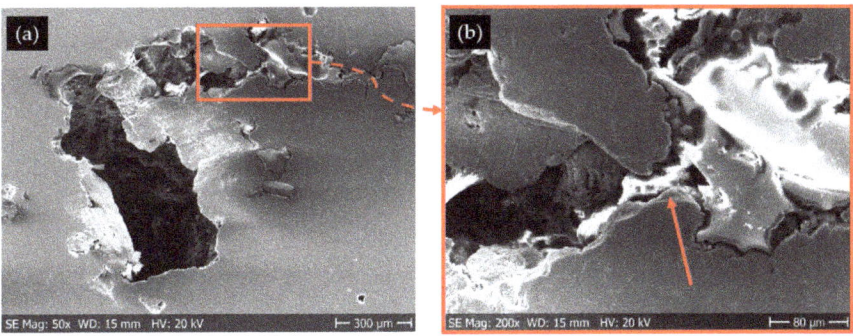

**Figure 7.** Some characteristic cavities identified on the metallographically prepared surface in section 5. (**a**) Complex cavity; (**b**) magnification of the area marked in (**a**).

Figure 7 shows a complex cavity partially filled with resin. The continuity gap extending below the plane of the section on the left side of Figure 7a forms a system with a thinner channel on the right. The meeting point of the smaller and larger cross-sectional cavity sections—the detail marked in Figure 7a—is shown enlarged in the image in Figure 7b. In the cavity system, a 15–20 µm constriction can be identified in the section marked with an arrow, the right side of which contains resin, while the left side does not. It can be assumed that during impregnation, the resin flowed into the cavity system through the cavity section on the right; however, for some reason, the filling of the left-hand section could not be completed. The constriction marked by the arrow must have increased the flow resistance, and thus the pressure difference or the time available was probably not sufficient to fill the cavity on the left.

Figure 8a shows the entire surface of cross-section 9 using LOM when the sample is under UV light. Due to the fluorescence of the impregnation resin, the discontinuities filled with resin glow in a bluish-white color, while the base matrix has a purple tint. The structure of the metallographic embedding resin is visible around the part and in the bore in Figure 8c,d.

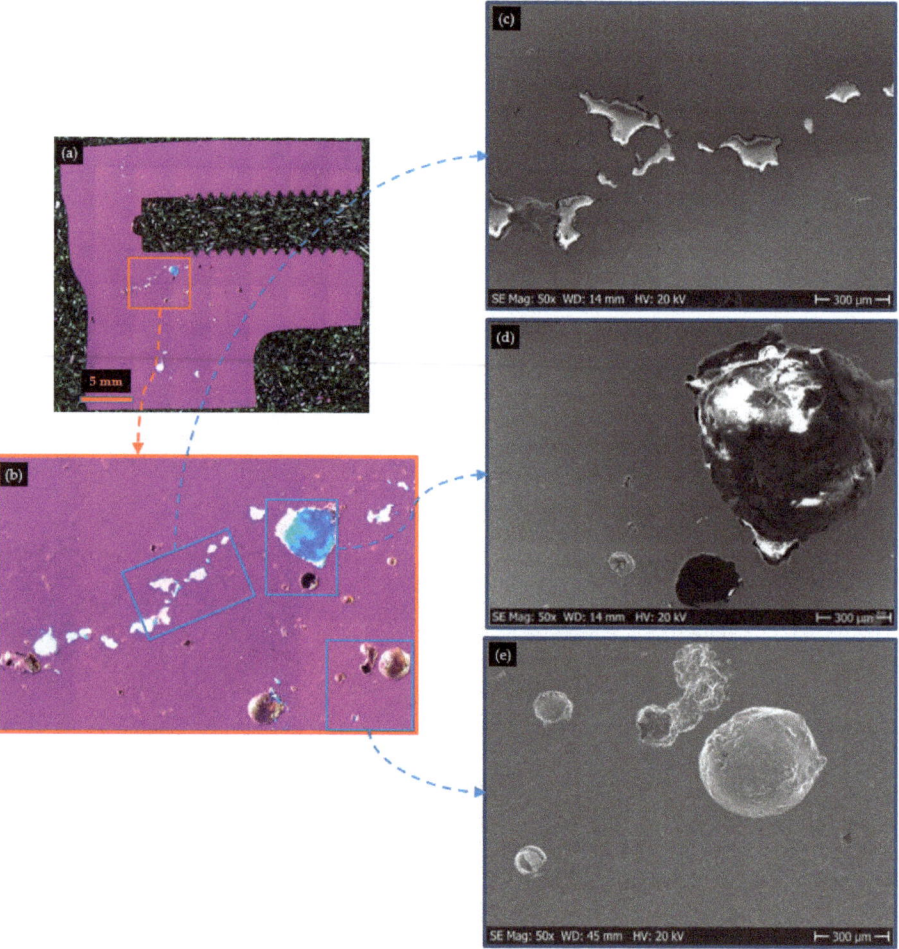

**Figure 8.** Material continuity defects identified in cross-section 9. (**a**) The entire surface under UV illumination (LOM); (**b**) the appearance of cavities under UV; (**c**) cavities filled with resin (SEM image); (**d**) a large cavity with resin (SEM image); (**e**) spherical cavities without resin (SEM image).

Due to the unidirectional UV illumination, the edges of the cavities are shimmering white, and thus automatic image analysis is limited for these images. Visual inspection of each area at higher magnifications reveals that the impregnating resin covers approximately half of the total cavity surface. The number of unsaturated cavities, mainly small ones, is significantly higher than the number of saturated cavities. The spherical continuity gaps, with a few exceptions, are generally free of resin and can probably be considered separated gas inclusions.

Figure 8b shows a larger magnification of the resin-filled cavity near the hole. Figure 8d shows a scanning electron micrograph of the larger cavity and its surroundings in the upper-right-third of the image. The large cavity clearly contains resin, and the surrounding smaller cavities, which are spherical and irregular in shape, do not. The presence of resin is also clearly visible in the scanning electron micrograph of the string-like shrinkage cavity row (Figure 8c), but the impregnating agent cannot be identified in the spherical cavity in Figure 8e (the cavities in the lower-right-third of Figure 8b). The UV-illumination optical microscopy inspection procedure clearly identifies cavities containing an impregnating resin on the metallographic section in our experience.

## 5. Conclusions

Die-cast aluminum alloy components with gas tightness requirements manufactured with high care in melt treatment and casting technology can still have leaking problems after impregnation, especially in castings with larger cross-sections. In order to map the nature and location of the possible leakage pathways, a detailed investigation of the material continuity defects and porosity of the casting is required. Leakage is likely to occur along contiguous leak paths with complex geometries of holes produced by gases and shrinkage cavities. The investigations presented in this study were carried out on a cast part, showing leakage. The inner structure of the cast part was analyzed by metallographic macro photography through an optical microscope and micro-CT images of the same planes in nine parallel cross-sectional positions. The optical microscopic and micro-CT procedures allowed resolutions of 3.5 and 30 μm, respectively.

The statistical evaluation of the cavities in the investigated planes (see Tables 1 and 2) shows that the defect density in the critical part of the casting is very high, exceeding 10 defects/mm$^2$ based on the optical microscopic examinations. However, the typical defect size is small; 80% of the defects are less than 30 μm. Overall, the ratio of the defected area is 1.3% for the surfaces tested. The defect density and defect rate, determined by the micro-CT examination, were significantly lower, which is a consequence of the lower resolution of this examination technique. For the defects larger than 60 microns, optical and micro-CT analyses yielded approximately the same statistical characteristics.

The cavity systems attached to the surface were partially saturated with resin during impregnation. In the case of the tested casting, the presence of impregnation resin in the cavity was not detectable by the micro-CT method. This is because the typical cavity size and cross-section are very small in relation to the wall thickness of the casting, and the resin comprises a small atomic number of elements. Therefore, there is minimal contrast difference between the resin-saturated and non-saturated cavities. Optical microscopy under UV illumination is suitable for the detection of impregnation resin in cavities.

**Author Contributions:** Conceptualization, M.R. and R.H.; methodology, M.R.; software, F.O., E.R.F. and T.B.; validation, M.R., J.G. and F.O.; formal analysis, M.R.; investigation, M.R.; resources, M.R.; writing—original draft preparation, M.R.; writing—review and editing, M.R. and R.H.; visualization, R.H. and F.O.; supervision, R.H. All authors have read and agreed to the published version of the manuscript.

**Funding:** This research was funded by the 2019-1.1.1-Market KFI-2019-00462 project and by the UNKP-22-3 New National Excellence Program of the Ministry of Human Capacities, Hungary.

**Data Availability Statement:** The data presented in this study are available on request from the corresponding author.

**Acknowledgments:** The authors would like to thank István Somogyi, Quality Engineer of Hanon Systems Auto Parts Hungary Ltd.; Viktor Nyeste, Technical Associate of Hanon Systems; and Sándor Orosz, Technical Manager of Euraseal Service Ltd. for providing the samples and technological information for the research; and Ádám Filep, Research Fellow of the Institute of Physical Metallurgy, Metalforming and Nanotechnology, University of Miskolc, for his assistance in the implementation of the CT scan.

**Conflicts of Interest:** The authors declare no conflict of interest. The funders had no role in the design of the study; in the collection, analyses, or interpretation of data; in the writing of the manuscript; or in the decision to publish the results.

## References

1. Anggraini, L. Analysis of Porosity Defects in Aluminum as Part Handle Motor Vehicle Lever Processed by High-pressure Die Casting IOP Conf. *Ser. Mater. Sci. Eng.* **2018**, *367*, 012039.
2. Felberbaum, M. Porosity in Aluminum Alloys. Ph.D Thesis, EPFL, Lausanne, Switzerland, 2010.
3. Hanxue, C.; Mengyao, H.; Chao, S.; Peng, L. The influence of different vacuum degree on the porosity and mechanical properties of aluminum die casting. *Vacuum* **2017**, *146*, 278–281.
4. Szalva, P.; Orbulov, I.N. The Effect of Vacuum on the Mechanical Properties of Die Cast Aluminum AlSi9Cu3(Fe) Alloy. *Int. J. Metalcast.* **2019**, *13*, 853–864. [CrossRef]
5. Lordan, E.; Zhang, Y.; Dou, K.; Jacot, A.; Tzileroglou, C.; Wang, S.; Wang, Y.; Patel, J.; Lazaro-Nebreda, J.; Zhou, X.; et al. High-Pressure Die Casting: A Review of Progress from the EPSRC Future LiME Hub. *Metals* **2022**, *12*, 1575. [CrossRef]
6. Campbell, J. *Complete Casting Handbook: Metal Casting Processes, Metallurgy, Techniques and Design*; Butterworth-Heinemann: Oxford, UK, 2015.
7. Carmignato, S. Computed tomography as a promising solution for industrial quality control and inspection of castings. *Metall. Sci. Technol.* **2012**, *1–30*, 5–14.
8. Fuchs, P.; Kröger, T.; Garbe Christoph, S. Defect detection in CT scans of cast aluminum parts: A machine vision perspective. *Neurocomputing* **2021**, *453*, 85–96. [CrossRef]
9. Jolly, M.R.; Prabhakar, A.; Sturzu, B.; Hollstein, K.; Singh, R.; Thomas, S.; Shaw, A. Review of non-destructive testing (NDT) techniques and their applicability to thick walled composites. *Procedia CIRP* **2015**, *38*, 129–136. [CrossRef]
10. Vásárhelyi, L.; Kónya, Z.; Kukovecz, Á.; Vajtai, R. Microcomputed tomography-based characterization of advanced materials: A review. *Mater. Today Adv.* **2020**, *8*, 1–13. [CrossRef]
11. Zhuang, L.; Shin, H.S.; Yeom, S.; Pham, C.N.; Kim, Y.J. A novel method for estimating subresolution porosity from CT images and its application to homogeneity evaluation of porous media. *Sci. Rep.* **2022**, *12*, 16229. [CrossRef]
12. Jaques, V.A.; Du Plessis, A.; Zemek, M.; Šalplachta, J.; Stubianová, Z.; Zikmund, T.; Kaiser, J. Review of porosity uncertainty estimation methods in computed tomography dataset. *Meas. Sci. Technol.* **2021**, *32*, 122001. [CrossRef]
13. Kokhan, V.; Grigoriev, M.; Buzmakov, A.; Uvarov, V.; Ingacheva, A.; Shvets, E. Segmentation criteria in the problem of porosity determination based on CT scans. *Twelfth Int. Conf. Mach. Vis.* **2020**, *11433*, 378–385. [CrossRef]
14. Reedy, C.L.; Reedy, C.L. High-resolution micro-CT with 3D image analysis for porosity characterization of historic bricks. *Herit. Sci.* **2022**, *10*, 83. [CrossRef]
15. Limodin, N.; El Bartali, A.; Wang, L.; Lachambre, J.; Buffiere, J.Y.; Charkaluk, E. Application of X-ray microtomography to study the influence of the casting microstructure upon the tensile behaviour of an Al–Si alloy. *Nucl. Instrum. Methods Phys. Res. Sect. B Beam Interact. Mater. At.* **2014**, *324*, 57–62. [CrossRef]
16. Garb, C.; Leitner, M.; Tauscher, M.; Weidt, M.; Brunner, R. Statistical analysis of micropore size distributions in Al–Si castings evaluated by X-ray computed tomography. *Int. J. Mater. Res.* **2018**, *109*, 889–899. [CrossRef]
17. Weidt, M.; Hardin, R.A.; Garb, C.; Rosc, J.; Brunner, R.; Beckermann, C. Prediction of porosity characteristics of aluminium castings based on X-ray CT measurements. *Int. J. Cast Met. Res.* **2018**, *31*, 289–307. [CrossRef]
18. Nicoletto, G.; Konečná, R.; Fintova, S. Characterization of microshrinkage casting defects of Al–Si alloys by X-ray computed tomography and metallography. *Int. J. Fatigue* **2012**, *41*, 39–46. [CrossRef]
19. Zhang, Y.; Majeed, A.; Muzamil, M.; Lv, J.; Peng, T.; Patel, V. Investigation for macro mechanical behavior explicitly for thin-walled parts of AlSi10Mg alloy using selective laser melting technique. *J. Manuf. Process.* **2021**, *66*, 269–280. [CrossRef]
20. Lloyd, G. Finding a Solution to the Eternal Problem of Porosity in Casting. *Die Cast. Eng.* **2012**, *40*, 1–7.
21. Soga, N.; Bandara, A.; Kan, K.; Koike, A.; Aoki, T. Micro-computed tomography to analyze industrial die-cast Al-alloys and examine impregnation polymer resin as a casting cavity sealant. *Prod. Eng.* **2021**, *15*, 885–896. [CrossRef]

**Disclaimer/Publisher's Note:** The statements, opinions and data contained in all publications are solely those of the individual author(s) and contributor(s) and not of MDPI and/or the editor(s). MDPI and/or the editor(s) disclaim responsibility for any injury to people or property resulting from any ideas, methods, instructions or products referred to in the content.

Article

# Grain Size Distribution of DP 600 Steel Using Single-Pass Asymmetrical Wedge Test

Urška Klančnik [1,*], Peter Fajfar [2], Jan Foder [3], Heinz Palkowski [4], Jaka Burja [5] and Grega Klančnik [6,*]

1   Valji d.o.o., Železarska Cesta 3, SI-3220 Štore, Slovenia
2   Department of Materials and Metallurgy, Faculty of Natural Sciences and Engineering, University of Ljubljana, Aškerčeva Cesta 12, SI-1000 Ljubljana, Slovenia
3   SIJ Acroni d.o.o., C. Borisa Kidriča 44, SI-4271 Jesenice, Slovenia
4   Institute of Metallurgy, Technical University of Clausthal, Robert-Koch-Straße 42, D-38678 Clausthal-Zellerfeld, Germany
5   Institute of Metals and Technology, Lepi Pot 11, SI-1000 Ljubljana, Slovenia
6   Pro Labor d.o.o., Podvin 20, SI-3310 Žalec, Slovenia
*   Correspondence: urska.klancnik@valji.si (U.K.); klancnik.grega@gmail.com (G.K.)

**Abstract:** Grain size distribution after the completion of a phase transformation was studied through the laboratory-controlled hot-plastic deformation of dual phase 600 (DP 600) steel using a specially prepared asymmetric single-pass hot-rolling wedge test with a refined reheating grain size instead of the usual coarse-grained starting microstructure observed in practice. The experiment was performed to reduce generally needed experimental trials to observe the microstructure development at elevated temperatures, where stable and unstable conditions could be observed as in the industrial hot-rolling practice. For this purpose, experimental stress–strain curves and softening behaviors were used concerning FEM simulations to reproduce in situ hot-rolling conditions to interpret the grain size distribution. The presented study revealed that the usual approach found in the literature for microstructure investigation and evolution with a hot-rolling wedge test was deficient concerning the observed field of interest. The degree of potential error concerning the implemented deformation per notch position, as well as the stress–strain rate and related mean flow stresses, were highly related to the geometry of the specimen and the material behavior itself, which could be defined by the actual hardening and softening kinetics (recrystallization and grain growth at elevated temperatures and longer interpass times). The grain size distribution at 1100–1070 °C was observed up to a 3.45 s$^{-1}$ strain rate and, based on its stable forming behavior according to the FEM simulations and the optimal refined grain size, the optimal deformation was positioned between $e = 0.2$ and $e = 0.5$.

**Keywords:** wedge test; hot-rolling; grain size distribution; dual-phase steel

## 1. Introduction

### 1.1. Dual Phase Steels

Dual-phase (DP) steel is highly interesting and is intended for many demanding applications, such as the automobile industry. Its high strength-to-weight ratio can be related to fuel economy as a part of the green transformation promoted by the EU. These compositions are similar to lean qualities, making them interesting for further study. The use of the wedge test for obtaining data on grain size evolution under various rolling parameters is demonstrated in this paper.

DP steels are considered common representatives of advanced high-strength steels (AHSS) for advanced safety components in the automotive industry due to their high tensile strength. The high strength is obtained by (higher carbon) martensite (M) content, and the optimal yield stress is provided by the ferrite (F) content. Therefore, their good cold-formability properties can be exploited for rather complex shapes. A complex microstructure was obtained during sheet processing; bainite (B), retained austenite, and carbides could also be identified alongside F and M. Usually, the martensite fraction is from 10 to 30 %. In the case of high-strength DP 1200 steel up to 38 vol.% or higher [1–4].

AHSS sheets are mainly delivered after cold rolling and are covered with metal coatings to protect the steel surface against corrosion [4–6]. They are also available in a hot-rolled state. For ultrafine (UF) refinement, two on-line methods are generally used: modern thermomechanical processing (which involves controlling the cooling rate during the $\gamma \rightarrow \alpha$ transformation, strain (–deformation)-induced ferrite nucleation (DIFT), etc.) and severe plastic deformation strategies at elevated temperatures, introducing low-temperature regimes, etc. [7]. An example of these two methods is when one part of the deformation is carried out in the roughing stage (R) for full austenite recrystallization, and the other part is carried out in the low-temperature regime with several large strain deformations. They are followed by inter-critical annealing (IA) and quenching for the production of ultrafine-grained DP steel [1]. Ultrafine ferrite grain (ULFG) can be produced by a single- or multi-pass deformation at IA temperatures by promoting the dynamic-strain-induced transformation of austenite to ferrite [8]. Under ordinary hot-rolling procedures, the DP steel's microstructure is composed of F and pearlite (P) bands, while microalloying additions (such as Nb) may cause additional B formation. Classical microstructure development using off-line reheating, for example, in thin strips, can be carried out by heating the products within the IA region between $A_{r1}$ and $A_{r3}$, followed by quenching [3]. These rolling schedules and heat treatments can, therefore, be adapted to the chemical composition, secondary precipitation, and potential influence of the solid drag (SD). In this study, a single laboratory "roughing pass" was conducted at a reasonably low roughing temperature of 1100 °C, which is close to the conditions of the final roughing passes used in the industry.

*1.2. The Wedge Test*

Non-standardized wedge-shaped specimens are used for research regarding metal forming, the workability of specific alloys, and to evaluate microstructure evolution. The tests of wedge specimen deformations can be used for hot forging, pressing, as well as (hot-) rolling tests. The rolling tests of wedge-shaped specimens (henceforth called wedge rolling tests) deform their former tapered geometry with a single pass into a thin sheet of equal thickness; however, different deformation conditions are applied regardless of the rolled length of the material under investigation. The direction of rolling begins from the lower height to fulfill the bite condition and gradually progresses toward the higher end. The wedge specimen can be designed to include a narrow part at the beginning (the so-called tongue) with a height smaller than the rolling gap. This in itself provides a good control sample, given that the tongue is not submitted to any material deformation. The wedge specimen must end with a parallel plane length of at least 1/3 of the total specimen's length to ensure a stable finish [9].

The wedge rolling test can be used to obtain information on the slab-to-plate or plate-to-strip reduction ratio, specifically concerning the grain size and microstructure evolution during deformation [10–14]. It can also be used for technological formability studies, such as observing crack formations under different technological parameters of melt preparation, casting, and hot rolling. This was demonstrated in the case of 25CrMo4 steel by extracting samples from different positions of blanks, blooms, and slabs. The height strain, $e$ (or $\varepsilon$), the force, and the strain rate ($\dot{\varepsilon}$) were close to linear with increasing values along the rolled length [15]. A possible mathematical solution for the evaluation of rolled wedge samples was proposed by Kubina et al. [16], where a macro image of the rolled sample was evaluated with a computer algorithm. The output was the mean strains (both true and relative), the mean strain rates, and the broadening of the sample along the sample's length. However, due to the dynamic nature of the wedge rolling test and its high dependency on the sample geometry, rolling stand, etc., the instructions for sampling and microstructural investigation should be given in a precise manner to ensure good comparability and repeatability.

In this study, the wedge rolling test was performed using DP 600 steel. The nature of the grain size evolution in grain refining and grain growth was studied at an elevated temperature and different length positions to ensure different thickness reductions with the corresponding strain rates ($\dot{\varepsilon}$). The main issue in the wedge-rolling tests was in the comparability of individual tests in terms of the sampling location significance and possible deformation-related influences that could affect the result.

## 2. Materials and Methods

### 2.1. Materials

The steel charge was melted using an open induction furnace. The steel (with $T_{liq}$ = 1520 °C) was cast into ingots at approx. 50 K of superheat and left to cool. No stress relieving was necessary due to the lean composition with poor hardenability. Ingots were used to prepare a tapered wedge sample with the dimensions shown in Figure 1a. The wedge sample consisted of three distinct areas: first, the thinnest starting part, referred to as the tongue (which also served as the reference point for microstructural evaluation), ensured a smoother start to the rolling; second, the tapered rising part enabled the dynamic test of varying strains and strain rates. Finally, the finishing thick area served to stabilize the rolling process and prevented dimensional post-rolling abnormalities [9]. Furthermore, notches were mechanically cut into the side of the wedge to signify the selected computed deformations (given in Table 1) for easier observation after rolling. A schematic representation of hot-deformation and cooling regimes to prevent excessive primary austenite grain (PAG) coarsening is shown in Figure 1b. Cooling was similar to offline heat treatments for DP steels. The cooling started from the full austenite region, crossing the ($\alpha + \gamma$) region by adjusting the cooling rate, and finally, with the intense transformation of undercooled (remaining) austenite (A) into a-thermal martensite (M), the test was completed [5].

**Table 1.** Engineering (relative) deformation per notch position (/) using a 3 mm rolling gap.

| Position | $e_1$ | $e_2$ | $e_3$ | $e_4$ | $e_5$ | $e_6$ | $e_7$ | $e_8$ | $e_9$ * |
|---|---|---|---|---|---|---|---|---|---|
| Target | 0.04 | 0.12 | 0.21 | 0.32 | 0.41 | 0.50 | 0.60 | 0.70 | 0.75 |
| Actual | 0.05 | 0.08 | 0.12 | 0.21 | 0.32 | 0.39 | 0.51 | 0.62 | 0.68 |
| Actual ratio ** | 1.05 | 1.08 | 1.13 | 1.27 | 1.46 | 1.64 | 2.04 | 2.62 | 3.12 |

\* Notch position $e_9$ marks the final flat end of the wedge sample. ** Measured on actual starting and achieved exit thickness per notch position. (Calculation of true strain from relative follows: $\varepsilon_h = -\ln(1-e)$).

### 2.2. Experimental Testing

The final rolled sheet was revealed to be rather flat, with no excessive edge waves or flatness issues. This could be an indication that due to a rather low rolling speed (and consequently $\dot{\varepsilon}$), the final microstructure along the sample length was cross-sectionally uniform (not isotropic) despite it being an asymmetric wedge test [17]. Figure 1c shows the location of the notch positions where with each increased number (from $e_1$ to $e_9$), an increase in the stored energy during rolling was expected. This means that, by increasing the deformation, the driving force for recrystallization also continuously increases [18]. The maximum ratio $\frac{h_0}{h_1}$, where $h_0$ signified the starting thickness and $h_1$ the final thickness, was set to 4:1. The ratio 3:1 was estimated to be the lowest ratio per notch position of the highest reduction for the given study and the wedge geometry setup. The wedge maximum thickness was sufficient due to the maximum allowable bite angle $\alpha_o$ concerning the friction coefficient. The steel wedge sample was reheated to the temperature of 1100 °C. The temperature of the sample after rolling was measured to be, on average, 1070 °C. For a single-run hot-rolling test, a laboratory two-high rolling mill with roll diameters (2$R$) of 294 mm each was used. The work rolls were also preheated to minimize heat extraction during rolling (a drop of 30 K was measured). The preheating of the rolls was conducted for 3 hours before the actual test was run to achieve quite a homogeneous surface temperature of approx. 80 °C. Based on the work roll's rotation frequency of $n = 8$ rpm, the average rolling speed $v$ was calculated as:

$$v = \omega \cdot R \tag{1}$$

where $\omega$ is the angular velocity. Therewith, the average rolling speed was calculated to be 0.12 m/s. The mean strain rate, $\dot{\varepsilon}$, based on the contact time of 2.5 s, was calculated to be up to 4.1 s$^{-1}$, and when estimated with the length of the deformation zone $L_d$, the maximum values were up to 3.98 s$^{-1}$ (Equations (2)–(4)) [19].

$$\dot{\varepsilon} = \frac{\varepsilon}{L_d} \cdot v \tag{2}$$

$$L_d = \sqrt{R \cdot (h_0 - h_1)} \tag{3}$$

$$\cos \alpha_0 = 1 - \frac{h_0 - h_1}{2R} \tag{4}$$

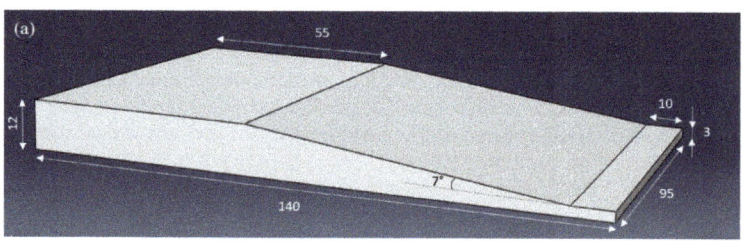

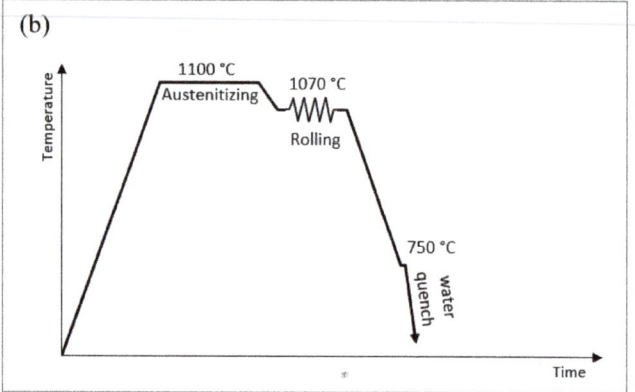

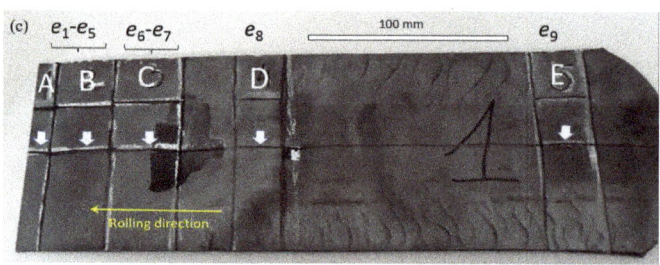

**Figure 1.** (**a**) Geometry of the test sample (sketch taken from the FEM simulation), (**b**) Test schedule, and (**c**) Final sample with locations selected for investigation (marked A to E). The arrows represent the position of the grain size evolution study (through the thickness cross-section). The edge integrity in rolling length is well preserved—the material acts ductile during rolling.

The first part of the moderate cooling of the nominally 3 mm thick sheets after rolling was conducted to promote ferrite nucleation in the ($\alpha + \gamma$) region down to approx. 750 °C, which was close to the usual IA. Further intense cooling was performed to promote $\gamma \rightarrow M$ transformation, similar to step quenching [4].

The samples were extracted for a light optical microstructure observation (OM, Zeiss Axio Imager 2, Francoforte, Germany) and field-emission scanning electron microscope investigation (FE-SEM, Zeiss Supra VP55, Francoforte, Germany), as seen in Figure 1c. They were viewed in the transverse direction according to the rolling direction. After grinding and polishing, the microstructure was observed after etching with 2% Nital. FE-SEM observations were performed on non-etched samples. Secondary electron (SE) images were used for grain size evaluation and were analyzed using the commercial ImageJ software (1.53k, open-source software, NIH, Bethesda, MD, USA) and by analyzing Feret's diameter.

Table 1 reveals the target and actually achieved engineering deformation per notch position for the performed wedge rolling test. The target deformations were chosen based on roll mill stiffness, safety during sheet production, and allowable roll mill displacement. The achieved true strain was in a range between $\varepsilon = 0.05$ and 0.96. The per-notch position deformation could be considered as having a cumulative (residual) deformation representing multiple passes with an insufficient delay between passes for softening (or achieving $t_{50}$). For example, in usual recrystallization rolling, at least 0.5 of the cumulative engineering strain was targeted for proper uniform grain size evolution. This was consistent up to notch $e_7$ regarding the achieved strain $e$. Here, we already had an obvious discrepancy between the target and achieved deformation based solely on the geometry of the hot-rolled flat sheet.

The per-notch positions for engineering strains $e_1$ to $e_3$ were regarded as limited regarding the metallurgical pass needed for the full cross-section PAG evolution. Additionally, secondary recrystallization was expected during post-rolling cooling. The latter was closer to broad sizing as seen in industrial practice, with weak first $\varepsilon$ passes, and were not considered as actual roughing or effective metallurgical passes [20]. Above $e_3$ and up to $e_5/e_6$ was the region where the last roughing pass before finishing in plates was considered in actual practice, and post-deformation SRX is regarded as the main softening kinetic. Above $e_5/e_6$, the $\varepsilon_T$ was already above the usual per pass engineering strain for the last roughing phase of most of the thin plate rolling productions' practice. This meant that the main softening mechanism in the last positions (between $e_6$ and $e_9$) on the wedge if $\varepsilon_c$ or $\varepsilon_p$ were also considered (latter as grain size dependent), was among the dynamic, static recovery (DRV, SRV) also fast post-deformation meta-dynamic recrystallization (MDRX). SRX could also be activated. The maximum peak obtained from the $\sigma$-$\varepsilon$ curve was evaluated at approximately $\varepsilon = 0.3$–0.4 with the softening above $\varepsilon = 0.4$ for 1100 °C at 1 s$^{-1}$ and 3 s$^{-1}$. The $\sigma$-$\varepsilon$ curves indicated a DRX phenomenon ($T > T_{nrx}$), as shown later in the manuscript. This was observed with the hot compression curve; reheating to 1200 °C $\rightarrow$ cooling $\rightarrow$ deformation at 1100 °C and $\varepsilon_T = 0.7$, quench. This meant that with and above $e_7$, the effect of starting PAG was complex, depending on the DRX kinetics. It also indicated that at least part of the production/reheating grain size history of the wedge should be minimized concerning the achieved final (originally preserved) PAG; it was known that under DRX, the recrystallization kinetics were independent of the starting (initial) PAG if $\sigma_{steady}$ was achieved. However, the post-deformed softening on nucleated DRX grains and grain size evolution was complex and depended on the status of individual grains concerning the preserved dislocation density and actual grain mobility.

Based on the results shown further in this paper, this also meant that a different PAG evolution per notch position was expected in the activated post-deformation process [21]. After deformation and based on the predicted CCT calculated using JMatPro, the approximate cooling rate that was needed to prevent pearlite formation was approx. 10 K/s, which is in agreement with [2] for similar compositions used for DP microstructure formation using the deformational dilatometry test.

## 2.3. Computer Modeling

Two commercial thermodynamic tools were used, namely Thermo-Calc (Thermo-Calc 2022a, TCFE11: Steels/Fe-Alloys v11.0, Thermo-Calc Software AB, Solna, Sweden) and JMatPro 6.1 (General Steel database). JMatPro 6.1 (Sente Software Ltd., Guildford, UK) was also used for the estimation of approximate cooling rates for $\gamma \rightarrow \alpha$ transformation before intense quenching in the cold water of approx. 18 °C. Additionally, the non-commercial model from Bhadeshia, Mucg83, was used for basic temperature determinations [22].

For a more detailed understanding of the hot working process when using a tapered wedge test, finite element (FEM) simulations were conducted using Abaqus software. The model was calculated as a 3D stress Dynamic/Explicit problem with the wedge being deformably extruded solid, and both rolls were 3D discrete rigid objects. The mesh of the wedge sample consisted of 2904 standard C3D8R linear brick elements, while each of the rolls was meshed with 1111 R3D4 and R3D3 quadrilateral and triangular elements, respectively. The mesh of the wedge sample was constructed with a bias and element seeds ranging from 4 mm to 2 mm. Figure 2 shows the mesh sensitivity analysis of the wedge sample, revealing that decreasing the element size below 4 mm gains produced neglectable little accuracy at the expense of a much longer computing time.

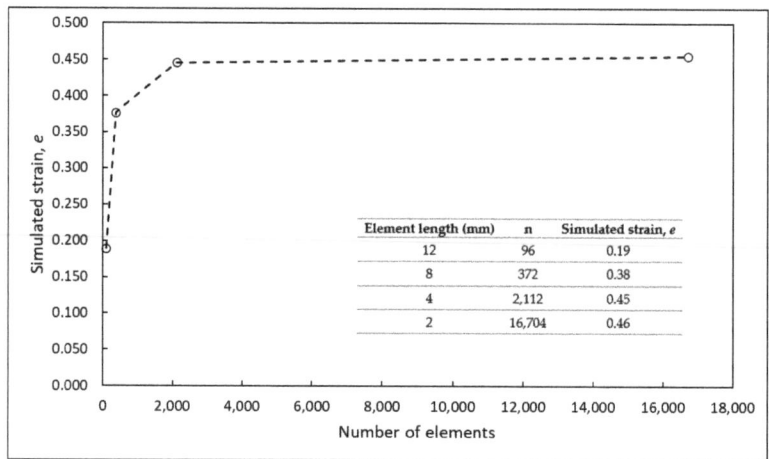

**Figure 2.** Mesh sensitivity analysis of the wedge sample mesh (C3D8R elements); a decreasing element size below 4 mm gained neglectable little accuracy at the expense of a significantly longer computing time.

In the setting of the simulation, the heat transfer between the wedge and the rolls was neglected, similar to the work of Parsa et al. [10]. The temperature effect was taken into account by employing the mechanical properties of the wedge as being temperature dependent. The data used for the temperature-dependent mechanical properties (Young's modulus, density, Poisson ratio) were obtained with the thermodynamic prediction software described above and the data shown in Table 2. The stress–strain curves used were taken from actual measurements performed on the Gleeble thermo-mechanical simulator, as an example shown in Figure 3. Kinematic sliding with the penalty was used as an interaction property; a friction coefficient of 0.3 was used based on the work of Edberg et al. [23] to calculate the flow stresses with the friction-hill model for a hot strip mill [18]

**Table 2.** Material properties used in the FEM calculation.

| Density (g/cm³) | Young's Modulus (GPa) | Poisson's Ratio |
|---|---|---|
| 7.485 | 98.85 | 0.36 |

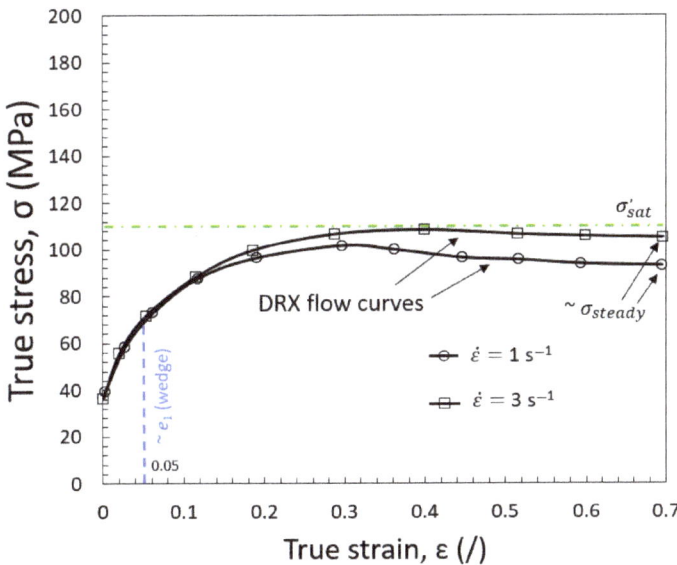

**Figure 3.** An example of experimental and fitted Gleeble hot compression true stress-true strain ($\sigma$-$\varepsilon$) diagram: DRX flow curves for both deformation rates (1 and 3 s$^{-1}$) at 1100 °C are presented for the C-Mn type of composition and were also used for DP 600 in this study. The estimated absolute error is +/−10 MPa (based on the Sellars constitutive equation, adapted after [3]). For the highest rate, a simplified hypothetical saturation stress limit ($\sigma'_{sat}$) was added (dotted line) to observe the softening due to DRX.

## 3. Results and Discussion

### 3.1. Chemical Composition

A basic composition, typical for lean (C-Mn) structural steels, was set (with low carbon, high manganese grade) where manganese defined the proportion of ferrite and pearlite under ordinary air-cooled conditions [24,25]. No precipitation-strengthening carbide or nitride-forming elements were intentionally added to the given study to allow full softening kinetics (for close to equiaxed austenite shape formation) concerning the kinematic parameters of wedge rolling. Additionally, due to the very thin sheet material, there was no need for Cu, Mo, or Ni to enhance hardenability. A similar basic composition could also be used for S690QL thin plate grades (<8 mm) using thermo-mechanical controlled processes (TMCP) and DP steels which were intended for bake hardening [2,17]. The added Al produced no precipitation hardening [24]. In Table 3, showing the measured chemical composition, it was obvious that a rather high nitrogen content was achieved due to the open induction melting procedures and the used pre-alloyed material. Soluble nitrogen influences the impact transition temperature; for this test, this was considered allowable. Therefore, the soluble nitrogen was not controlled, and grain refinement by HAGB and coarsening were observed concerning the rolling parameters. The alloys used in this case were highly pure, resulting in a low oligo-element content and cleanliness concerning non-metallic inclusions (NMI). Sulfur and phosphorous were both under 0.0015% and 0.01%, respectively. The material was prepared using remelts for the synthesis of low carbon, which was close to C-Mn type steel [26]. No boron or other microalloying additions were introduced for this purpose. The metal melt was deoxidized using SiMn and Al. The calculated carbon equivalent, based on the measured compositions of an ingot using the CEV (IIW) equation, was 0.42. This indicated that the material was weldable.

Table 3. Chemical composition of the as-cast ingot (in wt.%).

| C | Si | Mn | Cr | Ni | Mo | Al * | Cu | $N_{tot}$ | Ti + Nb + V | Fe | CEV [26] |
|---|---|---|---|---|---|---|---|---|---|---|---|
| 0.072 | 0.19 | 1.38 | 0.56 | <0.01 | <0.01 | 0.031 | <0.01 | 0.018 | 0.0125 | Bal. | 0.42 |

* Minimum 0.02 wt.% Al was considered for a fully killed grade. The ratio Al: N < 2.0 indicates the potential availability of soluble nitrogen in the austenitic matrix as no other nitrogen-binding elements were introduced.

Based on the chemical composition, $T_{nrx}$ was estimated to be at 841 and 864 °C based on the Boratto–Barbosa equation [27] and modified equation [28], respectively. Having a similar composition and microalloying additions, the $T_{nrx}$ was properly increased as in the work of Song et al. [8]. The $T_{nrx}$ was considered as a recrystallization stop temperature (RST). The recrystallization low temperature (RLT), despite the minimum solid drag and Zener force, if any, was considered to be 934 °C, which was close yet still approx. 140 °C under the used single-pass temperature. The $A_{e3}$ was predicted by Thermo-Calc and JMatPro 6.1 to be 834 °C and 830 °C, respectively. This meant that R and FRT were conducted in the full austenitic recrystallization range (Type-I, static recrystallization) according to Irvin et al. [24], and no substantial pancaking could be visible, even by direct online water quench.

The $M_s$ was predicted by JMatPro to be 438 °C, while the CCT diagram using KIN was predicted to be 477 °C (Figure 4). Additionally, based on the Bhadeshia model [22], the $M_s$ was predicted to be 463 °C. All information about the starting $M_s$ position revealed the possible self-tempering effect of prior M upon continuous cooling. The predicted [28] $A_{r3}$ was 764 °C and, based on KIN, $A_{c3}$ = 828 °C. The intense cooling should already be partially performed inside the IA region based on $A_{e3}$ and also $A_{r3}$ conditions. The delay time from FRT to the beginning of the water quench was sufficient for proper polygonal ferrite development under $A_{e3}$ [8]. In this case, the formed ferrite was not impinged after nucleation due to the lack of pinning particles and their subsequent deformation. Due to the rather fast cooling from FRT to the region of $A_{r3}$, the overall PAG coarsening was limited.

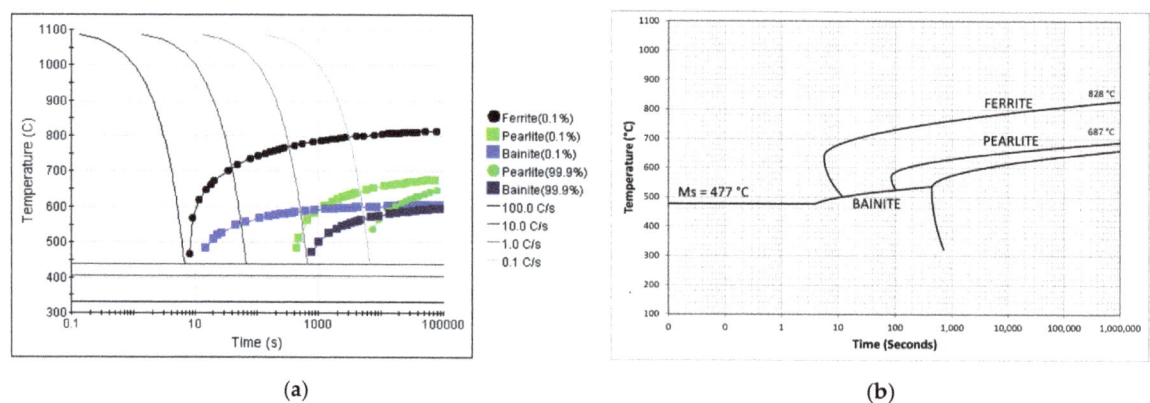

Figure 4. (a) Predicted CCT using JMatPro and (b) KIN with the internal material database.

### 3.2. FEM Simulations and Calculations

The wedge rolling test in itself is considered to be a dynamic test as every geometrically dependent rolling parameter ($\varepsilon$, $\dot{\varepsilon}$, $\alpha_o$) is both time and location dependent. Based on Equations (2)–(4), the different calculated mean parameters of the rolling in dependence of the per notch position are shown in Figure 5. Expectedly, an increasing trend can be noticed for all three parameters.

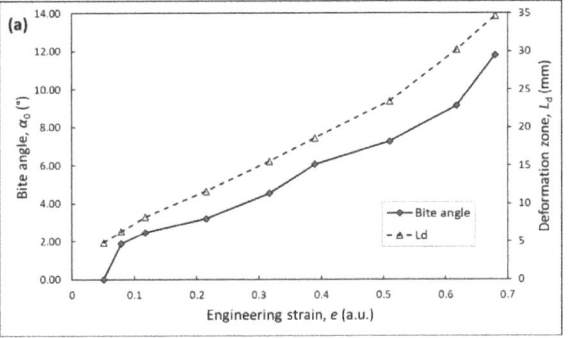 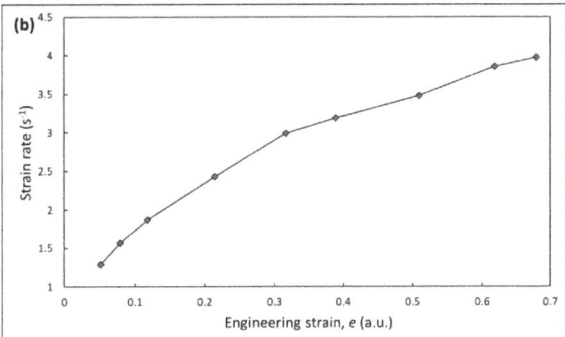

**Figure 5.** Calculated mean values of (**a**) the bite angle, deformation zone, and (**b**) strain rate with the per notch position of increasing mean strain, $e$. Each point on the diagram represents an increasing notch position (see Table 1 for reference).

Figure 6 represents the deformed wedge sample after rolling. The parameters, as described in the previous section, were taken from measurements during the actual test (i.e., the rotational speed of the rolls, velocity of the wedge sample, the rolling gap, etc.). When comparing the dimensions of the simulated sample to the actual one, it was observed that the simulation resulted in larger dimensions, even though the overall projection of the simulated sample was visually almost identical to the actual one (see Figure 1c). When comparing the simulation results to real experiments, a certain deviation was to be expected. In our case, the geometrical deviation most likely stemmed from the fact that shrinking during cooling was not included. Furthermore, the simulation was stopped after the sample left the rolling gap; therefore, the temperature drop and stress relaxation was not incorporated. The second reason was most likely linked to the fact that the simulation kept the end thickness fixed at 3 mm, while in reality, the thickness slightly increased toward the end of the wedge sample due to roll displacement (also evident from the actual ratio in Table 1).

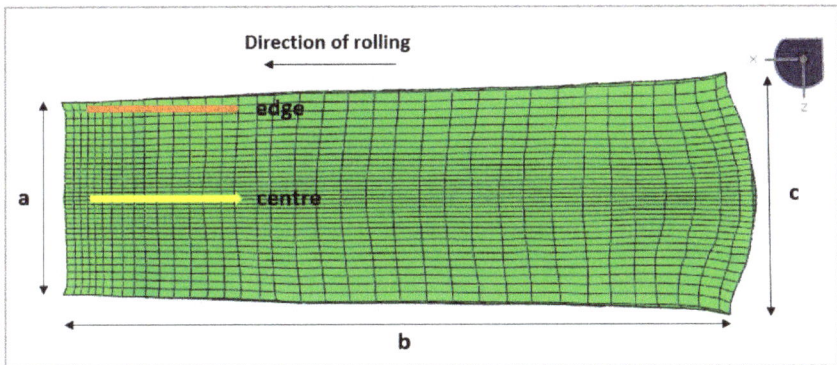

**Figure 6.** Image of the deformed wedge sample after post-rolling according to the simulation.

The dimensions of the wedge sample were marked with letters from $a$ to $c$; a comparison between the simulation and the actual test is given in Table 4. The center and edge directions are marked in the image with arrows (Figure 6) highlighting the two different areas for the strain and strain rate evaluation of the wedge sample.

**Table 4.** Wedge sample dimensions post-rolling: comparison between the actual test and the simulation (in mm).

|  | a | b | c | End Thickness |
|---|---|---|---|---|
| Actual test | 95.15 | 281.52 | 101.28 | 2.98–3.86 |
| Simulation | 95.86 | 322.58 | 112.80 | 3.00 |

The simulation of the true strain, $\varepsilon$, on the deformed wedge sample (Figure 7) showed an unequal distribution of the strain over the wedge's planar projection. The simulated strain was, as expected, lowest at the tongue part of the wedge sample and started increasing toward the thicker end. At first, the strain increased almost linearly over the entire width and localized with higher strain zones that started appearing in the middle of the wedge's width around notch $e_7$. This phenomenon continued throughout the rest of the wedge's length; lines of equal strain transformed from straight into almost parabolic (also visible by the deformation of elements), as seen in Figure 8a. The inequality of the predicted strain at the edge and in the center of the wedge sample was more emphasized, where the values calculated on the edge were increasingly lower compared to the values predicted in the center. This indicated a certain loss of strain and strain rate control during rolling.

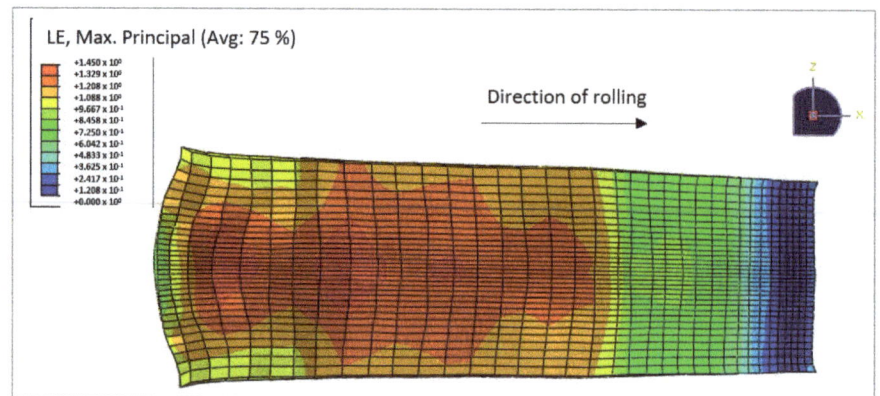

**Figure 7.** Simulated true strain, $\varepsilon$, of the deformed wedge sample.

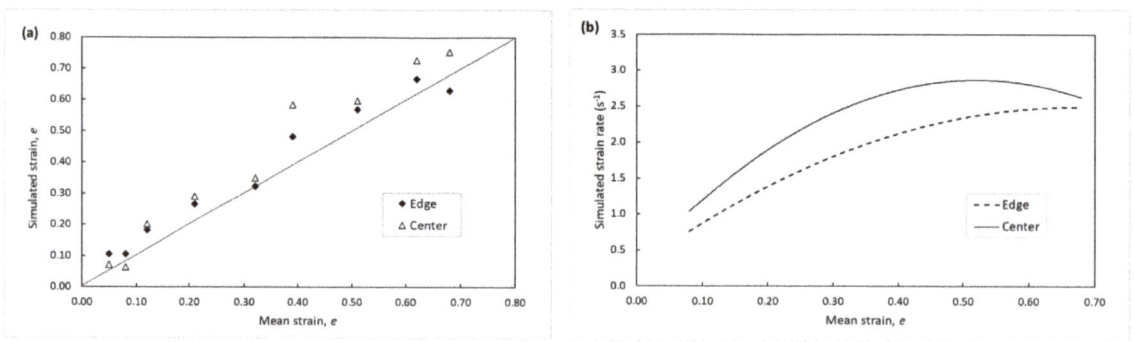

**Figure 8.** (a) Simulated relative strain $e$ (calculated from the software's output $\varepsilon$) to the mean relative strain based on the notches on the wedge sample (Table 1). (b) Simulated strain rate (given as an average trend) in correlation to the mean relative strain based on the notches on the wedge sample (dotted line—edge, solid line—center). Both simulated values are shown for the edge and the center of the wedge sample in the longitudinal direction, see Figure 6.

All the simulated strain values were predicted to be higher than the mean values from individual notch positions, according to Figure 8a. The simulation computation took into account the mutual interaction of individual elements representing the partial volume of the sample, which the theoretical calculation could not account for. Most likely, for this same reason, the strain in the center of the sample in the simulation was calculated to be higher than on the edge. The strain rate, calculated from the software's output strain $\varepsilon$ per notch position (simply as $\frac{\Delta \varepsilon}{\Delta t}$ for individual elements corresponding to a specific notch position), also showed a different rate between the edge and the center of the sample, see Figure 8b. Compared to the calculated mean strain rate, the simulated values were lower: the maximum simulated levels of the strain rate were 3.45 s$^{-1}$ and 2.93 s$^{-1}$ (center of the sample, notch $e_8$ and edge of the sample, notch $e_8$, respectively) while the maximum mean strain rate was calculated as 3.98 s$^{-1}$ (notch $e_9$).

The result of the predicted unequal strain across the planar projection as well as the differences predicted between the edge and the center of the wedge suggest that caution must be taken on how to sample the rolled wedge. The predicted notch positions might not be a sufficient marker for the achieved strain during the test, especially if the samples for metallographic investigation are taken from the center of the wedge. This, of course, depends on the chosen geometry of the wedge sample, as the result is highly dependent on the dimensions. Further tests are being performed to evaluate the impact of geometry variability on the changes in the strain and the strain rate distribution of the wedge sample.

*3.3. Grain Size Evolution*

Under similar hot rolling schedules and different starting PAGs, the starting difference in grain size evolution was expected if the per pass and total reduction with recrystallization were considered. Some mills produced coils of similar compositions, as used for DP, with cumulative $e$ = 0.82–0.88 to obtain proper final microstructure regardless of the starting PAG [18]. The importance of the starting grain size was already observed when comparing the shapes of single stress–strain curves with fine or coarse starting PAG at elevated temperatures obtained by torsion and hot compression tests [3]. The microstructure control over the wedge sample during intense reheating resulted in a fine starting grain size, which was achieved by cold charging. Intense reheating is also performed in the industry (where possible) for IA to take advantage of the uniform distribution of cementite. The cementite acts as a potential nucleation site of austenite [29]. When partial SRX is activated at sufficiently high temperatures, the starting new PAG can easily grow due to high HAGB mobility until sufficient roughing passes are introduced to limit/stop the HAGB mobility, and the continuous refining of PAG can again be observed with further passes [17,25]. By using a wedge rolling test, the notch positions are usually observed where sufficient $\varepsilon$ (or $e$) is introduced for an effective through-section deformation, achieving the uniform cross-section dislocation density and promoting a repeatable dislocation-free grain formation to minimize any microstructural cross-section variation (microstructural non-uniformity). However, unstable grain refining processes are also highly interesting, and other (presumably the starting notch) positions have also been considered. The grain size in this study was considered only by the high-angle boundaries.

The average measured values of anisotropy usually increase per position concerning temperature and, based on Figure 9a, are within the values of 1.3 for DP 600; these values are calculated as an average between the grain sizes determined in the longitudinal (L.G.) and transverse (T.G.) direction. If the ratio between L.G. and T.G. is unity or close to unity, then no anisotropy (transverse to rolling) is present after the completion of the test. Based on these results, the microstructure can be regarded as mainly equiaxed. Based on L.G./T.G., only modest anisotropy was expectedly interpreted, potentially due to the individual deformed coarser grains observed in the metallographic samples. It was concluded that recrystallization rolling was obtained and went well with the predicted $T_{nrx}$. Some texturing appeared very modest and more emphasized in the region of highest compressions, as expected in hot-rolled sheets or strips [5].

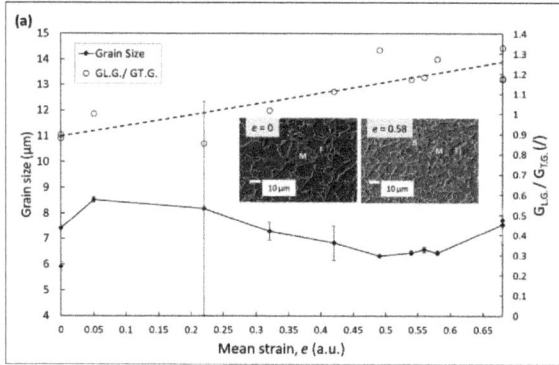
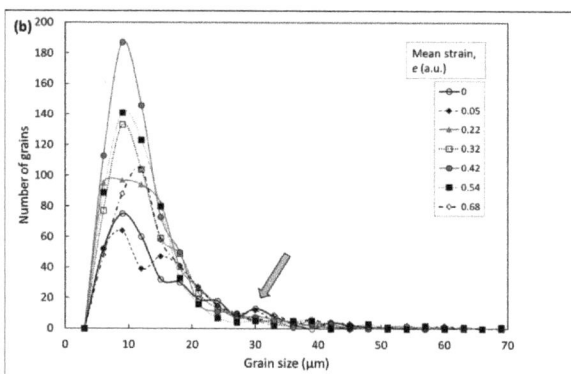

**Figure 9.** (a) Average grain size in correlation with the calculated mean strain, $e$, per specific notch position; measured on the metallographic samples taken from the rolled wedge sample. The secondary axis represents the ratio between the average grain size in the longitudinal direction (L.G.) and the transverse direction (T.G.) for the given deformation as an indication of modest (if any) pancaking at high $e$. (b) Mixed grain distribution using Feret's diameter and visible grain size multimodality with only a single pass. The results are based on SE micro-photographs taken at 2500× magnification.

Mixed grains were counted (partly as the fraction variation in phases is rather low): martensite (M), self-tempered martensite (SM), lower Bainite (LB), and ferrite (F) were included. Figure 9 shows that the final average grain size (F + M/self-tempered M/low B) observed on the finally cooled and transformed microstructure of the as-rolled sheet, regardless of the observed position, was in a range between 5 and 12 μm. This meant that we were achieving conventional to coarser grain sizes (CG) with F sizes on average of approx. 9–7 μm, yet no fine grains or ultra-fine grains were gained for DP steels (2–5 μm and <2 μm, respectively) as the process itself did not involve multi-forming operations, thermal cycling, etc. If these results are compared with classical high-temperature reheat and hot-rolled C-Mn steels, cooled under air with similar compositions, a rather refined structure was obtained in this work, indicating the importance of a proper low-temperature reheating temperature (of an ingot, slab, etc.), and a holding time adjustment in respect of the pre-existing state (quality of as-cast, pre-deformed state) to promote a fine transformed structure due to the initial fine and homogeneous PAG. The described fine-grained structure was observed from the first to the last notch position. This was achieved without using costly elements such as Nb, Mo, and similar. The degree of PAG evolution and a related transformed microstructure was successfully controlled by the reheat, roughing, and final rolling temperature, which introduced intense cooling as basic metallurgical tools for the minimization of grain coarsening [24,30]. Based on the coarsest observed transformed PAG within bi- and multi-modal peaks in Figure 9b, an estimation of the maximum PAG was set to be under 40 μm, which was consistent with similar values expected for the recrystallized grains of austenite in commercial grades [20].

Figure 9b reveals that, despite the relatively fine structure obtained on average, most grains were located between 3 and 20 μm. Locally rather coarse grains were also obtained, indicative of the anisotropy ratio. The localized coarse grains could exceed sizes of 40 μm up to 70 μm (related to transformed PAG into M/SM/B as an indicator). The local coarse grains were far from the fine-grain steel grade observed on average. The excessive transformed PAG size affected the ductility, as shown in [31]. Additionally, bi- or multimodality was enhanced at lower deformations ($e$ up to 0.22). This indicated the unstable recrystallization process in early per-notch positions in relation to deformation among the phase-related modality. The intensity of multimodality was, however, low and the curve resembled the asymmetric Gauss distribution regardless of the deformation.

PAG coarsening was observed on the last notch positions and at the maximum deformations achieved for the given test. The thickness of the final sheet was not completely equal along the entire length and thicker exit thickness was achieved on the last notch positions despite achieving a higher $\varepsilon_t$, as shown in Table 1. Therefore, slower cooling (longer times for grain growth) of the as-rolled structure was possible for these positions, partially due to the sheet manipulation and/or higher achieved thickness. However, based on the FEM simulations and material characteristics of DP 600, additional phenomena should be considered. Due to the starting fine structure, sufficiently high roughing temperature, sufficiently low strain rates, and achieved cumulative $\varepsilon$ based on the stress–strain curves, DRX could be activated on the last positions. Therefore, grain growth was possible during cooling based on the low (strain-related) incubation time, high HAGB mobility already under SRX, and the related lack of pinning particles to retard secondary recrystallization. Sudden grain growth was often observed in hot strip rolling when MDRX was activated below 6 mm of the exit thicknesses (based on [31]). As discussed, Figure 7 (based on FEM simulations) shows strain localizations above $e = 0.6$; hence, most representative sample positions for grain size interpretations should be under $e = 0.6$ by considering the constant temperature of the sample and the limited range of $\dot{\varepsilon}$.

In practice, contrary to a well-defined temperature regime under hot compression tests, the temperature uniformity using the wedge sample was more demanding, and the intensity of SRX, MDR/MDRX was, in some cases, also possibly related to non-uniform temperature distribution before and after completion of the test (as a part of adiabatic heating, variation in the roll chill per notch position, etc.).

Based on the results shown in Figure 9b and the laboratory setup of $\dot{\varepsilon}$ up to $3.45 \text{ s}^{-1}$, the optimum deformations for temperatures of 1100–1070 °C were obtained between $e = 0.2$ and $e = 0.5$ and went well with the overall strain uniformity achieved after the rolling test. Engineering strains $e$ were given from industrial practicality.

The characteristic of the grain-size curve visible in Figure 9a included only SRX (as PAG dependent) with no grain growth as a part of the secondary recrystallization at a close to constant temperature and, disregarding the obvious changes in $\dot{\varepsilon}$, the nature of this curve could be described based on the Beynon and Sellars type of equation [32], where $\varepsilon$ was considered from Table 1 for each notch position with the same starting PAG, which was written as $D_0$:

$$D_{SRX} = A \cdot D_0^B \cdot \varepsilon^{-1} \quad (5)$$

where $A$ and $B$ should be experimentally determined to calculate the achieved $SRX$ grain, $D_{SRX}$. The maximum transformed PAG (evaluated with the mode transition from IV to II, Figure 10) based on Equation (5) was at $e = 0.05$ of the deformation and was in relatively good agreement with the measured grain size distribution seen in Figure 9a.

The schematic representation of a potential PAG microstructure evolution (conditioning), as shown in Figure 10, was observed during a single pass by hot wedge rolling and was given for plain C-Mn-type steels, including low alloyed grades (as DP steels) as well as abrasion-resistant, high strength low alloyed steels (HSLA). This scheme showed a different PAG evolution above and under $T_{nrx}$ when various deformations per position at elevated reheating and rolling temperatures, cooling rates, and the overall changed hot-rolling schedules were implemented. The effects of the higher reheating (soaking) temperatures resulting in coarse starting PAG (mode I) or local PAG growth (mode IV and mode V) were also indicated, the latter due to the starting refined PAG and/or sufficiently high temperature for grain boundary mobility or post-rolling normal grain growth. Mode V was also related to the actual rolling speed as the flow curve (stress–strain) was related to temperature and the strain rate affecting the values of the Zener-Hollomon parameters and the hardening/softening of the material. It is visible that based on the scheme, we were able to produce DP 600 response within modes IV, II, and partially V due to the overall rather low $\dot{\varepsilon}$, and still observe an overall fine-grained structure per notch position. The modes of the grain size evolution presented in Figure 10 are shown to better understand the nature of refining and/or coarsening through a simple descriptive information methodology. Based on the results in Figures 9a and 10, the overall trend was observed and widely accepted; the total reduction that increased the overall refining was observed under SRX regardless of the starting PAG.

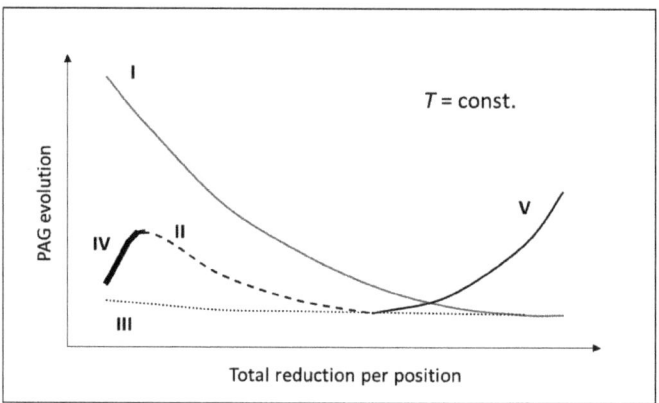

**Figure 10.** Based on the example of the literature review [3,17,20,25,33,34] and the current work, a schematic representation of grain size evolution was proposed concerning PAG, rolling schedule (geometry and related $\varepsilon_t$), temperature, and cooling regime considering the metallurgical per pass reduction. Here, Mode I represents continuous grain refining at higher starting roughing temperatures (coarse starting PAG); Mode II represents continuous grain refining at lower roughing temperatures and/or finer starting PAG; Mode III represents deformations usually conducted under $T_{nrx}$; Mode IV represents grain growth due to insufficient deformation during roughing, the pre-existing high density of HAGB and a lack of pinning particles; Mode V represents post-deformation grain growth after completion of SRX (with low $t_{50}$) or activated fast MDRX with the available post-deformation cooling rate and lack of second phase particles/SD or synergetic effect of decreased roll chill and adiabatic heating on the last notch positions in dependence on the roll length.

In Figure 11, only ferrite grains were determined for the grain size evolution study. Ferrite was considered a ductility holder and revealed a similar trend in the grain size evolution per notch position to the mixed grains shown previously in the text. The results show that controlling the final PAG also controlled the ferrite grains. Overall, no preferential orientation of ferrite was recognized due to a single pass run at an elevated temperature for the given test. The starting temperature for intense quenching was, however, under $A_{e3}$. Considering the equilibrium (lever) predictions conducted by using Thermo-Calc and JMatPro 6.1 and the results in Figure 11f, the temperature of the quench start was set to be between 770 °C and 780 °C. The measured surface temperature was, on average, 750 °C (under $T_{nrx}$) and had a reasonable agreement. The areal fraction of (self-tempered) martensite blocks and/or LB revealed that the per notch position and increased deformation reveal minor variations in martensite and ferrite content ($V_F = 1 - V_M$). It was recognized that if hard M/SM was considered a measure of tensile strength, with the increasing ratio of the original to final thickness for different wedge geometries under relatively low strain rates, the mechanical tensile properties were weakly changing for the current rolling schedule. Based on the increasing deformations of the hot-rolled wedge, ferrite fraction also changed moderately, indicating weak yet still measurable increased nucleation sites for F formation per $e$. The various F fractions could be obtained by manipulating the number of passes concerning the actual temperature of the deformations and IA temperature, etc. [4,8]. In this case, only polygonal ferrite was obtained. If the quench temperature of the sheet was in the region outside the formation of polygonal ferrite (PF), acicular ferrite (AF) could also be promoted. As $V_M$ varied with the chemical composition, the austenite grain size, the actual time available for the phase transformation of austenite into ferrite, the cooling rate ($V_M$ increases with cooling (quench) rate), etc., no considerable changes per position for the variation in the highly dislocated M content was considered for the wedge results [3,4]. The dislocation density of austenite was also estimated before the phase transformation due to the variation in $\varepsilon$ ($e$) and $\dot{\varepsilon}$ affecting the stress and force during rolling. The estimated dislocation density values were estimated from $1.0 \times 10^{10}$ m$^{-2}$ for under $e_1$ to $1.25 \times 10^{15}$ m$^{-2}$ for the last $e_9$ position (using the approach as in previous work [25]).

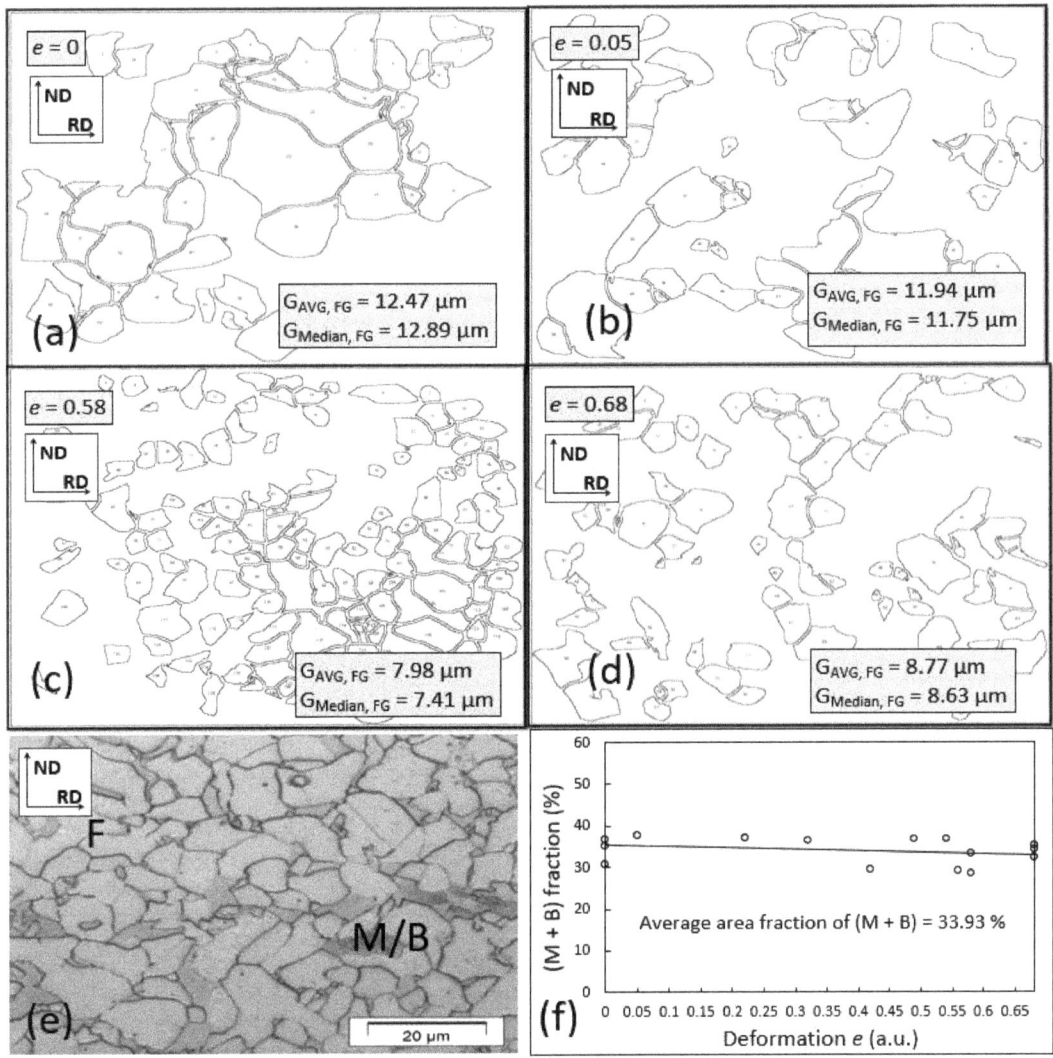

**Figure 11.** (**a**–**d**): An example of the graphical representation using the exploded distribution of the coarse grain (CG) to fine grain microstructure with a representation using solely the ferrite grain (FG) size. Enlargement: 2500×. ND and RD represent normal and rolling directions, respectively. (**e**) An example of OM with weak banding with M enrichment. Here, M and B also stand for SM and LB. Enlargement: 500×. (**f**) Minor deviation of M percentage on finally cooled per notch positions.

## 4. Conclusions

The non-standardized wedge rolling test of DP600, based on the above-presented results, was used in this case study for a single-pass schedule under isothermal conditions and could also be used for a multi-pass rolling schedule to observe the degree of grain refinement and grain growth in a post-cooled transformed microstructure by including the temperature profile. Based on the results, and regardless of the position, the main transformed microstructure consisted of martensite/self-tempered martensite and ferrite as a part of polygonal ferrite. No excessive formation of acicular ferrite was observed.

A single pass wedge rolling test at 1100–1070 °C was recognized to have similar asymmetric Gaussian grain size distributions between all per notch positions. A single pass could provide bi- or multi-modality, as seen in this study, despite it being modest. Partially the multi-modality originated from having a multi-phase microstructure; phase-related modality was also due to the unstable recrystallization on the first per notch positions. Local excessive coarse grains were observed up to mean $e$ = 0.4.

The FEM simulation revealed a discrepancy between the rolled edge and the mid-width positions studied in this case. This indicated that sample extraction should be carefully designed to correlate the grain size evolution to the correct local $\varepsilon$. These achieved true deformations were seemingly highly affected by the chosen material as well as by the wedge geometry and the used rolling parameters (roll displacement). This indicated that simply observing the positions of intentionally (mechanically) formed notches before and after rolling could add uncertainty concerning the target deformations and potentially increase the measurement error for the grades with a higher solid drag and/or pinning pressure on HAGB (this was avoided in this study due to the lean chemical composition). By controlling the rolling speed at a fixed wedge geometry, as in this study, the strain and strain rate distributions could be, to a certain degree, controlled satisfactorily in a way to obtain the grain size evolution for different strains per notch position at a rather minor variation in the strain rates per notch position (by considering the dynamic nature of the test). This meant that the experimental data obtained from the wedge were potentially more comparable to more regular and standardized methods for material hot behavior studies and the investigation obtained from Gleeble tests, etc. Single wedge tests could be used successfully for quantitative analysis closer to other methods instead of being solely qualitative, but only by introducing some of the aspects shown in the presented paper.

Based on the conclusion made on DP 600 steel, a rather refined structure (within 5 to 12 µm) was obtained starting from the intense reheating for fine initial PAG control in all the performed rolling variations along the wedge length. However, the aim was to observe and determine the optimal strain needed for stable grain size evolution at a single roughing temperature as a single-step deformation process to produce the finest possible microstructure instead of introducing additional rolling passes, adding special heat treatments, thermal cycling, etc., which are basic techniques to provide even finer grain sizes in all the phase constituents involved [4,35,36]. Based on the overall rolling parameters and cooling schedule, the optimum deformations per notch positions for the roughing stage of a 3 mm thick sheet between the temperature interval of 1100–1070 °C and up to 3.45 s$^{-1}$ strain rate were obtained between $e$ = 0.2 and $e$ = 0.5 as optimum grain size refining for balanced ductility and strength. The latter is the main target during DP production. The interval $e$ = 0.2 and $e$ = 0.5 agreed well based on FEM simulations, with the plastic deformation region of improved deformation stability. The FEM simulation therefore revealed, and this was considered relevant, that not all per-notch positions were used for the material study; this was based on the enhanced appearance of strain localization at higher starting thicknesses, as in this case. The work presented here sets a basic foundation for the potential further development of wedge geometry using C-Mn-type steels, including DP steels and others, for basic kinematic and material investigations.

**Author Contributions:** Conceptualization, G.K. and U.K.; methodology, G.K., U.K., P.F. and H.P.; software, U.K., G.K. and J.F.; validation, G.K., U.K. and P.F.; formal analysis, G.K., P.F. and U.K.; investigation, G.K. and J.F.; data curation, U.K. and G.K.; writing—original draft preparation, G.K. and U.K.; writing—review and editing, P.F., H.P., J.B. and J.F. All authors have read and agreed to the published version of the manuscript.

**Funding:** This research received no external funding.

**Data Availability Statement:** Data sharing is not applicable.

**Acknowledgments:** The authors would like to acknowledge Douglas Stalheim from DGS Metallurgical Solutions Inc. and Paul Lalley from CBMM for overall productive discussions about lean alloy steel grade topics. The authors would also like to acknowledge TU Clausthal for the execution of laboratory rolling and for helping with the numerical simulations.

**Conflicts of Interest:** The authors declare no conflict of interest.

## References

1. Calcagnotto, M.; Ponge, D.; Demir, E.; Raabe, D. Orientation gradients and geometrically necessary dislocations in ultrafine grained dual-phase steels studied by 2D and 3D EBSD. *Mater. Sci. Eng. A* **2010**, *527*, 2738–2746. [CrossRef]
2. Asadi, M.; Soliman, M.; Palkowski, H. Advanced High-Strength Steels: Bake Hardening. In *Encyclopedia of Iron, Steel, and Their Alloys*; Routledge: London, UK, 2016; pp. 16–45. [CrossRef]
3. Gladman, T. *The Physical Metallurgy of Microalloyed Steels*; The University of Leeds: Leeds, UK, 1997.
4. Nikkhah, S.; Mirzadeh, H.; Zamani, M. Fine tuning the mechanical properties of dual phase steel via thermomechanical processing of cold rolling and intercritical annealing. *Mater. Chem. Phys.* **2019**, *230*, 1–8. [CrossRef]
5. Tasan, C.C.; Diehl, M.; Yan, D.; Bechtold, M.; Roters, F.; Schemmann, L.; Zheng, C.; Peranio, N.; Ponge, D.; Koyama, M.; et al. An Overview of Dual-Phase Steels: Advances in Microstructure-Oriented Processing and Micromechanically Guided Design. *Annu. Rev. Mater. Res.* **2015**, *45*, 391–431. [CrossRef]
6. Available online: https://www.ssab.com/en/brands-and-products/docol/automotive-steel-grades/dual-phase-steel (accessed on 10 April 2023).
7. Song, R.; Ponge, D.; Raabe, D.; Speer, J.G.; Matlock, D.K. Overview of processing, microstructure and mechanical properties of ultrafine grained bcc steels. *Mater. Sci. Eng. A* **2006**, *441*, 1–17. [CrossRef]
8. Mandal, M.; Patra, S.; Anand, K.A.; Chakrabarti, D. An experimental and mathematical study on the evolution of ultrafine ferrite structure during isothermal deformation of metastable austenite. *Mater. Sci. Eng. A* **2018**, *731*, 423–437. [CrossRef]
9. Klug, M. Optimization of the Specimen Geometry for the Wedge Rolling Test. Bachelor's Thesis, Faculty of Natural Sciences and Technology, University of Ljubljana, Ljubljana, Slovenia, 2013. (In Slovene)
10. Parsa, M.H.; Ahmadabadi, M.N.; Shirazi, H.; Poorganji, B.; Pournia, P. Evaluation of microstructure change and hot workability of high nickel high strength steel using wedge test. *J. Mater. Process. Technol.* **2008**, *199*, 304–313. [CrossRef]
11. Beladi, H.; Kelly, G.L.; Hodgson, P.D. Formation of ultrafine grained structure in plain carbon steels through thermomechanical processing. *Mater. Trans.* **2004**, *45*, 2214–2218. [CrossRef]
12. Vodopivec, F.; Kosec, L.; Godec, M. Hot ductility of austenite stainless steel with a solidification structure. *Mater. Technol.* **2006**, *40*, 129–137.
13. Rusz, S.; Schindler, I.; Suchánek, P.; Turoňová, P.; Kubečka, P.; Heger, M.; Hlisníkovský, M.; Liška, M. Utilization of the hot wedge test in research of hot formability of free-cutting stainless steels. *Acta Metall. Slovaca* **2007**, *13*, 577–582.
14. Čížek, L.; Greger, M.; Kocich, R.; Rusz, S.; Juřička, I.; Dobrzański, L.A.; Tański, T. Structure characteristics after rolling of magnesium alloys. In Proceedings of the 13th International Conference on Achievements in Mechanical and Materials Engineering, Gliwice-Wisla, Poland, 16–19 May 2005.
15. Kvíčala, M.; Klimek, M.; Schindler, I. Study of Technological Formability of Low-Alloyed Steel 25CrMo4. *Hutnické Listy* **2009**, *6*, 13–15.
16. Kubina, T.; Schindler, I.; Turoňova, P.; Heger, M.; Franz, J.; Liška, M.; Hlisníkovsky, M. Mathematic simulation of the wedge rolling test and computer processing of laboratory results. *Acta Metall. Slovaca* **2006**, *12*, 469–476.
17. Stalheim, D. Metallurgical strategy for optimized production of QT high-strength and abrasion-resistant plate steels. In Proceedings of the AISTech2019 Iron and Steel Technology Conference, Pittsburgh, PA, USA, 6–9 May 2019; pp. 1881–1892.
18. Lissel, L. Modeling the Microstructural Evolution during Hot Working of C-Mn and of Nb Microalloyed Steels Using a Physically Based Model. Ph.D. Thesis, Royal Institute of Technology, School of Industrial Engineering and Management, Material Science and Engineering, Division of Mechanical Metallurgy, Stockholm, Sweden, 2006.
19. Fajfar, P. *Predelava Materialov*; Department of Materials and Metallurgy, Faculty of Natural Sciences and Technology, University of Ljubljana: Ljubljana, Slovenia, 2016. (In Slovene)
20. Stalheim, D.; Glodowski, R. Fundamentals of the generation of fine grain as-rolled structural steels. In Proceedings of the AIST International Symposium on the Recent Developments in Plate Steels, Winter Park, CO, USA, 19–22 June 2011; pp. 25–32.

21. Zenga, R.; Huanga, L.; Lia, J.; Lib, H.; Zhua, H.; Zhang, X. Quantification of multiple softening processes occurring during multi-stage thermoforming of high-strength steel. *Int. J. Plast.* **2019**, *120*, 64–87. [CrossRef]
22. Mucg83 Program. Available online: https://www.phase-trans.msm.cam.ac.uk/map/steel/programs/mucg83.html (accessed on 10 April 2023).
23. Edberg, J.; Lindgren, L.E.; Jarl, M. The wedge rolling test. *J. Mater. Process. Technol.* **1994**, *42*, 227–238. [CrossRef]
24. Irvin, K.J. *Strong Structural Steels, Symposium Low Alloy High Strength Steels, the Metallurg Companies*; Frese-Druck: Düsseldorf, Germany, 1970.
25. Klančnik, G.; Foder, J.; Bradaškja, B.; Kralj, M.; Klančnik, U.; Lalley, P.; Stalheim, D. Hot Deformation Behavior of C-Mn Steel with Incomplete Recrystallization during Roughing Phase with and without Nb Addition. *Metals* **2022**, *12*, 1597. [CrossRef]
26. EN 10025-2; Hot Rolled Products of Structural Steels-Part 2, Technical Delivery Conditions for Non-Alloy Structural Steels. European Committee for Standardization: Brussels, Belgium, 2019.
27. Barbosa, R.; Boratto, F.; Yue, S.; Jonas, J.J. *Processing, Microstructure and Properties of HSLA Steels*; DeArdo, A.J., Ed.; The Minerals, Metals & Materials Society AIME: Warrendale, PA, USA, 1988; p. 51.
28. Stalheim, D.; Wright, M.R. Fundamentals of developing fine grained structures in "as rolled" long products. In Proceedings of the 51th Rolling Seminar—Processes, Rolled and Coated Products, Foz do Iguaçu, Brazil, 28–31 October 2014.
29. Azizi-Alizamini, H.; Militzer, M.; Poole, W.J. Formation of Ultrafine Grained Dual Phase Steels through Rapid Heating. *ISIJ Int.* **2011**, *51*, 958–964. [CrossRef]
30. Wang, J.; Kang, Y.; Yang, C.; Wang, Y. Effect of Heating Temperature on the Grain Size and Titanium Solid-Solution of Titanium Microalloyed Steels. *Mater. Sci. Appl.* **2019**, *10*, 558–567. [CrossRef]
31. Stalheim, D.G.; Barbosa, R.A.N.M.; de Moura Bastos, F.M.M.; Gorni, A.A.; Rebellato, M.A. Basic Metallurgy/Processing Design concepts for optimized hot strip structural steel in yield strengths from 300–700 MPa. In Proceedings of the 53th Rolling Seminar, ABM Week, Rio de Janeiro, Brazil, 27–29 September 2016.
32. Beynon, J.H.; Sellars, C.M. Modelling Microstructure and Its Effects during Multipass Hot Rolling. *ISIJ Int.* **1992**, *32*, 359–367. [CrossRef]
33. Siwecki, T. Modelling of Microstructure Evolution during Recrystallization Controlled Rolling. *ISIJ Int.* **1992**, *32*, 368–376. [CrossRef]
34. Bombac, D.; Peet, M.J.; Zenitani, S.; Kimura, S.; Kurimura, T.; Bhadeshia, H.K.D.H. An integrated hot rolling and microstructure model for dual-phase steels. *Model. Simul. Mater. Sci. Eng.* **2014**, *22*, 045005. [CrossRef]
35. Calcagnotto, M.; Ponge, D.; Raabe, D. Microstructure Control during Fabrication of Ultrafine Grained Dual-phase Steel: Characterization and Effect of Intercritical Annealing Parameters. *ISIJ Int.* **2012**, *52*, 874–883. [CrossRef]
36. Calcagnotto, M.; Ponge, D.; Raabe, D. Effect of grain refinement to 1 μm on strength and toughness of dual-phase steels. *Mater. Sci. Eng.* **2010**, *A527*, 7832–7840. [CrossRef]

**Disclaimer/Publisher's Note:** The statements, opinions and data contained in all publications are solely those of the individual author(s) and contributor(s) and not of MDPI and/or the editor(s). MDPI and/or the editor(s) disclaim responsibility for any injury to people or property resulting from any ideas, methods, instructions or products referred to in the content.

Article

# Effect of Machine Pin-Manufacturing Process Parameters by Plasma Nitriding on Microstructure and Hardness of Working Surfaces

Włodzimierz Dudziński [1], Daniel Medyński [1,*] and Paweł Sacher [2]

[1] Department of Production Engineering and Logistics, Faculty of Technical and Economic Sciences, Sejmowa 5A, 59-220 Legnica, Poland; wlodzimierz.dudzinski@collegiumwitelona.pl
[2] Sacher Company, Ekonomiczna 6, 59-700 Bolesławiec, Poland
\* Correspondence: daniel.medynski@collegiumwitelona.pl

Citation: Dudziński, W.; Medyński, D.; Sacher, P. Effect of Machine Pin-Manufacturing Process Parameters by Plasma Nitriding on Microstructure and Hardness of Working Surfaces. *Crystals* 2023, 13, 1091. https://doi.org/10.3390/cryst13071091

Academic Editor: Mingyi Zheng

Received: 6 June 2023
Revised: 7 July 2023
Accepted: 11 July 2023
Published: 13 July 2023

Copyright: © 2023 by the authors. Licensee MDPI, Basel, Switzerland. This article is an open access article distributed under the terms and conditions of the Creative Commons Attribution (CC BY) license (https://creativecommons.org/licenses/by/4.0/).

**Abstract:** This work concerns two stages of research into plasma nitriding (change of nitriding steel and modification of nitriding parameters). In the first stage, pins obtained from currently used steel were compared with pins made of an alternative material available on the market, using the same nitriding process parameters. As a result of the metallographic tests carried out, in the first case, the presence of a thin, porous, and heterogeneous nitrided layer or its absence was found, with the core in its raw state and not thermally improved. In the second case, the presence of a nitrided layer of small thickness with noticeable porosity on the surface of the sample was found, but with a core after heat treatment (incorrect process parameters). Therefore, modification of the parameters of the nitriding process was proposed, in terms of a mixture of gases, currents, time, and temperature of the nitriding process. As a result, a satisfactory effective thickness of the nitrided layer was obtained, consisting of a white near-surface zone with $\varepsilon$ and $\varepsilon+\gamma'$-type nitrides with a thickness of 8.7 to 10.2 μm, and a dark zone of internal nitriding with $\gamma'$ nitrides. The nitrides layer was continuous, compact, and well adhered to the steel surface. In the core of the samples, the presence of a fine-needle tempering sorbite structure with a small amount of fine bainite, which is correct for the steel after heat treatment and nitriding, was found. The most favorable parameters of the ion nitriding process were gas flow rate (1.5 L/min N; 0.4 L/min H; 0.3 L/min Ar); currents (BIAS—410 V 4.0 A, SCREEN—320 V 4.0 A); time (26 h and 35 min); and temperature (550 °C).

**Keywords:** pins; wear; heat treatment; chemical treatment; plasma nitriding

## 1. Introduction

Both in everyday life and in industry, products that can make our lives easier, better and safer are very important. With this in mind, there is a constant need to improve the approach to production. One aspect of product improvement is the need for increasingly efficient and cost-effective manufacturing methods. To ensure cost-effective production, especially for large-scale industrial processing, this approach requires continuous process improvement. The key in this aspect is surface engineering, which includes the use of traditional and innovative technologies to obtain a product surface with better properties compared to the existing ones [1,2]. One of the methods of improving the surface properties of the product is the treatment of its surface after production, usually aimed at its hardening. The need to harden the surface of materials, especially metals, is now more in demand than ever. This is due to the systematically growing demand for metallic materials that, with high hardness, show high strength, resistance to abrasive wear, and, very importantly, are obtained at reduced production costs. Modern surface-treatment methods most often include chemical vapor deposition (CVD), physical vapor deposition (PVD), cladding, and plasma nitriding [1–6].

Nitriding is one such surface-treatment method that is critical to modern industry. It is used in many fields, including machine, automotive, and tool industries [3,7–15]. The nitriding process guarantees the improvement of the surface properties, primarily alloy steels intended for nitriding, while maintaining minimum dimensional tolerances and high abrasion resistance. In addition to increasing the surface hardness, wear resistance, and corrosion resistance, nitriding can lead to an improvement in the fatigue strength of the component. This is possible due to the formation of appropriate phases with increased volume, causing the occurrence of compressive stresses in the surface layer.

This is possible because iron in Fe–C alloys occurs in various allotropic forms, which, combined with appropriate alloy additions, favors the formation of phases strengthening their surface in the nitriding process [16–19]. Up to a temperature of 912 °C, the Feα variety occurs, crystallizing in the BCC lattice. Between the temperatures of 912 and 1394 °C there is a Feγ variety crystallizing in the FCC lattice, and in the temperature range of 1394 to 1538 °C, there is a high-temperature Feδ variety crystallizing in the BCC lattice. The phases occurring in the iron–nitrogen system are shown in the phase equilibrium diagram in Figure 1.

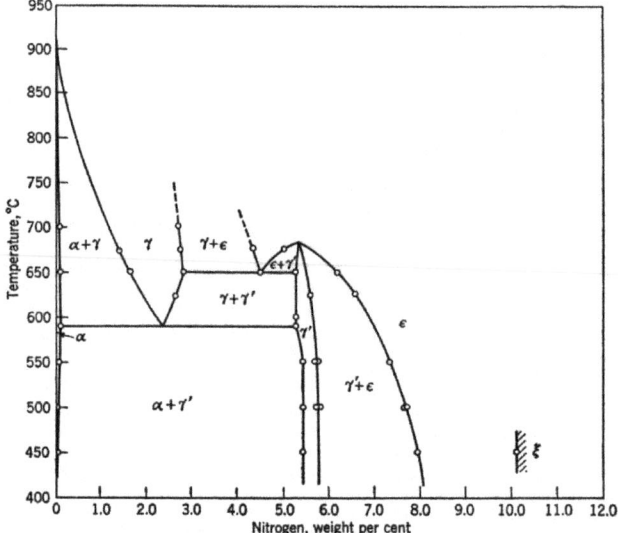

**Figure 1.** The Fe—N equilibrium phase diagram [20].

The graph shows that the solubility of nitrogen in Feα is low and reaches a maximum at 590 °C. At this temperature, the solubility is 0.10% and decreases rapidly with decreasing temperature, so that at 200 °C it is only 0.004%. This interstitial solution of nitrogen in the iron is called nitrogen ferrite.

There are two interstitial vacancies in the Feα lattice where nitrogen can be located. The radius of octahedral and tetrahedral vacancies is 0.019 nm and 0.052 nm, respectively [11]. Even though the radius of the N atom is 0.07 nm, it was found that it is in smaller octahedral vacancies in the Feα lattice, thus causing the expansion of the Feα lattice, an increase in internal stresses, and solution strengthening [20–23].

The solubility of nitrogen In Feγ is much higher, and this solution, called nitrogen austenite, is stable only at temperatures higher than 590 °C. The greatest solubility of nitrogen in Feγ is 2.80% at 650 °C, and then decreases to 2.35% at the eutectoid temperature. Nitrogen austenite with such a composition decomposes eutectoidly into a mixture of Feα(N) nitrogen ferrite and the γ′ phase. This eutectoid, formed during slow cooling, is called braunite. By rapid cooling, however, nitrogen austenite can be supercooled to temperatures at which nitrogen martensite is formed as a result of diffusion-free transformation.

The temperature Ms (martensite start) in this case is lower than in steels and, in the alloy with eutectoid composition, it is only 35 °C.

The $\gamma'$ phase is a solid solution based on iron nitride $Fe_4N$ with a narrow homogeneity range. This phase is stable up to 680 °C and then changes to the hexagonal $\varepsilon$ phase, which is a solid solution with a wide range of homogeneity. Like nitrogen austenite, the $\varepsilon$ phase containing 4.55% nitrogen decomposes at 650 °C into a eutectoid mixture of $\gamma$ and $\gamma'$ phases.

The iron nitrides show low durability and decompose at higher temperatures, which can be seen in the example of the $\gamma'$ phase. Due to this low stability, iron nitrides coagulate rapidly, even at low temperatures. The consequence of this is the low hardness of the surface of nitrided objects made of unalloyed steel because the dispersion of the formed nitrides quickly disappears.

More durable, and therefore less prone to coagulation, nitrides form nitrogen with transition metals that are components of alloy steels. These can be, for example, chromium nitrides CrN and $Cr_2N$ and molybdenum nitrides MoN, titanium nitrides TiN, or others. The most durable AlN nitrides are formed by aluminum and therefore are used in special grades of steel intended for nitriding. Apart from other alloying additives, this element is also used.

The structure of the diffusion layers formed depends primarily on the nitriding temperature, the chemical composition of the nitriding atmosphere, and the chemical composition of the material of the nitrided objects.

As already mentioned, the nitrided sample is characterized by increased surface hardness, increased wear resistance, high fatigue strength, increased corrosion resistance, and high dimensional stability [1,24–37]. This is possible because a sufficiently high concentration of nitrogen on the steel surface determines the formation of Fe–N phases. If the concentration of nitrogen increases sufficiently, a layer of these compounds with very high hardness forms on the surface. Below this layer, a diffusion zone is formed, consisting of nitrogen in the interstitial solid solution with fine precipitations of alloy nitrides [1,24]. This mechanism of precipitation hardening of the diffusion zone depends largely on the number of elements that favor the formation of nitrides and on the concentration of nitrogen itself [38]. A simplified cross-sectional diagram of a typical nitrided sample is shown in Figure 2. The diffusion zone consists of interstitial nitrogen dissolved in the Fe$\alpha$ lattice and Fe alloy carbonitrides [39–41]. It has been shown that the high surface hardness obtained as a result of nitriding results from the precipitation of fine-grained alloy carbonitrides [42–44]. The latter is a consequence of the presence of strongly nitride-forming elements in the steel substrate, such as Al, Cr, Mo, Ti, Mn, Si, and V [43–45]. The hardness in the diffusion zone depends on the type and number of alloying elements in the steel [38,46]. In addition, in high-alloy steels, there is a reduction in the hardening depth after nitriding treatment due to the precipitation of alloy nitrides, which additionally limits the diffusion of nitrogen to the substrate [47].

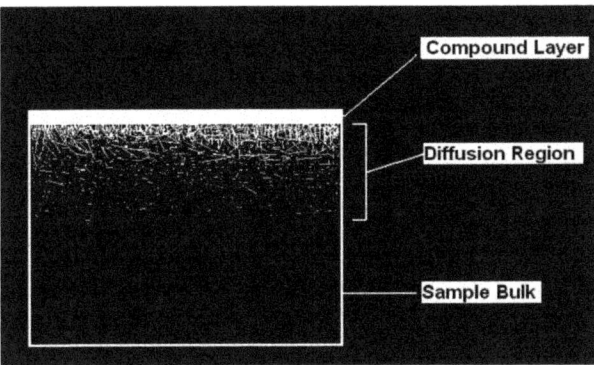

**Figure 2.** A diagram of a typical cross-sectional view of a nitrided component. Shown is the compound layer with the diffusion region beneath. Please note that this diagram is not to scale.

Plasma nitriding is one of the modern techniques of thermo–chemical treatment, which is usually carried out with the use of nitrogen and hydrogen [48–52]. The process is carried out under reduced pressure where a voltage is applied between the samples and the walls of the furnace. A glow discharge with a high level of ionization (plasma) is generated around the part [48,49]. In the area of the workpiece surface that is directly bombarded with ions, nitrogen-rich nitrides form and break down, releasing active nitrogen to the surface. The ability to adjust the degree of cathodic sputtering, the selection of the appropriate current–voltage characteristics, and the ability to increase or decrease the pressure in the working chamber enables the formation of the structure of nitrided layers [53]. Plasma nitriding makes it possible to modify the surface according to the desired properties [53–56]. By adjusting the gas mixture (proportions), it is possible to obtain tailored layers and hardness profiles, from a surface free of a layer of low nitrogen compounds and up to 20 microns thick, to a layer containing nitride compounds and a solid solution diffusion zone. The wide temperature range allows many applications beyond the possibilities of gas or salt bath processes.

Traditional nitriding techniques most often lead to the formation of defects, such as the formation of brittle surface layers, and the effect of non-uniform edges. To eliminate these problems, especially in the case of plasma nitriding, appropriate nitriding furnaces have been developed [57–59]. One of the recent innovations in this field was the introduction of the innovative technology of Active Screen Plasma Nitriding (ASPN) [56–59]. ASPN nitriding ensures the uniform formation of nitrides that reflect the shape of the workpiece. This effect is achieved by glowing nitrogen-activating plasma on the mesh as well as on the nitrided detail, which guarantees no damage or burns caused by oversized plasma discharges on the detail. This minimizes the risk of damage to the elements subjected to electric arc treatment and uneven heating of elements of different dimensions, and removes or minimizes the polarization voltage, which is applied to the treated elements in traditional plasma nitriding [59–62].

This work is the result of research and development activities carried out as part of the project entitled Regional Operational Program for the Lower Silesian Voivodeship for the years 2014–2020. As part of the project, actions were taken to modify the parameters of the technological process for the production of machine pins to improve the surface properties of the product. The research was carried out in two stages. In the first stage, they consisted of comparing the process from the point of view of the currently used steel with an alternative material, while maintaining the previously used parameters of the nitriding process. In the second stage, the parameters of the plasma nitriding process were modified.

## 2. Materials and Methods

The samples taken from the pins, which were plasma nitrided with Nitro-Tool technology in a vacuum furnace, using the ASPM method, were tested. Figure 3 shows a diagram of a vacuum furnace for nitriding using the ASPN method. During the tests, the ASPN nitriding method with active BIAS was used to obtain an even coverage with the nitrided layer.

In the first stage of the research, samples taken from pins currently manufactured by the Sacher Company, made of 38HMJ steel (designation according to the withdrawn PN—89/H84030.03 standard), in the state before and after the application of the plasma nitriding process, were tested. The parameters of the nitriding process typical for the company were used: gas flow rate (1.3 L/min N; 0.7 L/min H; 0.2 L/min Ar); currents (BIAS—420 V 5.0 A, SCREEN—320 V 3.0 A); time (26 h and 35 min); and temperature (510 °C).

As a result of the microscopic examination of samples subjected to heat–chemical treatment with the ion nitriding method, according to the parameters mentioned above, the presence of a thin, porous, and non-homogenous nitrided layer (or its absence) and a diffusion layer of small thickness was found on the surface of the sample with a raw, not thermally improved core.

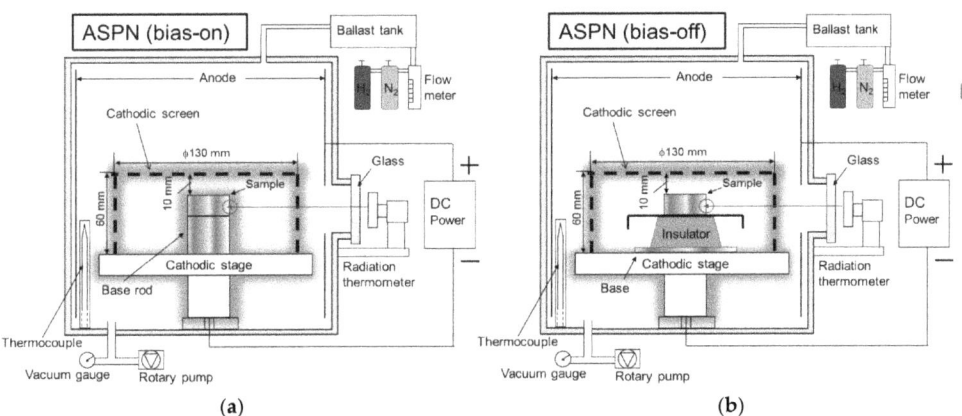

**Figure 3.** Schematic illustration of ASPN: (**a**) bias-on and (**b**) bias-off.

After analyzing the obtained results, the possibility of using modern grades of steel, e.g., 33CrMoV12—9 or 32CrAlMo7—10 (designation according to PN EN ISO 683—5:2021—10) was proposed as an alternative to the currently used steel used for the production of pins. Due to the unavailability of these materials, 42CrMo4 steel (designation according to EN ISO 683—1:2018—09) was used in the tests. After machining, the pins were subjected to the nitriding process, each time using the same process parameters as in the case of the first thermo–chemical treatment.

However, these actions did not bring the desired effect. Therefore, steps were taken to modify the parameters of the nitriding process. Thus, four test trials were carried out. Each time a different process parameter was changed (gas mixture, currents, time, and temperature), with the others returning to their initial values. Individual parameters were changed in the following ranges: gas flow rate (1.3–1.5 L/min N; 0.4–0.7 L/min H; 0.2–0.3 L/min Ar); currents (BIAS: 410–440 V 4.0–5.0 A, SCREEN: 320–330 V 3.0–4.0 A); time (from 26 h and 35 min to 30 h and 38 min); temperature (510–550 °C).

In each case, the pins (dimensions ⌀ 35 × 75) were produced using CNC machines: HASS SL—20 (Hass Automation Inc., Oxnard, CA, USA) and HASS VF5 (Hass Automation Inc., Oxnard, CA, USA). The pins were then subjected to ion plasma nitriding using a Nitro-Tool furnace (AMP Heat Treatment Technologies Company, Świebodzin, Poland).

Test samples were cut using a Secotom precision cutting machine (Struers, Copenhagen, Denmark) and then embedded. A grinding and polishing machine by LaboSystem (Struers, Copenhagen, Denmark) was used for grinding and polishing the specimens.

The metallographic tests consisted of the determination of the chemical composition using the spectral method using a GDS 750 QDP glow discharge analyzer. During the analysis, the following parameters were used to ionize the inert gas: U = 1250 V, I = 45 mA, 99(.9)% Ar (Leco, St. Joseph, MI, USA). The obtained results were the arithmetic mean of at least five measurements and the X-ray energy dispersion spectroscopy method (SEM Quanta 250 FEI with EDS detector by Oxford Instruments). The WDS by Oxford Instruments was also used for comparative purposes (for carbon and nitrogen content); measurement of the hardness distribution on the cross-section was carried out using a Vickers hardness tester MMT—X7B (Matsuzawa Co., Tokyo, Japan) under conditions compliant with PN EN ISO 6507—1:2018—05, as well as analysis of the microstructure of the core, nitrided layer and determining the thickness of the layer nitrided, using an Eclipse MA—200 metallographic microscope (Nikon, Tokyo, Japan), under conditions compliant with PN—82/H— 04550. Microscopic observations were carried out on samples non-etched and etched with nital-Mi1Fe (5 % alcohol solution $HNO_4$). Nital etching made it possible to visualize the microstructure of individual zones, i.e., the surface containing characteristic types of nitrides and the core without nitrides.

## 3. Results and Discussion

### 3.1. Testing of Currently Produced Pins—38HMJ Steel

In the first stage, metallographic tests were carried out on samples taken from pins made of 38HMJ steel (outdated designation according to the withdrawn PN—89/H84030.03 standard), delivered by the company before and after the nitriding process. The results of the analysis of chemical composition obtained by the GDS method of the base steel samples for nitriding are presented in Table 1.

**Table 1.** Results of chemical composition analysis of base steel samples intended for nitriding.

| Ordinal Number | Element | Chemical Composition of Steel (%$_{mass}$) |
|---|---|---|
| 1 | C | 0.392 |
| 2 | Cr | 1.270 |
| 3 | Al | 0.763 |
| 4 | Mn | 0.401 |
| 5 | Si | 0.308 |
| 6 | Mo | 0.166 |
| 7 | Ni | 0.086 |
| 8 | Ti | 0.004 |
| 9 | P | 0.006 |
| 10 | S | 0.001 |

Based on the analysis of the chemical composition, it was confirmed that the closest equivalent of the material from which the samples were made is 38HMJ steel (according to withdrawn PN—89/H84030.03).

#### 3.1.1. Microscopic Observations of Pins in Raw State

Microscopic observations of non-etched samples taken from pins in their raw state showed the presence of an increased number of non-metallic inclusions, mainly in the form of oxides and brittle silicates (Figure 4a). This indicates the low quality (metallurgical purity) of the tested material. After etching the reagent, a very finely dispersed, partially banded pearlitic structure with grains of free ferrite was found (Figure 4b).

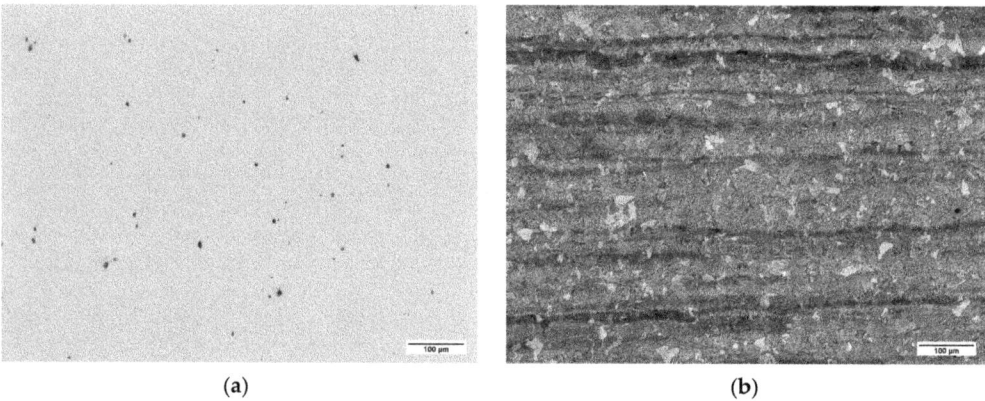

(a)            (b)

**Figure 4.** A sample taken from the pin before nitriding—38HMJ steel: (**a**) Increased number of non-metallic inclusions in the form of oxides and brittle silicates. Non-etched state; (**b**) Partially banded pearlitic-ferritic structure. Etched state.

This structure is typical for materials cooled from hot working temperatures (raw state) and is an incorrect structure, as it is the initial structure for steel before the nitriding process. It can cause a significant reduction in strength properties, and in particular cause cracks and chipping, and reduce the fatigue resistance of nitrided details.

In this case, it can be stated that no thermal improvement procedure was applied before the ion nitriding process.

### 3.1.2. Microscopic Observations of Pins after Nitriding

Based on the conducted microscopic examinations of nitral-etched samples after nitriding, the presence of a ferritic-pearlitic banded structure, typical of the raw state after hot plastic working, was found in the core (Figure 5). This indicates that the heat treatment required before the nitriding process was not applied. The nitrogen content on the surface of the samples after nitriding was an average of 8.50 %$_{mass}$ (EDS and WDS for comparison).

**Figure 5.** Sample after nitriding—core. Longitudinal section. Highly banded, heterogeneous ferritic structure with very finely dispersed pearlite. Etched state.

In the surface zone, a nitrided layer was found with a total metallographic thickness of about 0.2 mm (Figure 6), consisting of a very thin white near-surface layer with $\varepsilon$ and $\varepsilon+\gamma'$-type nitrides, 5 to 10 μm thick, and a dark nitriding zone with the $\gamma'$ nitride lattice forming a lattice on the former austenite grain boundaries.

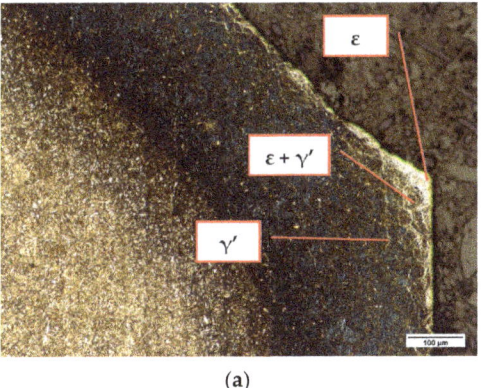
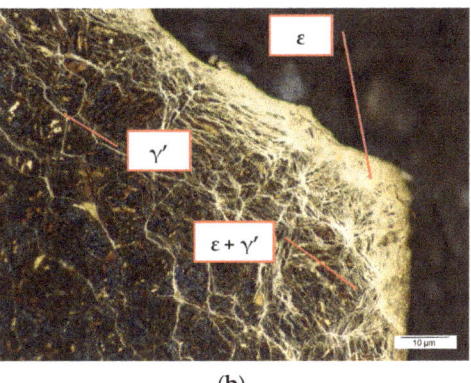

(a)      (b)

**Figure 6.** Nitrided layer in the corner of the sample after nitriding, etched state: (**a**) Visible bright, uneven, and very thin layer of $\varepsilon$, $\varepsilon+\gamma'$-type nitrides and a dark zone of internal nitriding with $\gamma'$ nitride precipitates forming a lattice on the former austenite grain boundaries; (**b**) Enlarged part of the area shown in Figure 7.

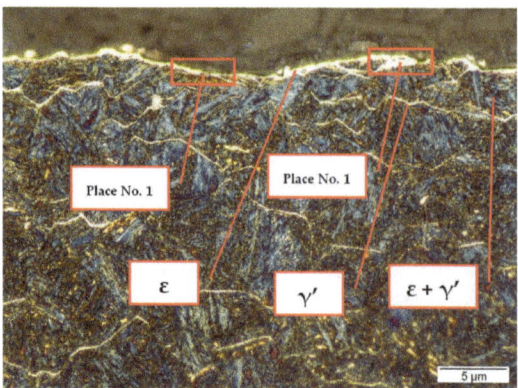

**Figure 7.** Nitrided layer on the sample surface after nitriding. Visible bright, uneven, and very thin layer of ε, ε+γ′-type nitrides and a dark zone of internal nitriding with γ′ nitride precipitates forming a lattice on the former austenite grain boundaries. Etched state.

Changes in the thickness of the nitride layer ε and ε+γ′ in the range from 2 to 5 μm and porosity features were observed locally (Figure 6a). In addition, a continuous network of nitrides distributed γ′ along the grain boundaries of the former austenite was found. The transition zone from the nitrided layer to the core has the correct structure. Summing up, the obtained structure is incorrect, showing a high tendency toward cracks and chipping of the surface layer during operation.

On the other hand, a dendritic bainitic structure was found in the core of the samples with precipitates of sorbite and troostite in the interdendritic spaces. This is an unacceptable defect, indicating a complete lack of application of thermal improvement after the plastic working process before the nitriding process or the use of incorrect technology in the thermal improvement process. It should be mentioned that the thickness of the nitrided layer should be from approx. 0.5 mm to approx. 0.7 mm. The small thickness of the nitrided layer requires a uniform, hard and resilient substrate, protecting the above-mentioned layer against its collapse and chipping under the action of local pressures.

In addition, it was found that in many places around the circumference of the sample, the nitrided layer was irregular, thicker in some places—place No. 1, thinner in some places—place No. 2 in others (Figure 7) or there is no such layer (Figure 8).

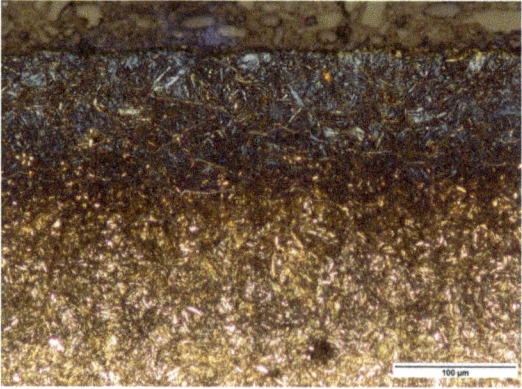

**Figure 8.** Nitrided layer on the surface of the sample after nitriding. Visible dark zone of internal nitriding with precipitates of γ′ nitrides forming a lattice on the boundaries of former austenite grains and no layer of ε, ε+γ′ nitrides on the surface. Etched state.

### 3.1.3. Pin Hardness Measurements before and after Nitriding

Hardness measurements were carried out using the Vickers method in conditions compliant with PN EN ISO 6507—1:2018 on the cross-section of the samples in the plane perpendicular to the nitrided surface in the cores of the samples taken from the pins before and after nitriding. A Matsuzawa hardness tester was used with a load of 1 kg (9807 N) operating for 15 s.

The average hardness of the base material in the core of the sample taken from the pin as supplied was found to be 255.8 HV1. The hardness in the core of the sample after the nitriding process does not show significant differences and amounts to 349.8 HV1, respectively.

In addition, measurements of the hardness distribution under the surface of the nitrided layer were made in two places on the sample after nitriding, using a Matsuzawa hardness tester with a load of 0.5 kg (4.904 N).

Determination of the thickness of the nitrided layer by measuring the hardness distribution was made in accordance with the PN—82/H—04550 standard. Hardness measurements were made along a line perpendicular to the surface of the sample at intervals of 50 μm. The measurement results are summarized in Tables 2 and 3.

**Table 2.** Measurement results of HV0.5 hardness distribution on the cross-section of the nitrided layer in the sample after nitriding—place No. 1.

| Distance from the Surface (μm) | 50 | 100 | 150 | 200 | 250 | 300 | 350 | 400 | 450 | 500 | 550 | 600 | 650 | 700 |
|---|---|---|---|---|---|---|---|---|---|---|---|---|---|---|
| Hardness HV0.5 | 732.0 | 663.9 | 442.7 | 373.1 | 369.9 | 364.7 | 372.8 | 369.4 | 365 | 384.7 | 376.4 | 378.1 | 377.9 | 366.4 |

**Table 3.** Measurement results of HV0.5 hardness distribution on the cross-section of the nitrided layer in the sample after nitriding—place No. 2.

| Distance from the Surface [μm] | 50 | 100 | 150 | 200 | 250 |
|---|---|---|---|---|---|
| Hardness HV0.5 | 761.2 | 729.9 | 567.9 | 396.9 | 390.1 |

As a result of hardness measurements, it was found that in the sample after nitriding in place No. 1, the effective thickness of the nitrided layer calculated up to the hardness of 500 HV0.5 is 0.13 mm (130 μm), and the maximum hardness of 732 HV0.5 in place No. 1 occurs at a distance of up to 0.05 mm (50 μm) below the sample surface. The hardness at a distance of approx. 0.2 mm from the surface is 373 HV0.5.

In the sample after nitriding in place No. 2, the effective thickness of the nitrided layer calculated up to the hardness of 500 HV0.5 is 0.18 mm (180 μm), and the maximum hardness of 761.2 HV0.5 also occurs at a distance of up to 0.05 mm (50 μm) below the surface of the sample. The hardness at a distance of about 0.2 mm from the surface is 396 HV0.5.

The obtained results indicate that 38HMJ steel should not be used for nitriding. It was proposed to use modern steel grades intended for nitriding, e.g., 33CrMoV12—9 or 32CrAlMo7—10 (designation according to PN EN ISO 683-5:2021—10), and in the absence of their availability, the use of 42CrMo4 steel (designation according to EN ISO 683-1: 2018—09).

### 3.2. Study of Pins Manufactured from Alternative Steel—42CrMo4

Due to the unavailability of 33CrMoV12—9 and 32CrAlMo7—10 steel, 42CrMo4 steel was used.

### 3.2.1. Pins before Nitriding

As a result of microscopic examinations carried out in the longitudinal and transverse direction to the direction of machining of the pins in the non-etched state, a small number of non-metallic inclusions was found, mainly in the form of point oxides, unevenly distributed (Figure 9a). After etching, a partially banded bainite structure of a dendritic character was found, with precipitates of sorbitol and troostite in the interdendritic spaces (Figure 9b).

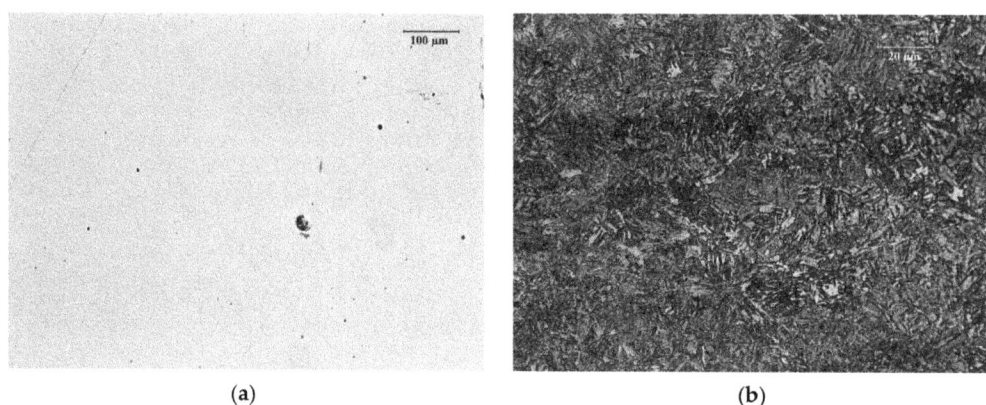

**Figure 9.** Sample before nitriding, steel 42CrMo4: (**a**) Visible small number of non-metallic inclusions, mainly in the form of point oxides, unevenly distributed. Non-etched state; (**b**) Visible partially banded bainite structure of dendritic character with precipitates of sorbitol and troostite in interdendritic spaces. Etched state.

### 3.2.2. Pin Surface after Nitriding

The analysis of the results of EDS (WDS analysis was used for comparison purposes for carbon and nitrogen) measurements made on the cross-section of the correctly nitrided samples showed the highest nitrogen saturation on their surface and a systematic decrease in the concentration of nitrogen towards the core, unlike the other elements (Figure 10).

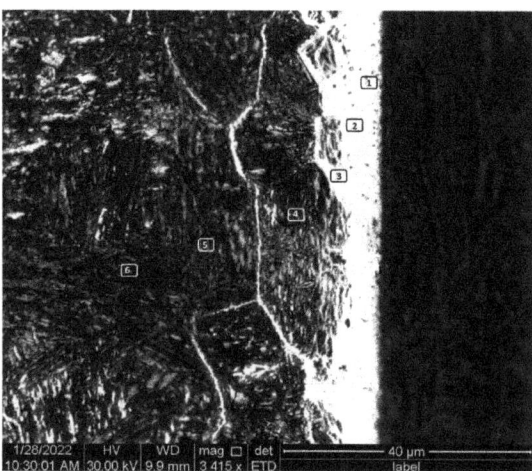

| | Element distribution ($\%_{mass}$) | | | | | |
|---|---|---|---|---|---|---|
| | Measurement No. | | | | | |
| | 1 | 2 | 3 | 4 | 5 | 6 |
| N | 8.51 | 7.33 | 6.71 | 4.55 | 3.10 | 2.05 |
| C | 0.17 | 0.21 | 0.25 | 0.31 | 0.35 | 0.37 |
| Cr | 1.04 | 1.14 | 1.17 | 1.20 | 1.22 | 1.26 |
| Al | 0.73 | 0.75 | 0.74 | 0.75 | 0.74 | 0.76 |
| Mn | 0.24 | 0.25 | 0.28 | 0.31 | 0.33 | 0.39 |
| Si | 0.27 | 0.28 | 0.29 | 0.28 | 0.30 | 0.30 |
| Mo | 0.11 | 0.10 | 0.12 | 0.13 | 0.12 | 0.15 |
| Fe | Rest | | | | | |

**Figure 10.** EDS analysis of the elemental distribution of the sample after nitriding (cross-section). Etching state.

Based on the conducted microscopic examinations, the presence of a nitrided layer with a metallographic thickness of approx. 0.3 mm was found (Figure 11). This layer consists of a white near-surface zone with nitrides of the ε and ε+γ′-type with a thickness of 10 to 19 µm, and a dark zone of internal nitriding with a network of γ′ nitrides forming a mesh on the boundaries of former austenite grains.

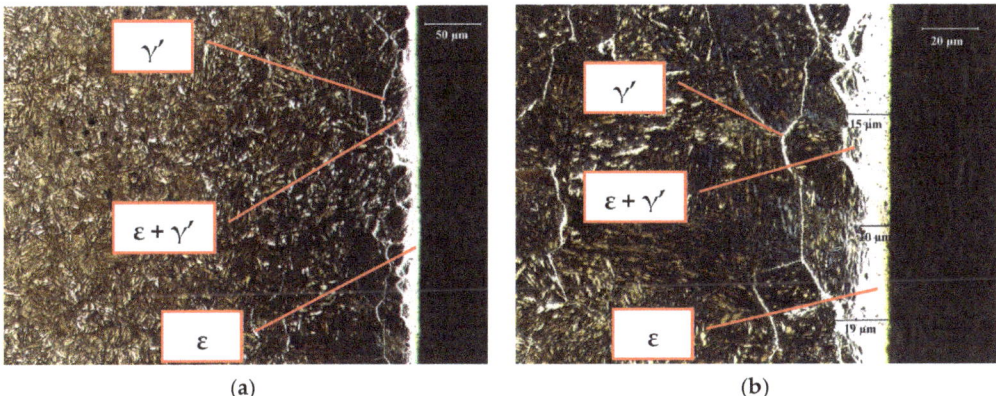

**Figure 11.** Nitrided layer: (**a**) Visible bright layer of ε, ε+γ′-type nitrides and a dark zone of internal nitriding with γ′ nitride precipitates forming a lattice on former austenite grain boundaries; (**b**) Magnified fragment of the same area. Etched state.

In addition, it was found that in some places around the circumference of the sample, the nitrided layer was non-uniform and porous (Figures 11a and 12). Changes in the thickness of the nitride layer ε and ε+γ′ in the range of 4 to 7 µm and porosity were observed locally. A lattice of γ′ nitrides distributed along the grain boundaries of the former austenite was also found. Moreover, the lack of an internal nitriding transition zone was observed. Locally, the hardness of the core was low and amounted to 320 HV1. It is an incorrect structure, showing a high tendency towards the cracking and chipping of the surface layer during operation.

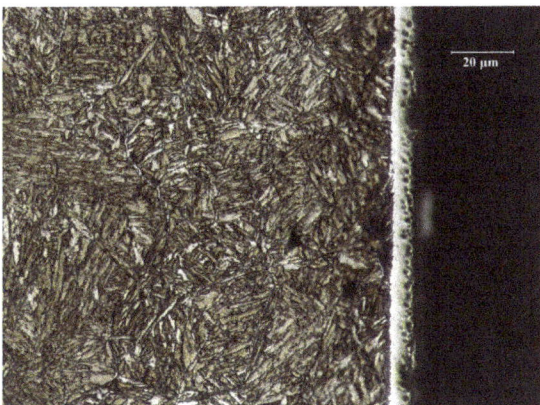

**Figure 12.** Nitrided layer. Visible uneven and porous bright layer of nitrides of the ε and ε+γ′-type, about 5 µm thick. A continuous grid of γ′ nitrides located along the grain boundaries of the former austenite is visible. No internal nitriding zone. Etched state.

In the core of the sample, a dendritic bainitic structure was found with precipitates of sorbite and troostite in the interdendritic spaces, typical for materials cooled quickly from hot working temperatures (typical for the raw state QT). This is an unacceptable defect, indicating a complete lack of application of the thermal improvement treatment after the plastic working process before the nitriding process or the use of an incorrect technology in the thermal improvement process.

As mentioned earlier, the thickness of the nitrided layer should be from about 0.5 mm to about 0.7 mm. The small thickness of the nitrided layer requires a uniform, hard and resilient substrate, protecting the above-mentioned layer against its collapse and chipping under the action of local pressures.

### 3.2.3. Measurements of Hardness of Samples Core after Nitriding

Core hardness measurements were made on the cross-section of the sample. The tests were carried out using the Vickers method in conditions compliant with PN EN ISO 6507—1:1999. As before, a Matsuzawa hardness tester with a load of 1 kg (9.807 N) operating for 15 s was used. As a result of the hardness measurements of the material in the core of the sample after the nitriding process, it was found that the average hardness of five measurements is 436.5 HV1.

### 3.2.4. Determination of Nitrided Layer Thickness

Determination of the thickness of the nitrided layer by measuring the hardness distribution was made in accordance with the PN—82/H—04550 standard. Hardness measurements were carried out using the Vickers method in conditions compliant with PN EN ISO 6507—1:2018 on the cross-section of the sample in the plane perpendicular to the nitrided surface. A Matsuzawa hardness tester was used with a load of 0.5 kg (4.904 N) operating for 15 s. Hardness measurements were made along a line perpendicular to the surface at intervals of 50 μm.

As a result of the hardness measurements, it was found that the effective thickness of the nitrided layer calculated to a hardness of 500 HV0.5 is 0.35 mm. The results of hardness measurements are summarized in Table 4 and additionally in the form of a graph in Figure 13.

**Table 4.** Measurement results of HV0.5 hardness distribution on the cross-section of the nitrided layer in the sample after nitriding.

| Distance from the Surface (μm) | 50 | 100 | 150 | 200 | 250 | 300 | 350 | 400 | 450 | 500 | 550 | 600 | 650 | 700 |
|---|---|---|---|---|---|---|---|---|---|---|---|---|---|---|
| Hardness HV 0.5 | 614.0 | 619.4 | 556.1 | 526.1 | 526.0 | 512.2 | 492.2 | 485.1 | 468.4 | 468.6 | 425.5 | 447.2 | 436.8 | 442.0 |

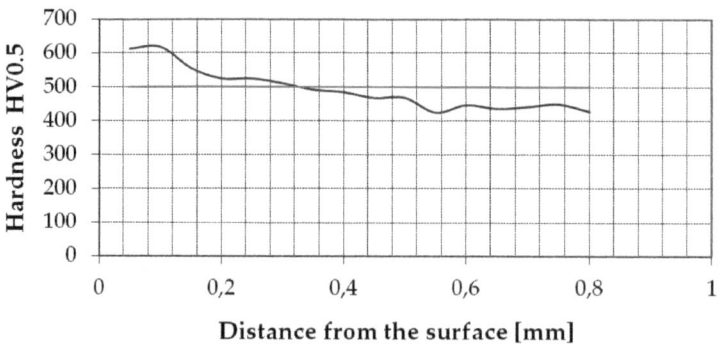

**Figure 13.** Hardness distribution in the nitrided layer with an HV load of 0.5 kg (4.904 N).

Referring to Section 3.2.2, it was found that reducing the nitrogen content from the surface towards to core of the samples caused changes in the microstructure from the formation of a white surface layer of ε-type nitrides, through a diffusion layer of ε+γ'-type nitrides, to the dark zone of γ'-type nitrides. This, in turn, resulted in a successive decrease in hardness, caused by a change in the types of nitrides in individual zones from the surface to the pin core.

As a result of the analyses carried out, it was found that the change in the material of 38HMJ steel to 42CrMo4 steel brought some improvement, but not as much as could be obtained using modern grades of steel, specially dedicated to the nitriding process.

To obtain a greater thickness of the nitrided layer, changing the parameters of the ion nitriding process was proposed, e.g., extending the nitriding time, changing the chemical composition of the nitriding atmosphere in the furnace, changing the BIAS potential, or possibly increasing the nitriding temperature. Before nitriding, it was necessary to carry out heat treatment, i.e., hardening from 880 °C in oil and tempering at 560 °C/1 h.

*3.3. Testing of Samples Taken from Pins Made of Nitrided 42CrMo4 Steel at Various Process Parameters*

Before nitriding, the pins were subjected to heat treatment, i.e., hardening from 880 °C in oil and tempering at 560 °C/1 h. Then, samples taken from the pins after the nitriding process carried out in accordance with the parameters given in Table 5 were examined.

**Table 5.** Parameters of the nitriding process.

| Samples No | Gas Flow Rate (L/min) | | | Currents | | Time (h/min) | Temperature (°C) |
|---|---|---|---|---|---|---|---|
| | N | H | Ar | BIAS | SCREEN | | |
| 1 | 1.5 | 0.4 | 0.3 | 410V 4.0A | 320V 3.0A | 26/35 | 530 |
| 2 | 1.3 | 0.7 | 0.2 | 440V 6.0A | 330V 3.0A | 26/35 | 510 |
| 3 | 1.3 | 0.7 | 0.2 | 410V 4.0A | 330V 3.0A | 30/38 | 510 |
| 4 | 1.3 | 0.7 | 0.2 | 410V 4.0A | 330V 4.0A | 26/35 | 550 |

As a result of the conducted microscopic observations of the samples etched with nital, it was found that in the core of all samples, there is a structure of fine-needle tempered sorbite with a small amount of fine bainite with a hardness of 330 to 366 HV0.5. This structure is typical for 42CrMo4 steel in the heat-treated state after the ion nitriding process and is correct (Figures 14–17).

3.3.1. Sample No. 1—Microscopic Observations

In sample No. 1, a nitrided layer with a metallographic thickness of approx. 0.3 mm was found, consisting of a white near-surface zone with ε and ε + γ' nitrides with a thickness of 8.7 to 10.2 μm and a dark internal nitriding zone with γ' nitrides. The layer of nitrides ε and ε + γ' is continuous, compact and adheres well to the steel substrate.

In the core of the sample, there is a structure typical of fine-needle sorbite tempering, typical of 42CrMo4 grade steel in the tempered state (Figure 14). The structure of the heat–chemically treated layer is typical for glow-nitrided layers and is correct.

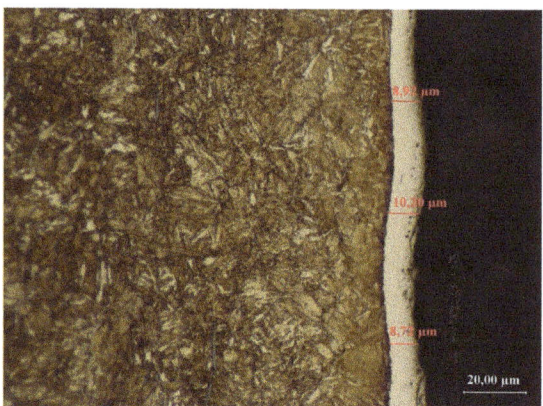

**Figure 14.** Nitrided layer in sample No. 1. Visible bright layer of $\varepsilon$, $\varepsilon + \gamma'$-type nitrides and a dark internal nitriding zone with $\gamma'$ nitride precipitates. Etched state.

3.3.2. Sample No. 2—Microscopic Observations

In sample No. 2, the presence of a nitrided layer with a metallographic thickness of approx. 0.1 mm was found, consisting of a white near-surface zone with $\varepsilon$ and $\varepsilon + \gamma'$ nitrides with a thickness of 4.2 to 5.6 µm and a dark zone of internal nitriding with $\gamma'$ nitrides. The layer of nitrides $\varepsilon$ and $\varepsilon + \gamma'$ is continuous, compact and adheres well to the steel substrate.

In the core of the sample, there is a typical structure of fine-needle tempered sorbite, typical for steel grade 40HM in the QT state. The structure of the heat–chemically treated layer is typical for glow-nitrided layers and is correct. The microstructure of the nitrided layer in sample No. 2 is shown in Figure 15.

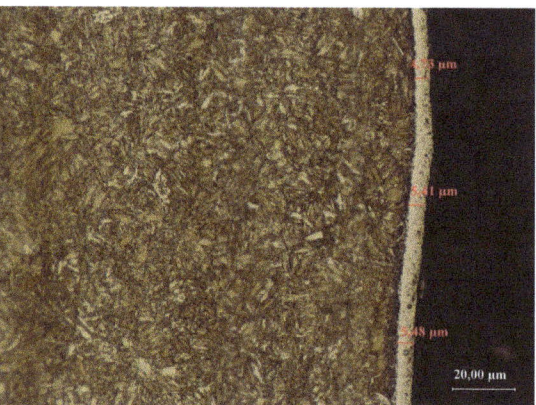

**Figure 15.** Nitrided layer in sample No. 2. Visible bright layer of $\varepsilon$, $\varepsilon + \gamma'$-type nitrides and a dark internal nitriding zone with $\gamma'$ nitride precipitates. Etched with Mi1Fe.

3.3.3. Sample No. 3—Microscopic Observations

In sample no. 3, a nitrided layer with a metallographic thickness of about 0.15 mm was found, consisting of a white near-surface zone with $\varepsilon$ and $\varepsilon + \gamma'$ nitrides with a thickness of 5.6 to 6.2 µm and a dark zone of internal nitriding with $\gamma'$ nitrides. The layer of nitrides $\varepsilon$ and $\varepsilon + \gamma'$ is continuous, compact and adheres well to the steel substrate.

In the core of the sample, there is a typical structure of fine-needle tempered sorbite, typical for steel grade 40 HM in the QT state. The structure of the heat–chemically treated layer is typical for glow-nitrided layers and is correct. The microstructure of the nitrided layer in sample no. 3 is shown in Figure 16.

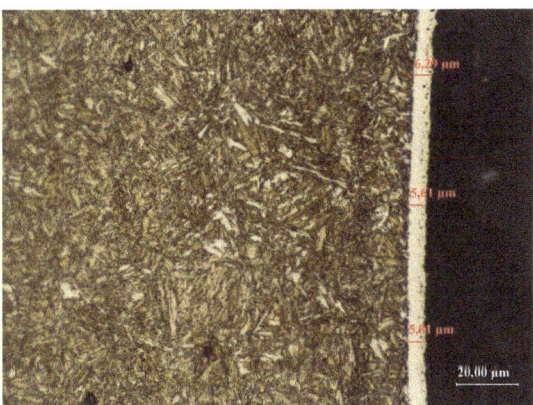

**Figure 16.** Nitrided layer in sample No. 3. Visible bright layer of ε, ε + γ′-type nitrides and a dark internal nitriding zone with γ′ nitride precipitates. Etched state.

3.3.4. Sample No. 4—Microscopic Observations

In sample no. 4, a nitrided layer with a metallographic thickness of approx. 0.15 mm was found, consisting of a white near-surface zone with ε and ε + γ′ nitrides with a thickness of 4.3 to 5.0 μm and a dark zone of internal nitriding with γ′ nitrides. The layer of nitrides ε and ε + γ′ is continuous, compact and adheres well to the steel substrate.

In the core of the sample, there is a typical structure of fine-needle tempered sorbite, typical for steel grade 40 HM in the QT state. The structure of the heat–chemically treated layer is typical for glow-nitrided layers and is correct. The microstructure of the nitrided layer in sample No. 4 is shown in Figure 17.

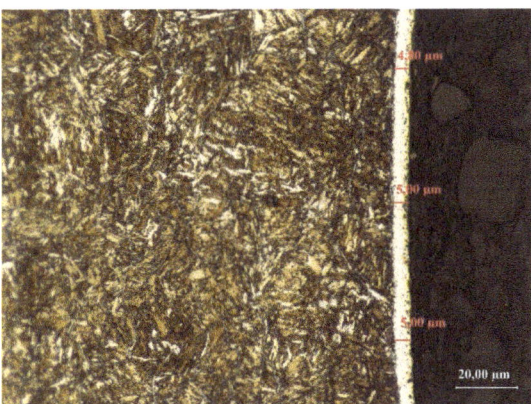

**Figure 17.** Nitrided layer in sample No. 4. Visible bright layer of ε, ε + γ′-type nitrides and a dark internal nitriding zone with γ′ nitride precipitates. Etched state.

3.3.5. Samples from 1 to 4-Hardness Measurements

Determination of the thickness of the nitrided layer by measuring the hardness distribution was made in accordance with the PN—82/H—04550 standard. Hardness measurements were carried out using the Vickers method in conditions compliant with PN EN ISO 6507—1:2018 on the cross-section of the sample in the plane perpendicular to the nitrided surface. During the measurements, a Matsuzawa hardness tester with a load of 0.5 kg (4.904 N) operating for 15 s was used. Hardness measurements were made along a line perpendicular to the surface at intervals of 50 μm. The results of hardness measurements are summarized in Table 6.

**Table 6.** Results of hardness distribution measurements on the cross-section of sample No. 1.

| | Sample No. 1 | | | | | | | | | |
|---|---|---|---|---|---|---|---|---|---|---|
| Distance from the surface [μm] | 50 | 100 | 150 | 200 | 250 | 300 | 350 | 400 | 450 | 500 |
| Hardness HV 0.5 | 621.0 | 638.7 | 605.2 | 539.1 | 512.1 | 478.7 | 444.9 | 419.3 | 416.8 | 366.7 |
| | Sample No. 2 | | | | | | | | | |
| Distance from the surface [μm] | 50 | 100 | 150 | 200 | 250 | 300 | 350 | 400 | 450 | 500 |
| Hardness HV 0.5 | 529.7 | 499.3 | 461.9 | 438.9 | 420.0 | 386.1 | 375.6 | 359.5 | 359.0 | 330.3 |
| | Sample No. 3 | | | | | | | | | |
| Distance from the surface [μm] | 50 | 100 | 150 | 200 | 250 | 300 | 350 | 400 | 450 | 500 |
| Hardness HV 0.5 | 568.5 | 529.9 | 483.7 | 434.3 | 409.7 | 383.5 | 356.6 | 347.6 | 335.5 | 341.7 |
| | Sample No. 4 | | | | | | | | | |
| Distance from the surface [μm] | 50 | 100 | 150 | 200 | 250 | 300 | 350 | 400 | 450 | 500 |
| Hardness HV 0.5 | 554.0 | 527.8 | 481.5 | 454.0 | 421.8 | 417.6 | 398.7 | 366.7 | 355.2 | 337.0 |

As a result of hardness measurements, it was found that in sample No. 1, the effective thickness of the nitrided layer calculated to hardness 500 HV0.5 was 0.3 mm, in sample No. 2 the effective thickness of the nitrided layer was 0.1 mm, and in samples No. 3 and 4 the effective thickness of the nitrided layers were 0.15 mm. It can be concluded that the change in the parameters of the ion nitriding process of the 42CrMo4 steel material after thermal improvement resulted in an improvement in the quality of the nitrided layers, in particular in terms of their uniformity, thickness, and hardness. The most favorable set of parameters of the plasma nitriding process was obtained for sample No.1 (Table 5). Changing the parameters of the ion nitriding process improved the diffusion penetration of active nitrogen atoms toward the core of the nitrided samples. A higher concentration of active nitrogen atoms (not exceeding the limit value) favored the formation of a thicker layer of ε-type nitrides, a diffusion zone containing ε + γ′-type nitrides, and a dark zone containing γ′-type nitrides. This, in turn, resulted in a higher surface hardness of the samples and a thicker hardened zone in the cross-section.

## 4. Conclusions

Tests have shown that 38HMJ steel in its raw state is not suitable for use on pins obtained by plasma nitriding. In its structure, the presence of many non-metallic inclusions with uneven distribution and inhomogeneous band structure of ferrite with finely dispersed pearlite was found. This shows the low quality of the material. Nitriding resulted in obtaining a thin, porous layer of ε-type nitrides and in some cases no layer at all.

At the same time, it was found that reducing the nitrogen content from the surface toward the core of the samples caused changes in the microstructure, most often from the formation of a white surface layer of ε-type nitrides, through a diffusion layer of ε + γ′-type nitrides, to the dark zone of γ′-type nitrides. This, in turn, resulted in a successive decrease in hardness, caused by a change in the types of nitrides in individual zones from the surface to the pin core.

The use of 42CrMo4 steel in its raw state brought slightly better results. However, the obtained nitride layer showed uneven thickness and signs of porosity.

To obtain a greater thickness and uniformity of the layer of nitrided 42CrMo4 steel, initial heat treatment was carried out (quenching from 880 °C in oil and tempering at 560 °C/1 h) and modification of the parameters of the plasma nitriding process.

This brought the intended effect, and the most favorable set of parameters for the plasma nitriding process was gas flow rate (1.5 L/min N; 0.4 L/min H; 0.3 L/min Ar); currents (BIAS—410 V 4.0 A, SCREEN—320 V 4.0 A); time (26 h and 35 min); and temperature (550 °C). This facilitated the diffusion penetration of active nitrogen atoms from the surface towards the core of the nitrided samples. It was found that the concentration of nitrogen in the cross-section of the nitrided samples decreased towards the core, in contrast with the other elements. This allowed the obtaining of a continuous and compact layer of nitrides that was well adhered to the steel surface. A relatively thick and continuous white layer of ε-type nitrides appeared on the surface of the nitrided sample, guaranteeing the highest hardness, then a layer of ε+γ′-type nitrides of lower hardness and a dark zone of internal nitriding with γ′-type nitrides of the lowest hardness. On the other hand, in the core of the nitrided samples, the presence of the correct structure of fine-needle hardened sorbite with a small amount of fine bainite was found.

**Author Contributions:** Conceptualization, D.M., W.D. and P.S.; methodology, D.M. and W.D.; software, P.S.; validation, D.M. and W.D.; formal analysis, D.M. and W.D.; investigation, D.M. and W.D.; resources, P.S.; data curation, D.M. and W.D.; writing—original draft preparation, D.M. and W.D.; writing—review and editing, D.M. and W.D.; visualization, D.M.; supervision, W.D.; project administration, D.M.; funding acquisition, P.S. All authors have read and agreed to the published version of the manuscript.

**Funding:** Regional Operational Program for the Lower Silesian Voivodeship for the years 2014–2020. Priority axis 1. Enterprises and innovations. Action 1.2. Innovative enterprises; Sub-measure 1.2.1, Innovative enterprises-horizontal competition. Witelon Collegium State University-Voucher for Innovation—Sacher.

**Conflicts of Interest:** The authors declare no conflict of interest. The funders had no role in the design of the study; in the collection, analyses, or interpretation of data; in the writing of the manuscript; or in the decision to publish the results.

# References

1. Dan, J.; Boving, H.; Hintermann, H. Hard coatings. *HAL Open Sci. J. Phys. IV Proc.* **1993**, *3*, 933–941. [CrossRef]
2. Bell, T. Surface engineering: Its current and future impact on tribology. *J. Phys. D Appl. Phys.* **1992**, *25*, A297. [CrossRef]
3. Bell, T. *Engineering the Surface to Combat Wear*; Tribology Series; Elsevier: Amsterdam, The Netherlands, 1993; Volume 25, pp. 27–37.
4. Ren, C.X.; Wang, Q.; Zhang, Z.J.; Zhang, P.; Zhang, Z.F. Surface strengthening behaviors of four structural steels processed by surface spinning strengthening. *Mater. Sci. Eng. A* **2017**, *704*, 262–273. [CrossRef]
5. Śliwa, A.; Mikuła, J.; Gołombek, K.; Tański, T.; Kwaśny, W.; Bonek, M.; Brytan, Z. Prediction of the properties of PVD/CVD coatings with the use of FEM analysis. *Appl. Surf. Sci.* **2016**, *388*, 281–287. [CrossRef]
6. Kıvak, T. Optimization of surface roughness and flank wear using the Taguchi method in milling of Hadfield steel with PVD and CVD coated inserts. *Measurement* **2014**, *50*, 19–28. [CrossRef]
7. Vetter, J.; Barbezat, G.; Crummenauer, J.; Avissar, J. Surface treatment selections for automotive applications. *Surf. Coat. Technol.* **2005**, *200*, 1962–1968. [CrossRef]
8. Ferreira, A.A.; Silva, F.J.; Pinto, A.G.; Sousa, V.F. Characterization of thin chromium coatings produced by PVD sputtering for optical applications. *Coatings* **2021**, *11*, 215. [CrossRef]
9. Frohn-Sörensen, P.; Cislo, C.; Paschke, H.; Stockinger, M.; Engel, B. Dry friction under pressure variation of PACVD TiN surfaces on selected automotive sheet metals for the application in unlubricated metal forming. *Wear* **2021**, *476*, 203750. [CrossRef]
10. Baptista, A.; Pinto, G.; Silva, F.J.; Ferreira, A.A.; Pinto, A.G.; Sousa, V.F. Wear characterization of chromium PVD coatings on polymeric substrate for automotive optical components. *Coatings* **2021**, *11*, 555. [CrossRef]

11. ChandraYadaw, R.; Singh, S.K.; Chattopadhyaya, S.; Kumar, S.; Singh, R.C. Comparative study of surface coating for automotive application—A review. *Int. J. Appl. Eng. Res.* **2018**, *13*, 17–22.
12. Ginting, A.; Skein, R.; Cuaca, D.; Masyithah, Z. The characteristics of CVD-and PVD-coated carbide tools in hard turning of AISI 4340. *Measurement* **2018**, *129*, 548–557. [CrossRef]
13. Martinho, R.P.; Silva, F.J.G.; Martins, C.; Lopes, H. Comparative study of PVD and CVD cutting tools performance in milling of duplex stainless steel. *Int. J. Adv. Manuf. Technol.* **2019**, *102*, 2423–2439. [CrossRef]
14. He, Q.; Paiva, J.M.; Kohlscheen, J.; Beake, B.D.; Veldhuis, S.C. An integrative approach to coating/carbide substrate design of CVD and PVD coated cutting tools during the machining of austenitic stainless steel. *Ceram. Int.* **2020**, *46*, 5149–5158. [CrossRef]
15. Thakur, A.; Gangopadhyay, S. Influence of tribological properties on the performance of uncoated, CVD and PVD coated tools in machining of Incoloy 825. *Tribol. Int.* **2016**, *102*, 198–212. [CrossRef]
16. Besley, N.A.; Johnston, R.L.; Stace, A.J.; Uppenbrink, J. Theoretical study of the structures and stabilities of iron clusters. *J. Mol. Struct.* **1995**, *341*, 75–90. [CrossRef]
17. Friehling, P.B.; Poulsen, F.W.; Somers, M.A. Nucleation of iron nitrides during gaseous nitriding of iron; effect of a preoxidation treatment. *Int. J. Mater. Res.* **2022**, *92*, 589–595.
18. Shakhnazarov, K.Y. Chernov's iron–carbon diagram, the structure and properties of steel. *Met. Sci. Heat Treat.* **2009**, *51*, 3–6. [CrossRef]
19. Davydov, S.V. Phase Equilibria in the Carbide Region of Iron–Carbon Phase Diagram. *Steel Transl.* **2020**, *50*, 888–896. [CrossRef]
20. Shohoji, N. *Uncracked Gaseous Ammonia $NH_3$ as a Powerful Nitriding Medium*; Lambert Academic Publishing: London, UK, 2017.
21. Cahn, R.W.; Haasen, P. *Physical Metallurgy*, 4th ed.; Elsevier Science B.V.: Amsterdam, The Netherlands, 1996.
22. Brandes, E.A.; Brook, G.B. *Smithells Metals Reference Book*, 7th ed.; Butterworth Heinemann: Oxford, UK, 1999.
23. Kovacs, W.; Russell, W. Ion Nitriding. In Proceedings of the an International Conference, Cleveland, OH, USA, 15–17 September 1986.
24. Jack, D.H.; Jack, K.H. Invited review: Carbides and nitrides in steel. *Mater. Sci. Eng.* **1973**, *11*, 1–27. [CrossRef]
25. Menthe, E.; Bulak, A.; Olfe, J.; Zimmermann, A.; Rie, K.T. Improvement of the mechanical properties of austenitic stainless steel after plasma nitriding. *Surf. Coat. Technol.* **2000**, *133*, 259–263. [CrossRef]
26. Chang, Y.Y.; Amrutwar, S. Effect of plasma nitriding pretreatment on the mechanical properties of AlCrSiN-coated tool steels. *Materials* **2019**, *12*, 795. [CrossRef]
27. Hernandez, M.H.S.M.; Staia, M.H.; Puchi-Cabrera, E.S. Evaluation of microstructure and mechanical properties of nitrided steels. *Surf. Coat. Technol.* **2008**, *202*, 1935–1943. [CrossRef]
28. Raj, V.R.; Vijayarannuathb, B.; Ramanan, N.; Ponshanmugakumar, A. Mechanical and corrosion resistant properties of nitrided low carbon steel. *Mater. Today Proc.* **2021**, *46*, 3288–3291. [CrossRef]
29. Özkan, D.; Yilmaz, M.A.; Karakurt, D.; Szala, M.; Walczak, M.; Bakdemir, S.A.; Türküz, C.; Sulukan, E. Effect of AISI H13 Steel Substrate Nitriding on AlCrN, ZrN, TiSiN, and TiCrN Multilayer PVD Coatings Wear and Friction Behaviors at a Different Temperature Level. *Materials* **2023**, *16*, 1594. [CrossRef]
30. Lin, N.; Liu, Q.; Zou, J.; Guo, J.; Li, D.; Yuan, S.; Ma, Y.; Wang, Z.; Wang, Z.; Tang, B. Surface texturing-plasma nitriding duplex treatment for improving tribological performance of AISI 316 stainless steel. *Materials* **2016**, *9*, 875. [CrossRef]
31. Adachi, S.; Ueda, N. Wear and corrosion properties of cold-sprayed AISI 316L coatings treated by combined plasma carburizing and nitriding at low temperature. *Coatings* **2018**, *8*, 456. [CrossRef]
32. Mateescu, A.O.; Mateescu, G.; Balan, A.; Ceaus, C.; Stamatin, I.; Cristea, D.; Samoila, C.; Ursutiu, D. Stainless steel surface nitriding in open atmosphere cold plasma: Improved mechanical, corrosion and wear resistance properties. *Materials* **2021**, *14*, 4836. [CrossRef]
33. Xu, M.; Kang, S.; Lu, J.; Yan, X.; Chen, T.; Wang, Z. Properties of a plasma-nitrided coating and a CrN x coating on the stainless steel bipolar plate of PEMFC. *Coatings* **2020**, *10*, 183. [CrossRef]
34. Fan, Y.; Yang, H.; Fan, H.; Liu, Q.; Lv, C.; Zhao, X.; Tang, M.; Wu, J.; Cao, X. Corrosion resistance of modified hexagonal boron nitride (h-BN) nanosheets doped acrylic acid coating on hot-dip galvanized steel. *Materials* **2020**, *13*, 2340. [CrossRef]
35. Sumiya, K.; Tokuyama, S.; Nishimoto, A.; Fukui, J.; Nishiyama, A. Application of active-screen plasma nitriding to an austenitic stainless steel small-diameter thin pipe. *Metals* **2021**, *11*, 366. [CrossRef]
36. Liu, Y.; Liu, D.; Zhang, X.; Li, W.; Ma, A.; Fan, K.; Xing, W. Effect of Alloying Elements and Low Temperature Plasma Nitriding on Corrosion Resistance of Stainless Steel. *Materials* **2022**, *15*, 6575. [CrossRef]
37. Edenhofer, B. The Ionitriding Process-Thermochemical Treatment of Steel and Cast Iron Materials. *Metall. Mater. Technol.* **1976**, *8*, 421–426.
38. D'Haen, J.; Quaeyhaegens, C.; Knuyt, G.; D'Olieslaeger, M.; Stals, L.M. Structure analysis of plasma-nitrided pure iron. *Surf. Coat. Technol.* **1995**, *74–75*, 405. [CrossRef]
39. Lengauer, W.; Binder, S.; Aigner, K.; Ettmayer, P.; Guillou, A.; Debuigne, J.; Groboth, G. Solid state properties of group IVb carbonitrides. *J. Alloys Compd.* **1995**, *217*, 137–147. [CrossRef]
40. Taneike, M.; Sawada, K.; Abe, F. Effect of carbon concentration on precipitation behavior of $M_{23}C_6$ carbides and MX carbonitrides in martensitic 9Cr steel during heat treatment. *Metall. Mater. Trans. A* **2004**, *35*, 1255–1262. [CrossRef]
41. Thelning, K.E. *Steel and Its Heat Treatment*; Butterworth: London, UK, 1984.
42. Akhlaghi, M.; Meka, S.R.; Jägle, E.A.; Kurz, S.J.; Bischoff, E.; Mittemeijer, E.J. Formation mechanisms of alloying element nitrides in recrystallized and deformed ferritic Fe-Cr-Al alloy. *Metall. Mater. Trans. A* **2016**, *47*, 4578–4593. [CrossRef]

43. Dossett, J.; Totten, G.E. Fundamentals of nitriding and nitrocarburizing. In *ASM Handbook: Steel Heat Treating Fundamentals and Processes*; ASM International: Materials Park, OH, USA, 2013; Volume 619.
44. Morita, Z.; Tanaka, T.; Yanai, T. Equilibria of nitride forming reactions in liquid iron alloys. *Metall. Trans. B* **1987**, *18*, 195–202. [CrossRef]
45. Babaskin, Y.Z.; Shipitsyn, S.Y. Microalloying of structural steel with nitride-forming elements. *Steel Transl.* **2009**, *39*, 1119–1121. [CrossRef]
46. Bell, T.; Birch, B.J.; Korotchenko, V.; Evans, S.P. Controlled nitriding in ammonia-hydrogen mixtures. In *Heat Treatment'73*; The Metals Society: London, UK, 1975.
47. Georges, J. Nitriding Process and Nitriding Furnace Therefore. Plasma Metal SA. U.S. Patent No. 5,989,363, 23 November 1999.
48. Malvos, H.; Ricard, A.; Szekely, J.; Michel, H.; Gantois, M.; Ablitzer, D. Modelling of a microwave postdischarge nitriding reactor. *Surf. Coat. Technol.* **1993**, *59*, 59–66. [CrossRef]
49. Aghajani, H.; Behrangi, S. *Plasma Nitriding of Steels*; Springer International Publishing: Cham, Switzerland, 2017.
50. Georges, J.; Cleugh, D. Active screen plasma nitriding. In *Stainless Steel 2000*; CRC Press: Boca Raton, FL, USA, 2020; pp. 377–387.
51. Ohtsu, N.; Miura, K.; Hirano, M.; Kodama, K. Investigation of admixed gas effect on plasma nitriding of AISI316L austenitic stainless steel. *Vacuum* **2021**, *193*, 110545. [CrossRef]
52. Priest, J.M.; Baldwin, M.J.; Fewell, M.P. The action of hydrogen in low-pressure rf-plasma nitriding. *Surf. Coat. Technol.* **2001**, *145*, 152–163. [CrossRef]
53. Zdravecká, E.; Slota, J.; Solfronk, P.; Kolnerová, M. Evaluation of the effect of different plasma-nitriding parameters on the properties of low-alloy steel. *J. Mater. Eng. Perform.* **2017**, *26*, 3588–3596. [CrossRef]
54. Kovacı, H.; Baran, Ö.; Yetim, A.F.; Bozkurt, Y.B.; Kara, L.; Çelik, A. The friction and wear performance of DLC coatings deposited on plasma nitrided AISI 4140 steel by magnetron sputtering under air and vacuum conditions. *Surf. Coat. Technol.* **2018**, *349*, 969–979. [CrossRef]
55. Jeong, B.Y.; Kim, M.H. Effects of the process parameters on the layer formation behavior of plasma nitrided steels. *Surf. Coat. Technol.* **2001**, *141*, 182–186. [CrossRef]
56. Yazdani, S.; Mahboubi, F. Comparison between microstructure, wear behavior, and corrosion resistance of plasma-nitrided and vacuum heat-treated electroless Ni–B coating. *J. Bio-Tribo-Corros.* **2019**, *5*, 69. [CrossRef]
57. Ricard, A.; Czerwiec, T.; Belmonte, T.; Bockel, S.; Michel, H. Detection by emission spectroscopy of active species in plasma–surface processes. *Thin Solid Film.* **1999**, *341*, 1–8. [CrossRef]
58. Zhao, C.; Li, C.X.; Dong, H.; Bell, T. Study on the active screen plasma nitriding and its nitriding mechanism. *Surf. Coat. Technol.* **2006**, *201*, 2320–2325. [CrossRef]
59. Li, C.X. Active screen plasma nitriding—An overview. *Surf. Eng.* **2010**, *26*, 135–141. [CrossRef]
60. Nishimoto, A.; Nagatsuka, K.; Narita, R.; Nii, H.; Akamatsu, K. Effect of the distance between screen and sample on active screen plasma nitriding properties. *Surf. Coat. Technol.* **2010**, *205*, S365–S368. [CrossRef]
61. Li, Y.; Wang, L.; Zhang, D.; Shen, L. Influence of bias voltage on the formation and properties of iron-based nitrides produced by plasma nitriding. *J. Alloys Compd.* **2010**, *497*, 285–289. [CrossRef]
62. Ahangarani, S.; Mahboubi, F.; Sabour, A.R. Effects of various nitriding parameters on active screen plasma nitriding behavior of a low-alloy steel. *Vacuum* **2006**, *80*, 1032–1037. [CrossRef]

**Disclaimer/Publisher's Note:** The statements, opinions and data contained in all publications are solely those of the individual author(s) and contributor(s) and not of MDPI and/or the editor(s). MDPI and/or the editor(s) disclaim responsibility for any injury to people or property resulting from any ideas, methods, instructions or products referred to in the content.

Article

# Study on Efficient Dephosphorization in Converter Based on Thermodynamic Calculation

Zhong-Liang Wang [1], Tian-Le Song [1], Li-Hua Zhao [2,*] and Yan-Ping Bao [1,*]

1. State Key Laboratory of Advanced Metallurgy, University of Science and Technology Beijing, Beijing 100083, China
2. School of Metallurgical and Ecological Engineering, University of Science and Technology Beijing, Beijing 100083, China
* Correspondence: zhaolihua@metall.ustb.edu.cn (L.-H.Z.); baoyp@ustb.edu.cn (Y.-P.B.)

**Abstract:** Given the accelerating depletion of iron ore resources, there is growing concern within the steel industry regarding the availability of high-phosphorus iron ore. However, it is important to note that the utilization of high-phosphorus iron ore may result in elevated phosphorus content and notable fluctuations in molten iron, thereby imposing additional challenges on the dephosphorization process in steelmaking. The most urgent issue in the process of converter steelmaking is how to achieve efficient dephosphorization. In this study, the influence of various factors on the logarithm of the phosphorus balance distribution ratio ($\lg L_P$), the logarithm of the $P_2O_5$ activity coefficient ($\lg \gamma_{P_2O_5}$), and the logarithm of the phosphorus capacity ($\lg C_p$) were examined through thermodynamic calculations. The impact of each factor on dephosphorization was analyzed, and the optimal conditions for the dephosphorization stage of the converter were determined. Furthermore, the influence of basicity and $Fe_tO$ content on the form of phosphorus in the slag was analyzed using FactSage 7.2 software, and the precipitation rules of the slag phases were explored. The thermodynamic calculation results indicated that increasing the basicity of the dephosphorization slag was beneficial for dephosphorization, but it should be maintained below 3. The best dephosphorization effect was achieved when the $Fe_tO$ content was around 20%. The reaction temperature during the dephosphorization stage should be kept low, as the dephosphorization efficiency decreased sharply with the increasing temperature. In dephosphorization slag, $Ca_3(PO_4)_2$ usually formed a solid solution with $Ca_2SiO_4$, so the form of phosphorus in the slag was mainly determined by the precipitation form and content of $Ca_2SiO_4$. The phases in the dephosphorization slag mainly consisted of a phosphorus-rich phase, an iron-rich phase, and a matrix phase. The results of scanning electron microscopy and X-ray diffraction analyses were consistent with the thermodynamic calculation results.

**Keywords:** steelmaking; dephosphorization; thermodynamics; precipitation

**Citation:** Wang, Z.-L.; Song, T.-L.; Zhao, L.-H.; Bao, Y.-P. Study on Efficient Dephosphorization in Converter Based on Thermodynamic Calculation. *Crystals* **2023**, *13*, 1132. https://doi.org/10.3390/cryst13071132

Academic Editors: Daniel Medyński, Grzegorz Lesiuk and Anna Burduk

Received: 7 June 2023
Revised: 18 July 2023
Accepted: 19 July 2023
Published: 20 July 2023

**Copyright:** © 2023 by the authors. Licensee MDPI, Basel, Switzerland. This article is an open access article distributed under the terms and conditions of the Creative Commons Attribution (CC BY) license (https://creativecommons.org/licenses/by/4.0/).

## 1. Introduction

As the main raw material of the steel industry, iron ore is an important strategic resource for a country [1–4]. With the rapid development of iron and steel metallurgy technology, the demand for iron ore is increasing [5]. Iron ore resources are relatively abundant and widely distributed worldwide, but rich iron ore resources will gradually become depleted, so it is very important to develop and utilize iron ore with high phosphorus content [6]. During the sintering and blast furnace smelting process, the phosphorus in the ore will enter the sinter and hot metal [7–9]. Currently, steel companies have increasingly higher requirements for the quality of iron concentrates and strict limits on the phosphorus content, with the phosphorus content of iron concentrates required to be less than 0.024% [10]. There are many steel companies in China, and the differences in raw material conditions such as the phosphorus content of hot metal are significant among various steel mills. Only a few large steel companies in China can use high-quality iron ore with low phosphorus content [11]. Many steel mills produce hot metal with a phosphorus content between 0.1 and 0.25%, and the phosphorus content in hot metal will further increase in the future [12].

The phosphorus in steel mainly comes from raw steelmaking materials such as hot metal, steel scrap, slagging materials, and deoxidizing alloys [13–15]. Phosphorus dissolves in the ferrite contained in steel [16]. It has a strong solid solution strengthening effect, significantly improving the strength and hardness of steel at room temperature. The strengthening effect of phosphorus is second only to carbon, which can significantly increase the yield strength and yield-to-tensile ratio of steel, but it also causes a significant decrease in plasticity and toughness (especially low-temperature toughness) and a sharp increase in the transition temperature of steel [17–19]. For most steel grades, phosphorus is a harmful element, so it is required to minimize the phosphorus content in the steel [20]. Dephosphorization is one of the most important functions in the converter steelmaking process, and the steel slag is the only place where phosphorus is removed from the steel. Converter dephosphorization mainly uses oxidation, which oxidizes the phosphorus in the steel to the slag phase under an oxidizing atmosphere to remove it [21]. The higher the phosphorus content in the hot metal, the greater the difficulty and technical requirements of the converter smelting process. Different conditions of phosphorus content in hot metal directly affect many key indicators such as the production rhythm of converter steelmaking, auxiliary material consumption, iron and steel material consumption, endpoint oxygen content, and endpoint accuracy rate [22–24]. As users' requirements for phosphorus content in steel become increasingly demanding, efficient dephosphorization in the converter has become the goal pursued by metallurgists. A large amount of research has been carried out at home and abroad on the dephosphorization reaction of steel, and a series of empirical and semi-empirical formulas have been obtained, which have certain guiding significance for actual production [25].

The removal trajectory of phosphorus involves the distribution between the steel and slag, as well as the distribution among different phases of slag. Therefore, to comprehensively understand the removal pattern of phosphorus in the converter and to obtain reasonable control conditions for the dephosphorization stage, it is necessary to consider both the reactions in the slag and the phase separation of dephosphorization slag. In this study, firstly, by using the thermodynamic empirical formulas summarized by previous researchers, the influence of different slag compositions and reaction temperatures on dephosphorization during slag–steel reactions was analyzed, and favorable control conditions for the dephosphorization stage of the converter were obtained. Then, based on the previous research, the form of phosphorus in the dephosphorization slag was analyzed, and the phase separation pattern of the dephosphorization slag was investigated. Industrial experiments were conducted to verify the thermodynamic calculation results, providing a research foundation for efficient dephosphorization in converters.

## 2. Research Method

The main factors affecting the dephosphorization effect of steel slag include the basicity of the slag; the contents of $Fe_tO$, $MgO$, $MnO$; and the reaction temperature. Therefore, it is necessary to study the impact and degree of each factor on the dephosphorization ability of steel slag. The main parameters for evaluating the dephosphorization reaction were the phosphorus balance distribution ratio, the $P_2O_5$ activity coefficient, and the phosphorus capacity. Theoretical calculations were carried out based on thermodynamic formulas to analyze the influence of the main components and temperature of dephosphorization slag on the dephosphorization effect and to provide guidance for optimizing the converter dephosphorization process [26]. The contents of slag components in the dephosphorization stage of the converter used for calculation are shown in Table 1. By fixing the content of other components and adjusting the contents of CaO and $SiO_2$ in the slag while keeping the basicity of the slag unchanged, the impact of a single factor on the dephosphorization effect was studied. Additionally, since the dephosphorization reaction was an exothermic process, the temperature was also an important factor affecting the dephosphorization efficiency. By examining the changes in the logarithm of the phosphorus balance distribution ratio (lg$L_p$), the logarithm of the P$_2$O$_5$ activity coefficient (lg$\gamma_{P_2O_5}$), and the logarithm of the phosphorus capacity (lg$C_p$) under the influence of various factors, the impact of each factor on dephosphorization was determined, and the optimal conditions for the converter dephosphorization stage were analyzed. To avoid any misunderstandings, we wanted to clarify the nature of the logarithmic expressions used in thermodynamic calculations. These expressions did not originate from fundamental laws or principles but, rather, were the result of previous researchers' summarizations and inductions. The logarithmic functions served as mathematical tools in our thermodynamic calculations, allowing us to quantitatively evaluate the efficiency of phosphorus removal from slag. It is important to emphasize that these expressions should not be interpreted as representing inherent laws but rather to facilitate the analysis and interpretation of the computational results.

Table 1. Components of converter slag for calculation (%).

| Composition | CaO | SiO$_2$ | Fe$_t$O | MnO | P$_2$O$_5$ | MgO |
|---|---|---|---|---|---|---|
| Content | 38 | 19 | 25 | 5 | 3 | 10 |

2.1. Phosphorus Balance Distribution Ratio

The phosphorus distribution ratio between the slag and steel could well reflect the dephosphorization degree in the converter steelmaking process. The larger the phosphorus distribution ratio was, the better the dephosphorization effect was. As shown in Equation (1), this study cited the thermodynamic empirical formula for dephosphorization experiments in a CaO − Fe$_t$O − SiO$_2$ − MgO − MnO slag system summarized by Ide and Fruehan. to calculate the phosphorus distribution ratio [27]. In the equation, [%X] represents the content of X element in the steel liquid, in %, and (%X) represents the content of X element in the slag phase, in %. The expression for the phosphorus distribution ratio in the slag and steel was derived from Equation (1), as shown in Equation (2). By substituting the known composition of the furnace slag in Table 1 into Equation (2), the theoretical phosphorus balance distribution ratio under this condition can be calculated. The $L_p$ was affected by factors other than the slag properties. Firstly, the activity coefficient of phosphorus in molten steel played a significant role in the distribution of phosphorus between the steel and slag. The activity coefficient was a parameter that described the activity of components in a molten system and was influenced by the concentration of phosphorus in the steel, the concentration of phosphorus in the slag, and the temperature. If the activity coefficient of phosphorus in the steel was relatively low compared to that in the slag, phosphorus tended to remain in the steel rather than combine with the slag. This led to a decrease in the $L_p$. Additionally, the oxygen potential (redox conditions) also affected the $L_p$. A higher oxygen potential promoted the formation of phosphorus oxide, while a lower oxygen potential facilitated the reduction of phosphorus, releasing it from the slag into the steel. Thus, variations in the oxygen potential altered the distribution of phosphorus between the steel and slag, thereby affecting the $L_p$. Other researchers have also summarized some empirical formulas for Lp, which were listed in Table 2. The main difference lay in the diverse calculation methods used for different slag systems. However, for the same slag system, it was recommended to choose an empirical formula that had been validated within the compositional range of the slag.

$$\lg \frac{(\%P)}{[\%P](\%T.Fe)^{2.5}} = 0.072[(\%CaO) + 0.15(\%MgO) + 0.6(\%P_2O_5) + 0.6(\%MnO)] + 11,570/T - 10.520 \qquad (1)$$

$$\lg L_P = \lg \frac{(\%P)}{[\%P]} = 0.072[(\%CaO) + 0.15(\%MgO) + 0.6(\%P_2O_5) + 0.6(\%MnO)] + 11,570/T - 10.520 + 2.5\lg(\%T.Fe) \qquad (2)$$

Table 2. Common empirical formulas for $L_p$ [27–33].

| $L_p$ Representation | $L_p$ Calculation Formula |
|---|---|
| $\dfrac{(\%P_2O_5)}{[\%P]}$ | $\lg L_P = 5.9(\%CaO) + 2.5\lg(\%FeO) + 0.5\lg(\%P_2O_5) - 0.5C - 0.36$<br>$\lg L_P = 0.072[(\%CaO) + 0.3(\%MgO) + 0.6(\%P_2O_5) + 0.2(\%MnO) + 1.2(\%CaF_2) - 0.5(\%Al_2O_3)]$<br>$+ 2.5\lg(\%T.Fe) + 11,570/T - 10.52$<br>$\lg L_P = \frac{1}{T}[162(\%CaO) + 127.5(\%MgO) + 28.5(\%MnO)] + \frac{11,000}{T} - 6.28 \times 10^{-4}(\%SiO_2) - 10.4$<br>$+ 2.5\lg(\%Fe_tO)$ |
| $\dfrac{(\%P)}{[\%P]}$ | $\lg L_P = \frac{22,350}{T} - 16 + 0.08(\%CaO) + 2.5\lg(\%T.Fe)$<br>$\lg L_P = 0.065[(\%Cao) + 0.55(\%MgO)] + 2.5\lg(\%T.Fe) + 12,230/T - 10.8$<br>$\lg L_P = 0.071[(\%Cao) + 0.1(\%MgO)] + 2.5\lg(\%T.Fe) + 8260/T - 8.56$<br>$\lg L_P = 0.073[(\%CaO) + 0.148(\%MgO) + 0.96(\%P_2O_5) + 0.144(\%SiO_2) + 0.22(\%Al_2O_3)] + 11,570/T$<br>$-10.46 + 2.5\lg(\%T.Fe) \pm 0.1$ |

### 2.2. P$_2$O$_5$ Activity Coefficient

The activity coefficient of the slag component was one of the main factors affecting the thermodynamic properties of the slag. Phosphorus in the steel was oxidized and entered the steel slag in the converter, and the activity coefficient of P$_2$O$_5$ in the slag could significantly affect the dephosphorization effect of the steel slag. In this study, Equation (3) was cited, which assumed that phosphorus and oxygen in the steel liquid obeyed Henry's Law, and their activity coefficients were calculated as 1 [34]. Based on a large amount of experimental data, a linear relationship between the P$_2$O$_5$ activity coefficient ($\gamma_{P_2O_5}$) and the slag composition was regressed. Here, $x_i$ represents the mole fraction of oxide $i$ in the slag. By substituting the known slag composition in Table 1 into Equation (3), the P$_2$O$_5$ activity coefficient in the slag under the given conditions can be calculated.

$$\lg \gamma_{P_2O_5} = -22.44 x_{CaO} - 15.3 x_{MgO} - 13.26 x_{MnO} - 12.24 x_{Fe_tO} + 2.04 x_{SiO_2} - 42,000/T + 23.58 \quad (3)$$

### 2.3. Phosphorus Capacity

Phosphate capacity (hereinafter referred to as P capacity) reflected the potential ability of steel slag to remove phosphorus. Since Wagner proposed the concept of P capacity and phosphide capacity, researchers had conducted extensive studies on the relationship between P capacity and slag composition. Many studies suggested that there was a good correlation between the P capacity of steel slag and its optical basicity, so P capacity could be used as a representation of slag basicity. This study cited the relationship between P capacity and optical basicity of various complex slags summarized by Bergman, as shown in Equation (4) [35]. The optical basicity $\Lambda$ of the slag composed of various types of oxides was calculated using Equation (5), where $m_i$ is the number of oxygen atoms in the oxide $i$ and $\Lambda_i$ is the theoretical optical basicity of the oxide, as shown in Table 3 [36]. The $C_p$ referred to the ability of the slag to adsorb and combine with phosphorus at a given temperature. It was primarily determined by the properties of the slag, including its chemical composition, melting point, and structure. Therefore, at a fixed temperature, the $C_p$ in steel was solely dependent on the properties of the slag and was not influenced by other factors.

$$\lg C_P = 21.55 \Lambda + 32,912/T - 27.90 \quad (4)$$

$$\Lambda = \sum \frac{m_i x_i}{\sum m_i x_i} \Lambda_i \quad (5)$$

Table 3. Theoretical optical basicity of some oxides.

| Oxide | CaO | SiO$_2$ | FeO | MnO | P$_2$O$_5$ | MgO |
|---|---|---|---|---|---|---|
| Optical basicity | 1 | 0.47 | 0.72 | 0.95 | 0.38 | 0.92 |

## 2.4. Phase Precipitated Calculation

The high phosphorus distribution among the slag phases was a favorable driving condition for dephosphorization in the converter process. Through morphology and phase analysis of the slag, it was found that the slag mainly consisted of phosphorus-rich phase, iron-rich phase, and matrix phase. Phosphorus in the slag mainly existed in the phosphorus-rich phase and the matrix phase, and the enrichment of phosphorus in the phosphorus-rich phase had an important impact on the efficiency of dephosphorization in the converter process. Based on the composition of the slag during the converter process, this study used the Equilib module in the FactSage 7.2 thermodynamic software to calculate the phase precipitation process of the slag during the solidification process to investigate the influence of slag composition on phase precipitation in different slag systems. To ensure accurate calculations, we followed the recommended procedures provided by FactSage 7.2 for handling phosphorus-containing systems. This involved specifying the slag's elemental compositions, considering the interactions and phase stability of phosphorus-containing compounds, and utilizing the appropriate thermodynamic database within FactSage 7.2. This software has the advantages of having powerful calculation functions, rich database content, and a simple operating interface and has been widely used in thermodynamic simulation and calculation in the metallurgical field, achieving satisfactory results. The slag composition used for the theoretical calculation of phase precipitation is shown in Table 4.

**Table 4.** Components of converter slag for phase precipitation calculation.

| No. | CaO | SiO$_2$ | Fe$_t$O | MnO | P$_2$O$_5$ | MgO |
|---|---|---|---|---|---|---|
| 1 | 28.5 | 28.5 | 25 | 5 | 3 | 10 |
| 2 | 38 | 19 | 25 | 5 | 3 | 10 |
| 3 | 42.75 | 14.25 | 25 | 5 | 3 | 10 |
| 4 | 48 | 24 | 10 | 5 | 3 | 10 |
| 5 | 44.67 | 22.33 | 15 | 5 | 3 | 10 |
| 6 | 41.33 | 20.67 | 20 | 5 | 3 | 10 |

## 3. Results and Discussion

### 3.1. Effect of Composition and Temperature on Dephosphorization

#### 3.1.1. Basicity

The representation method for binary basicity in slag was R = (%CaO)/(%SiO$_2$), and it had an important impact on the ability of slag to remove phosphorus. By fixing the content of other components in the slag, the basicity range of the slag was set to 1–4, and the impact of slag basicity on various dephosphorization indicators was analyzed. Figure 1 shows the effect of slag basicity on the three dephosphorization indicators $\lg L_p$, $\lg \gamma_{P_2O_5}$, and $\lg C_p$. As the slag basicity increased, $\lg L_p$ and $\lg C_p$ increased, while $\lg \gamma_{P_2O_5}$ decreased. CaO was a reactant in the dephosphorization reaction. Generally, when the slag basicity increased, the CaO content in the slag also increased. The increase in CaO content not only increased the concentration of the reactant in the dephosphorization reaction but also reduced $\gamma_{P_2O_5}$. Both effects increased the phosphorus distribution ratio, which was beneficial for dephosphorization. At the same time, as shown in Figure 1, as the basicity increased, the growth rate of $\lg L_p$ and $\lg C_p$ slowed down, and the decreasing trend of $\lg \gamma_{P_2O_5}$ also slowed down. Therefore, although increasing the basicity of the converter slag was beneficial for dephosphorization, it inevitably increased lime consumption. If the basicity was too high, the effect of dephosphorization would not be significantly increased, and it could also lead to lime waste. Overall, when the slag basicity was less than 3, $\lg L_p$ and $\lg C_p$ increased rapidly, while $\lg \gamma_{P_2O_5}$ decreased rapidly as the slag basicity increased. When the slag basicity exceeded 3, the growth rate of $\lg L_p$ and $\lg C_p$ slowed down, and the decreasing rate of $\lg \gamma_{P_2O_5}$ also slowed down as the basicity increased. During the converter dephosphorization stage, if a large amount of lime was added to maintain a high slag basicity, a large amount of un-melted lime accumulated in the furnace, which not only worsened the dephosphorization kinetic conditions but also did not contribute to dephosphorization due to the short time spent in the dephosphorization stage, and the smelting cost thus increased. Therefore, considering the comprehensive dephosphorization effect and steelmaking cost, the basicity of the converter slag during the dephosphorization stage should not exceed 3, and the effects of other factors on dephosphorization in this article were also calculated based on a basicity of no more than 3.

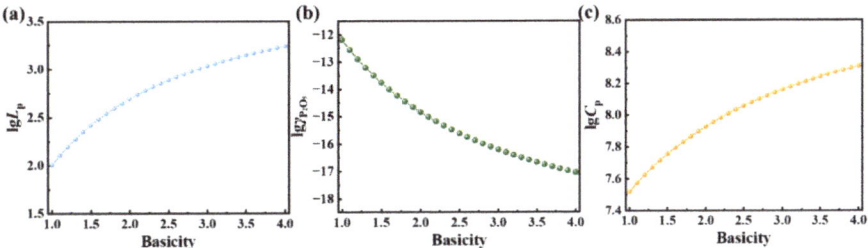

**Figure 1.** Effect of basicity on (**a**) $\lg L_P$, (**b**) $\lg \gamma_{P_2O_5}$, and (**c**) $\lg C_P$ in slag.

### 3.1.2. $Fe_tO$ Content

The principle of dephosphorization in the converter was oxidation. The $Fe_tO$ content in the slag had an important influence on dephosphorization and was a significant indicator of the oxidation capacity of the slag. The main method used to study the effect of $Fe_tO$ content in the slag during the dephosphorization stage in the converter was to vary the $Fe_tO$ content in the slag between 10 and 30% while keeping the other components in the slag constant. At the same time, the basicity of the slag was fixed at 1, 2, and 3, and the effect of $Fe_tO$ content on dephosphorization in the slag at different basicity levels was analyzed. Figure 2 shows the effect of $Fe_tO$ content in the slag on three dephosphorization indicators, $\lg L_P$, $\lg \gamma_{P_2O_5}$, and $\lg C_P$. It could be seen that as the $Fe_tO$ content in the slag increased, $\lg \gamma_{P_2O_5}$ tended to increase linearly, and the higher the basicity, the more obvious the increase. $\lg C_P$ tended to decrease linearly, and the higher the basicity, the more obvious the decrease. Overall, when the $Fe_tO$ content in the slag was around 20%, $\lg L_P$ was the largest, and both too high and too low $Fe_tO$ contents are not conducive to dephosphorization. $Fe_tO$ was a reactant in the dephosphorization reaction, and increasing its content could increase the oxygen potential in the slag, which was beneficial to dephosphorization. However, increasing the $Fe_tO$ content in the slag would dilute the concentration of alkaline oxides in the slag, especially the concentration of CaO, and the higher the basicity, the more obvious the dilution effect, ultimately resulting in a decrease in $\lg C_P$ and an increase in $\lg \gamma_{P_2O_5}$.

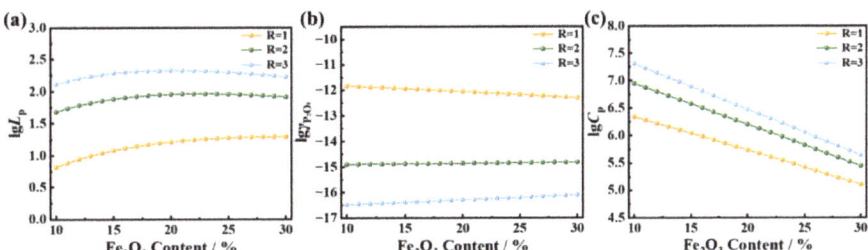

**Figure 2.** Effect of $Fe_tO$ content on (**a**) $\lg L_P$, (**b**) $\lg \gamma_{P_2O_5}$, and (**c**) $\lg C_P$ in slag.

### 3.1.3. MgO Content

MgO was the main component for improving the refractoriness of slag. The main purpose of adding MgO to converter slag was to enhance the effect of slag splashing protection and maintain the condition of the converter. However, MgO in the slag also had certain ability to fix phosphorus. As MgO could combine with $P_2O_5$ to form phosphate, it had some impact on the dephosphorization ability of the slag. In order to analyze the influence of the MgO content in the slag on the dephosphorization effect, the same method as above was adopted. The MgO content in the slag was varied between 1% and 11%, while the basicity and other component contents were kept constant. At the same time, in order to study the effect of MgO content on dephosphorization under different basicity levels, the basicity values were fixed at 1, 2, and 3. Figure 3 showed the influence of MgO content in the slag on three dephosphorization indicators, $\lg L_p$, $\lg \gamma_{P_2O_5}$, and $\lg C_p$. It could be seen that as the MgO content in the slag increased, $\lg L_p$ and $\lg \gamma_{P_2O_5}$ decreased linearly, while $\lg C_p$ increased linearly, and the change became more obvious with higher basicity. MgO itself was a weak alkaline substance, and its ability to fix phosphorus was not as good as CaO. CaO was the most important factor influencing $\lg L_p$, $\lg \gamma_{P_2O_5}$, and $\lg C_p$. The introduction of higher MgO content in the slag led to a consequent dilution of the concentration of CaO. As a result, the ability of CaO to effectively capture phosphorus decreased, leading to a reduction in $\lg L_p$ and $\lg \gamma_{P_2O_5}$. However, owing to its inherent basicity, the increase in MgO content caused $\lg C_p$ to increase. Adding a certain amount of MgO to the slag is necessary for dephosphorization and furnace lining protection, as it decreases the activity coefficient of $P_2O_5$, thereby enhancing the dephosphorization process.

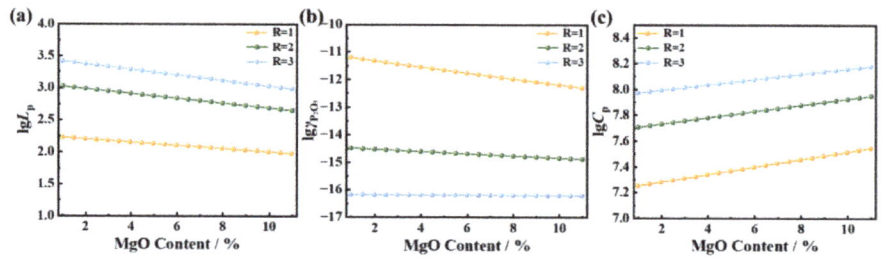

**Figure 3.** Effect of MgO content on (**a**) $\lg L_p$, (**b**) $\lg \gamma_{P_2O_5}$, and (**c**) $\lg C_p$ in slag.

### 3.1.4. MnO Content

The presence of MnO had a certain impact on the dephosphorization in the converter. In order to study its effect on dephosphorization, the same method as above was adopted to fix the content of other components in the slag and vary the MnO content in the slag within the range of 1–11%. At the same time, in order to study the effect of MnO content on dephosphorization under different levels of basicity, the basicity was fixed at 1, 2, and 3. Figure 4 showed the effect of MnO content in the slag on the three dephosphorization indicators $\lg L_p$, $\lg \gamma_{P_2O_5}$, and $\lg C_p$. It could be seen that as the MnO content in the slag increased, $\lg L_p$ and $\lg \gamma_{P_2O_5}$ decreased linearly, while $\lg C_p$ tended to increase linearly. The variation trend was similar to that of MgO. Because MnO was a weak alkaline substance, its ability to fix phosphorus was inferior to that of CaO. CaO was the most important influencing factor for $\lg L_p$, $\lg \gamma_{P_2O_5}$, and $\lg C_p$. The increase in the MnO content also diluted the concentration of CaO in the slag, resulting in a decrease in the phosphorus-fixing ability of CaO, leading to a decrease in $\lg L_p$ and $\lg \gamma_{P_2O_5}$ and an increase in $\lg C_p$. Therefore, the increase in the MnO content would reduce the dephosphorization ability of the slag, but its basicity would increase the phosphorus capacity of the slag. There would inevitably be a certain amount of elemental manganese in the molten iron, and the MnO content in the slag was mainly determined by the conditions of the molten iron.

### 3.1.5. Temperature

The dephosphorization reaction was a strongly exothermic reaction, so temperature had an important influence on dephosphorization. In order to study the effect of temperature on dephosphorization, the composition of the slag was kept constant, and the temperature was varied between 1200 °C and 1600 °C. In addition, in order to study the effect of temperature on dephosphorization under different levels of basicity, the basicity was fixed at 1, 2, and 3. Figure 5 shows the effect of temperature on the three dephosphorization indicators $\lg L_p$, $\lg \gamma_{P_2O_5}$, and $\lg C_p$. As can be seen from Figure 5, as the temperature increased, $\lg L_p$ and $\lg C_p$ decreased significantly, while $\lg \gamma_{P_2O_5}$ increased sharply. Therefore, from a thermodynamic point of view, the reaction temperature during the dephosphorization stage of the converter should be reduced as much as possible. However, one of the main functions of the converter was to raise the temperature. If the temperature control during the dephosphorization stage was too low, the burden of raising the temperature during the decarbonization stage would be too heavy, which could not meet the requirements of steelmaking. Therefore, the temperature during the dephosphorization stage should be reduced as much as possible under the premise of meeting the requirements of steelmaking.

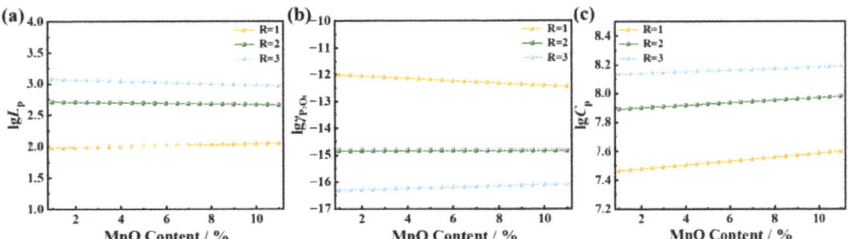

**Figure 4.** Effect of MnO content on (**a**) $\lg L_p$, (**b**) $\lg \gamma_{P_2O_5}$, and (**c**) $\lg C_p$ in slag.

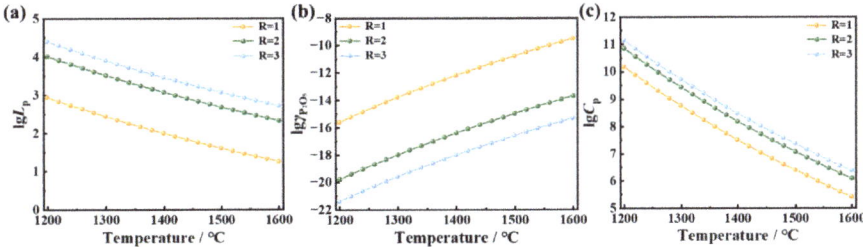

**Figure 5.** Effect of temperature on (**a**) $\lg L_p$, (**b**) $\lg \gamma_{P_2O_5}$, and (**c**) $\lg C_p$ in slag.

*3.2. Effect of Steel Slag Composition on Precipitated Phase*

3.2.1. Basicity

It was found from the above research that the most significant factors affecting the ability of dephosphorization from converter steel slag were basicity, $Fe_tO$ content, and temperature. Therefore, FactSage 7.2 software was used to further analyze the effects of basicity and $Fe_tO$ content on the phosphorus forms in the slag and explore the precipitation behavior of slag phases. Figure 6 shows the phase precipitation process of slag samples No. 1, No. 2, and No. 3 in Table 4 during the solidification process from 1700 °C to 700 °C. Figure 6a shows the phase precipitation of converter slag with a basicity of 1. The spinel phase began to precipitate at around 1350 °C and peaked at 1250 °C. At this time, the cordierite phase began to precipitate, which gradually decreased after peaking at 1180 °C. This was mainly due to $Fe_tO$ entering the cordierite phase and forming $Ca_3Fe_2Si_3O_{12}$, the precipitation amount of which continued to increase with decreasing temperature until it reached the maximum value at 950 °C. The phosphorus-containing $Ca_3P_2O_8$ phase began to appear at 1090 °C, quickly reached the maximum value, and remained stable. Figure 6b shows the phase precipitation of converter slag with a basicity of 2. The $Ca_2Fe_2O_5$ phase began to precipitate at 1360 °C, and the $Ca_7MgSi_4O_{16}$ phase began to precipitate at 1270 °C, with its precipitation amount gradually increasing as the temperature decreased. When the temperature dropped to 1160 °C, the $Ca_7P_2Si_2O_{16}$ phase precipitated, and in the continued cooling process, this phase gradually decomposed to form a new phase, $Ca_5P_2SiO_{12}$, which remained stable. Figure 6c showed the phase precipitation of converter slag with basicity of 3, which was basically the same as that of basicity 2. However, due to the lower $SiO_2$ content in the slag, the phosphorus-containing phase that finally formed was $Ca_4P_2O_9$.

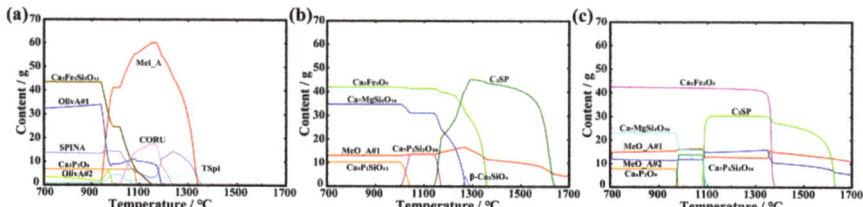

**Figure 6.** Calculation of phase precipitation in slag system: (**a**) No. 1, (**b**) No. 2, (**c**) No. 3.

3.2.2. $Fe_tO$ Content

Figure 7 shows the phase precipitation process of slag samples No. 4, No. 5, and No. 6 in Table 4 during the solidification process from 1700 °C to 700 °C. Figure 7a presents the case of converter slag with 10% $Fe_tO$ content. The $\alpha$-$Ca_2SiO_4$ phase began to precipitate at 1640 °C, and then underwent a crystal phase transition at 1400 °C, where $\alpha$-$Ca_2SiO_4$ disappeared and $\beta$-$Ca_2SiO_4$ began to precipitate. At the same time, the $Ca_7MgSi_4O_{16}$ phase started to precipitate in the slag at 1340 °C, and the precipitation amount continued to increase as the temperature decreased, reaching a maximum at 1100 °C. $Ca_2Fe_2O_5$ phase began to precipitate at 1260 °C. When the temperature dropped to 1190 °C, $Ca_7P_2Si_2O_{16}$ containing phosphorus began to precipitate. During the continued cooling process, this phase gradually decomposed and formed a new phase, $Ca_5P_2SiO_{12}$, which remained stable. Figure 7b presents the case of converter slag with 15% $Fe_tO$ content. The $\alpha$-$Ca_2SiO_4$ phase began to precipitate at 1520 °C and then underwent a crystal phase transition at 1390 °C, where $\alpha$-$Ca_2SiO_4$ disappeared and $\beta$-$Ca_2SiO_4$ began to precipitate, but the precipitation amount was less than that of slag No. 4. Until 1350 °C, the precipitated $\beta$-$Ca_2SiO_4$ gradually reacted with the slag, causing the precipitation amount of $Ca_7MgSi_4O_{16}$ phase to gradually increase. The transformation of the phosphorus-containing phase was basically the same as that of slag No. 4, and eventually became the $Ca_5P_2SiO_{12}$ phase. Figure 7c presents the case of converter slag with 20% $Fe_tO$ content. In this slag, the precipitation amount of $\alpha$-$Ca_2SiO_4$ was small, and no crystal phase transition occurred. The $Ca_3MgSi_2O_8$ phase began to precipitate at 1400 °C, and reached a maximum at 1350 °C, before quickly transforming into the $Ca_7MgSi_4O_{16}$ phase. The phosphorus-containing phase that finally stabilized in the slag was also $Ca_5P_2SiO_{12}$.

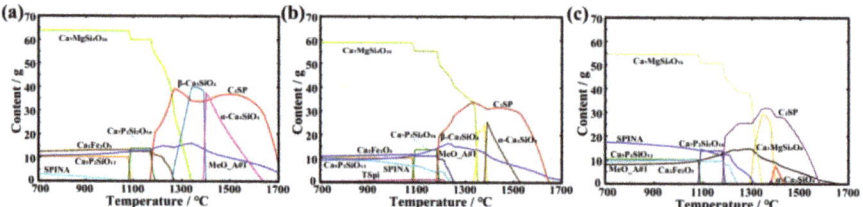

**Figure 7.** Calculation of phase precipitation in slag system: (**a**) No. 4, (**b**) No. 5, (**c**) No. 6.

*3.3. Industrial Experiment Verification*

In pursuit of faster production rhythms, the converter smelting process was usually operated continuously, as shown in Figure 8. Scrap steel was loaded as the starting point for smelting, followed by molten iron. Then, rapid blowing was carried out using a porous oxygen lance, and slag-making auxiliary materials were added. The oxygen supply time varied depending on the size of the converter but was generally around 20 min. The blowing lance was generally operated in a low–medium–high–low mode and adjusted according to the slag formation inside the furnace during production. Then, the steel was tapped into a ladle for further treatment. After pouring the steel, nitrogen was used for slag splashing to protect the furnace, and finally, the slag was tapped out. The most pressing issue in actual production is how to remove phosphorus from the molten iron and fix it in the slag within a short amount of time. Based on the previous theoretical calculations, it is necessary to ensure a relatively high basicity of the slag during the phosphorus removal stage in the converter while maintaining sufficient fluidity. Additionally, the content of $Fe_tO$, MgO, and MnO in the steel slag also has a certain influence on phosphorus removal. Therefore, in the smelting process using the single-slag method, controlling the basicity of the converter slag at around 2.5 and the $Fe_tO$ content at approximately 20% achieves the optimal phosphorus removal effect. The composition of the slag at this stage is shown in Table 5, and micro-area morphology observation and XRD phase analysis were performed. Each slag sample is quenched in water at approximately 1000 °C to ensure consistency in the cooling process. The results presented are the average values obtained from these three samples, aiming to enhance the reliability of the findings despite the constraints in obtaining a larger quantity of samples in an industrial setting.

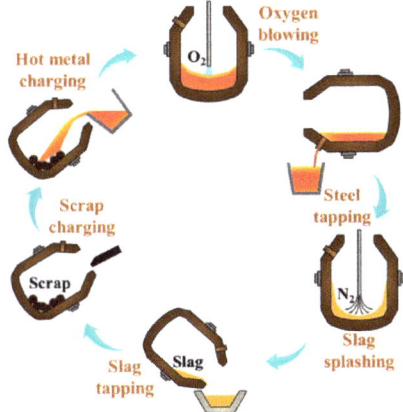

**Figure 8.** Production process of converter steelmaking process.

**Table 5.** Components of converter slag sampled on site.

| Composition | CaO | $SiO_2$ | $Fe_tO$ | MnO | $P_2O_5$ | MgO |
| --- | --- | --- | --- | --- | --- | --- |
| Content | 39.1 | 15.5 | 21.2 | 6.8 | 3.3 | 7.7 |

Figure 9a,b are backscattered electron micrographs of the dephosphorization slag which clearly showed the presence of three distinct phases: the phosphorus-rich phase, the iron-rich phase, and the matrix phase. Typically, the phosphorus-rich phase in the converter slag predominantly exists as the $Ca_5P_2SiO_{12}$ phase, and phosphorus mainly resides in this phase, which appears as a dark gray color in the micrograph. The iron-rich phase in the slag typically appears as white in scanning electron microscopy and is mainly composed of iron oxide or iron–manganese oxide. The matrix phase appears as gray and contains elements that had not fully entered either the phosphorus-rich or iron-rich phases. Figure 9c presents the XRD pattern of the experimental slag, which mainly consists of five phases: $Ca_5P_2SiO_{12}$, $Ca_2Fe_2O_5$, FeS, $(MnO)_{0.165}(CaO)_{0.835}$, and FeO. These phases are in good agreement with the calculated results of the precipitates in Figure 7c, with the final phosphorus-containing phase being $Ca_5P_2SiO_{12}$.

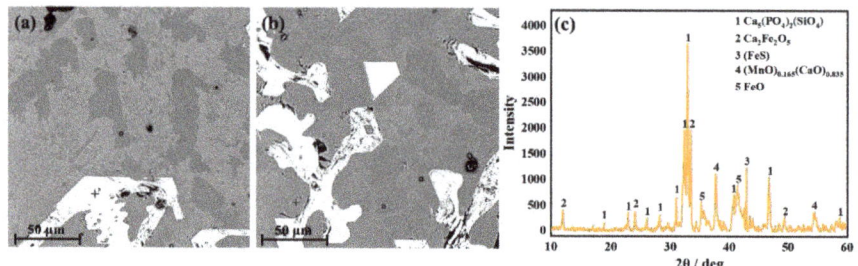

**Figure 9.** (**a**,**b**) Microscopic morphology and (**c**) XRD pattern of slag in converter dephosphorization stage.

In order to investigate the phosphorus distribution ratio in practical production scenarios, we conducted compositional measurements on 15 sets of steel slag samples and their corresponding steel liquid compositions in Table 6. These samples were directly obtained from the production site, ensuring their relevance to real-world conditions. By analyzing the collected data and building upon the existing Equation (2), we were able to summarize and derive a new Equation (6). The computations were performed using the SPSS 25.0 software (Statistical Package for the Social Sciences), a widely used tool for statistical analysis. The $p$-values for each coefficient were calculated using Spearman's correlation analysis and are presented in Table 7. All coefficients demonstrated statistical significance with $p$-values less than 0.05, passing the significance test. Additionally, the equation achieved a high $R^2$ value of 0.905, indicating a strong fit and excellent predictive performance. This revised equation provides an accurate prediction of the phosphorus distribution ratio between the slag and steel liquid during the dephosphorization process. Our equation is primarily developed for a specific research object, namely a 100-ton converter in a particular steel plant, and therefore, it does not possess universal applicability. It is important to note that different equipment conditions and operational processes have a significant impact on phosphorus partitioning ratios, as these parameters influence the extent of the phosphorus removal reactions. Furthermore, it should be emphasized that the derived equation does not represent the phosphorus partitioning at long-term equilibrium. In our specific study, the overall duration of the converter process is approximately 30 min, with a phosphorus removal reaction time of less than 10 min for the molten steel. Consequently, the equation we have derived reflects the non-equilibrium conditions prevailing during the relatively short time span of the converter process. The equation has been validated extensively within a suitable range of basicity, which spans from 1.3 to 2.4, and a temperature range of 1300 to 1450 °C. The accuracy of the equation has been thoroughly verified within these specified ranges. Therefore, we can ensure the reliability of the equation within these validated conditions. Figure 10 presents the comparison between the measured phosphorus partitioning ratios and the predictions obtained using different equations. It can be observed that the predictions obtained using Equation (6) exhibit a small deviation from the measured values, indicating a good agreement. However, the predictions obtained from the other three equations deviate significantly from the measured values. These discrepancies can be attributed to the non-equilibrium state during the phosphorus removal process in the converter and the incomplete phosphorus removal due to the insufficient effectiveness of lime in the slag.

$$\lg L_P = \lg \frac{(\%P)}{[\%P]} = 0.072[0.8(\%CaO) + 0.35(\%MgO) + 0.7(\%P_2O_5) + 0.8(\%MnO)] + 11,570/T - 10.620 + 2.5\lg(\%T.Fe) \tag{6}$$

**Table 6.** Common empirical formulas for $L_p$.

| No. | Main Components of Dephosphorization Slag/% | | | | | | [P] Content in Molten Steel/% | Temperature/°C |
|---|---|---|---|---|---|---|---|---|
| | CaO | SiO$_2$ | Fe$_t$O | MnO | P$_2$O$_5$ | MgO | | |
| 1 | 32.6 | 18.1 | 25.3 | 5.4 | 3.7 | 7.7 | 0.137 | 1332 |
| 2 | 31.8 | 19.5 | 27.7 | 5.7 | 3.8 | 5 | 0.143 | 1400 |
| 3 | 37.2 | 21.2 | 18.9 | 6.1 | 4.3 | 5.3 | 0.144 | 1375 |
| 4 | 35.8 | 18.8 | 22.5 | 3.8 | 4.9 | 7.1 | 0.139 | 1386 |
| 5 | 34.8 | 20.7 | 24.4 | 6.1 | 4.2 | 3.9 | 0.138 | 1367 |
| 6 | 32.8 | 17 | 29.4 | 3.9 | 3.6 | 6.8 | 0.134 | 1393 |
| 7 | 36.8 | 17.2 | 23.7 | 4.2 | 4.8 | 6.5 | 0.142 | 1377 |
| 8 | 37.9 | 19.4 | 25.3 | 2.9 | 4.6 | 4.1 | 0.143 | 1368 |
| 9 | 34.2 | 19 | 26.9 | 3.4 | 4 | 5.8 | 0.15 | 1378 |
| 10 | 42 | 18.1 | 18.2 | 4.1 | 4.4 | 5.3 | 0.143 | 1354 |
| 11 | 33.6 | 15.8 | 26 | 5.3 | 4.3 | 6.9 | 0.142 | 1380 |
| 12 | 38.2 | 20.3 | 16 | 7.3 | 4.3 | 7.4 | 0.138 | 1327 |
| 13 | 42.3 | 22.3 | 14.6 | 4.9 | 2.8 | 6.3 | 0.143 | 1352 |
| 14 | 32.2 | 23.6 | 23.3 | 4.4 | 3.4 | 4.6 | 0.145 | 1368 |
| 15 | 31.3 | 15.7 | 28.5 | 5.6 | 3.6 | 6.7 | 0.149 | 1383 |

**Table 7.** Spearman correlation analysis results for coefficients.

| Coefficient | (%CaO) | (%MgO) | (%P$_2$O$_5$) | (%MnO) | 1/T | lg(%T.Fe) |
|---|---|---|---|---|---|---|
| $p$-value | 0.000 | 0.011 | 0.000 | 0.003 | 0.006 | 0.000 |

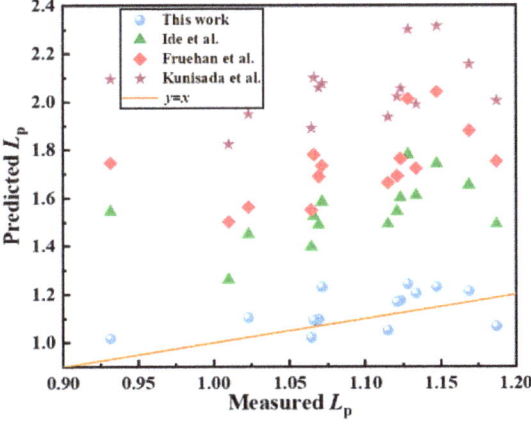

**Figure 10.** Measured and predicted phosphorus partitioning ratios using different equations [27,32,33].

## 4. Conclusions

This study investigated the behavior of steel slag during the dephosphorization stage in the converter process using thermodynamic calculations and sampling analysis. The optimal thermodynamic conditions for this stage were obtained and provide a certain reference for practical production. The following conclusions were drawn:

1. The results of thermodynamic formula calculations showed that increasing the basicity of the dephosphorization slag was beneficial for dephosphorization, but it should be kept below 3. When the Fe$_t$O content was around 20%, the dephosphorization effect was the best, and a too high or too low Fe$_t$O content was not conducive to dephosphorization. The contents of MgO, MnO, and other components had little effect on the dephosphorization effect. The reaction temperature should be kept relatively low during the dephosphorization stage, and the dephosphorization efficiency sharply decreased with the increase in temperature.

2. During the solidification process of the dephosphorization slag, phosphorus existed in the form of $Ca_3(PO_4)_2$. When the slag basicity and $Fe_tO$ content changed, $Ca_3(PO_4)_2$ only changed its precipitation form, but the precipitation amount remained unchanged. $Ca_3(PO_4)_2$ in the slag usually formed a solid solution with $Ca_2SiO_4$, so the form of phosphorus in the slag was mainly determined by the precipitation form and content of $Ca_2SiO_4$.
3. The phases in the dephosphorization slag were mainly composed of phosphorus-rich phases, iron-rich phases, and the matrix phase. This was consistent with the results of scanning electron microscopy observation and XRD analysis. The developed prediction model accurately estimates the phosphorus distribution ratio between the slag and steel liquid.

**Author Contributions:** Conceptualization, Z.-L.W. and T.-L.S.; methodology, L.-H.Z.; software, T.-L.S.; validation, Z.-L.W., Y.-P.B. and L.-H.Z.; funding acquisition, Y.-P.B. All authors have read and agreed to the published version of the manuscript.

**Funding:** This research was funded by the National Natural Science Foundation of China, grant number 51574019.

**Data Availability Statement:** Not applicable.

**Conflicts of Interest:** The authors declare no conflict of interest.

# References

1. Hao, X.; An, H. Comparative study on transmission mechanism of supply shortage risk in the international trade of iron ore, pig iron and crude steel. *Resour. Policy* **2022**, *79*, 103022. [CrossRef]
2. Ma, S.; Wen, Z.; Chen, J.; Wen, Z. Mode of circular economy in China's iron and steel industry: A case study in Wu'an city. *J. Clean. Prod.* **2014**, *64*, 505–512. [CrossRef]
3. Zhang, Q.; Sun, Y.; Han, Y.; Li, Y.; Gao, P. Review on coal-based reduction and magnetic separation for refractory iron-bearing resources. *Int. J. Miner. Metall. Mater.* **2022**, *29*, 2087–2105. [CrossRef]
4. Wang, Z.; Bao, Y.; Wang, D.; Wang, M. Effective removal of phosphorus from high phosphorus steel slag using carbonized rice husk. *J. Environ. Sci.* **2023**, *124*, 156–164. [CrossRef] [PubMed]
5. Lawrence, K.; Nehring, M. Market structure differences impacting Australian iron ore and metallurgical coal industries. *Minerals* **2015**, *5*, 473–487. [CrossRef]
6. Sun, Y.; Han, Y.; Gao, P.; Wang, Z.; Ren, D. Recovery of iron from high phosphorus oolitic iron ore using coal-based reduction followed by magnetic separation. *Int. J. Miner. Metall. Mater.* **2013**, *20*, 411–419. [CrossRef]
7. Liu, X.; Feng, H.; Chen, L.; Qin, X.; Sun, F. Hot metal yield optimization of a blast furnace based on constructal theory. *Energy* **2016**, *104*, 33–41. [CrossRef]
8. Zhu, D.; Yang, C.; Pan, J.; Lu, L.; Guo, Z.; Liu, X. An integrated approach for production of stainless steel master alloy from a low grade chromite concentrate. *Powder Technol.* **2018**, *335*, 103–113. [CrossRef]
9. Chen, Y.; Zuo, H. Gasification behavior of phosphorus during pre-reduction sintering of medium-high phosphorus iron ore. *ISIJ Int.* **2021**, *61*, 1459–1468. [CrossRef]
10. Kubo, H.; Matsubae-Yokoyama, K.; Nagasaka, T. Magnetic separation of phosphorus enriched phase from multiphase dephosphorization slag. *ISIJ Int.* **2010**, *50*, 59–64. [CrossRef]
11. Zhang, X.; Han, Y.; Sun, Y.; Li, Y. Innovative utilization of refractory iron ore via suspension magnetization roasting: A pilot-scale study. *Powder Technol.* **2019**, *352*, 16–24. [CrossRef]
12. Lughofer, E.; Pollak, R.; Feilmayr, C.; Schatzl, M.; Saminger-Platz, S. Prediction and explanation models for hot metal temperature, silicon concentration, and cooling capacity in ironmaking blast furnaces. *Steel Res. Int.* **2021**, *92*, 2100078. [CrossRef]
13. Jeong, Y.; Matsubae-Yokoyama, K.; Kubo, H.; Pak, J.; Nagasaka, T. Substance flow analysis of phosphorus and manganese correlated with South Korean steel industry. *Resour. Conserv. Recycl.* **2009**, *53*, 479–489. [CrossRef]
14. Dippenaar, R. Industrial uses of slag (the use and re-use of iron and steelmaking slags). *Ironmak. Steelmak.* **2005**, *32*, 35–46. [CrossRef]
15. Wang, L.; Guo, P.; Kong, L.; Zhao, P. Industrial application prospects and key issues of the pure-hydrogen reduction process. *Int. J. Miner. Metall. Mater.* **2022**, *29*, 1922–1931. [CrossRef]
16. Liu, Z.; Kobayashi, Y.; Yin, F.; Kuwabara, M.; Nagai, K. Nucleation of acicular ferrite on sulfide inclusion during rapid solidification of low carbon steel. *ISIJ Int.* **2007**, *47*, 1781–1788. [CrossRef]
17. Gibbs, P.J.; De Moor, E.; Merwin, M.J.; Clausen, B.; Speer, J.G.; Matlock, D.K. Austenite stability effects on tensile behavior of manganese-enriched-austenite transformation-induced plasticity steel. *Metall. Mater. Trans. A* **2011**, *42*, 3691–3702. [CrossRef]

18. Zurutuza, I.; Isasti, N.; Detemple, E.; Schwinn, V.; Mohrbacher, H.; Uranga, P. Effect of Nb and Mo additions in the microstructure/tensile property relationship in high strength quenched and quenched and tempered boron steels. *Metals* **2020**, *11*, 29. [CrossRef]
19. Yao, X.; Huang, J.; Qiao, Y.; Sun, M.; Wang, B.; Xu, B. Precipitation behavior of carbides and its effect on the microstructure and mechanical properties of 15CrNi$_3$MoV Steel. *Metals* **2022**, *12*, 1758. [CrossRef]
20. Tian, Z.; Li, B.; Zhang, X.; Jiang, Z. Double slag operation dephosphorization in BOF for producing low phosphorus steel. *J. Iron Steel Res. Int.* **2009**, *16*, 6–14. [CrossRef]
21. Lv, M.; Zhu, R.; Yang, L. High efficiency dephosphorization by mixed injection during steelmaking process. *Steel Res. Int.* **2019**, *90*, 1800454. [CrossRef]
22. Lin, L.; Zeng, J. Consideration of green intelligent steel processes and narrow window stability control technology on steel quality. *Int. J. Miner. Metall. Mater.* **2021**, *28*, 1264–1273. [CrossRef]
23. Qian, Q.; Dong, Q.; Xu, J.; Zhao, W.; Li, M. A metallurgical dynamics-based method for production state characterization and end-point time prediction of basic oxygen furnace steelmaking. *Metals* **2022**, *13*, 2. [CrossRef]
24. Li, H.; Li, X.; Liu, X.; Bu, X.; Li, H.; Lyu, Q. Industrial internet platforms: Applications in BF ironmaking. *Ironmak. Steelmak.* **2022**, *49*, 905–916. [CrossRef]
25. Huss, J.; Berg, M.; Kojola, N. Experimental study on phosphorus partitions between liquid iron and liquid slags based on DRI. *Metall. Mater. Trans. B* **2020**, *51*, 786–794. [CrossRef]
26. Wang, Z.; Bao, Y.; Wang, D.; Gu, C.; Wang, M. Study on the effect of different factors on the change of the phosphorus-rich phase in high phosphorus steel slag. *Crystals* **2022**, *12*, 1030. [CrossRef]
27. Ide, K.; Fruehan, R.J. Evaluation of phosphorus reaction equilibrium in steelmaking. *Iron Steelmak.* **2000**, *27*, 65–70.
28. Balajiva, K.; Quarrell, A.G.; Vajragupta, P. A laboratory investigation of the phosphorus reaction in the basic steeling porcess. *J. Iron Steel Inst.* **1946**, *153*, 115–150.
29. Suito, H.; Inoue, R. Thermodynamic assessment of hot metal and steel dephosphorization with MnO-containing BOF slags. *ISIJ Int.* **1995**, *35*, 258–265. [CrossRef]
30. Turkdogan, E.T. Assessment of P$_2$O$_5$ activity coefficients in molten slags. *ISIJ Int.* **2000**, *40*, 964–970. [CrossRef]
31. Sen, N. Studies on dephosphorisation of steel in induction furnace. *Steel Res. Int.* **2006**, *77*, 242–249. [CrossRef]
32. Assis, A.N.; Tayeb, M.A.; Sridhar, S.; Fruehan, R.J. Phosphorus equilibrium between liquid iron and CaO-SiO$_2$-MgO-Al$_2$O$_3$-FeO-P$_2$O$_5$ slags: EAF slags, the effect of alumina and new correlation. *Metals* **2019**, *9*, 116. [CrossRef]
33. Kunisada, K.; Iwai, H. Effect of Na$_2$O on phosphorus distribution between liquid iron and CaO-based slags. *ISIJ Int.* **2007**, *27*, 263–269. [CrossRef]
34. Basu, S.; Lahiri, A.K.; Seetharaman, S. A model for activity coefficient of P2O5 in BOF slag and phosphorus distribution between liquid steel and slag. *ISIJ Int.* **2007**, *47*, 1236–1238. [CrossRef]
35. Bergman, Å. Representation of phosphorus and vanadium equilibria between liquid iron and complex steelmaking type slags. *ISIJ Int.* **1988**, *28*, 945–951. [CrossRef]
36. Ray, H.S.; Pal, S. Simple method for theoretical estimation of viscosity of oxide melts using optical basicity. *Ironmak. Steelmak.* **2004**, *31*, 125–130. [CrossRef]

**Disclaimer/Publisher's Note:** The statements, opinions and data contained in all publications are solely those of the individual author(s) and contributor(s) and not of MDPI and/or the editor(s). MDPI and/or the editor(s) disclaim responsibility for any injury to people or property resulting from any ideas, methods, instructions or products referred to in the content.

Article

# In-Situ Study of Temperature- and Magnetic-Field-Induced Incomplete Martensitic Transformation in Fe-Mn-Ga

Xiaoming Sun [1,2,*], Jingyi Cui [1,2], Shaofu Li [1,2], Zhiyuan Ma [3,4], Klaus-Dieter Liss [5,6], Runguang Li [7] and Zhen Chen [8,*]

1. State Key Laboratory of Multiphase Complex Systems, Institute of Process Engineering, Chinese Academy of Sciences, Beijing 100190, China
2. School of Chemical Engineering, University of Chinese Academy of Sciences, Beijing 100049, China
3. Department of Materials Science and Engineering, China University of Petroleum-Beijing, Beijing 102249, China
4. X-ray Science Division, Argonne National Laboratory, Argonne, IL 60439, USA
5. School of Mechanical, Materials, Mechatronic and Biomedical Engineering, University of Wollongong, Wollongong, NSW 2522, Australia
6. Australian Nuclear Science and Technology Organisation, Lucas Heights, NSW 2234, Australia
7. Department of Civil and Mechanical Engineering, Technical University of Denmark, DK-2800 Kgs. Lyngby, Denmark
8. Xi'an Rare Metal Materials Institute Co., Ltd., Xi'an 710000, China
* Correspondence: xmsun19@ipe.ac.cn (X.S.); cz0521shu@126.com (Z.C.)

**Abstract:** Significant interest in the stoichiometric and off-stoichiometric $Fe_2MnGa$ alloys is based on their complex phase transition behavior and potential application. In this study, temperature- and magnetic-field-induced phase transformations in the $Fe_{41.5}Mn_{28}Ga_{30.5}$ magnetic shape memory alloy were investigated by in situ synchrotron high-energy X-ray diffraction and in situ neutron diffraction techniques. It was found that incomplete phase transformation and phase coexistence behavior are always observed while applying and removing fields in $Fe_{41.5}Mn_{28}Ga_{30.5}$. Typically, even at 4 K and under 0 T, or increasing the magnetic field to 11 T at 250 K, it can be directly detected that the martensite and austenite are in competition, making the phase transition incomplete. TEM observations at 300 K and 150 K indicate that the anti-phase boundaries and B2 precipitates may lead to field-induced incomplete phase transformation behavior collectively. The present study may enrich the understanding of field-induced martensitic transformation in the Fe-Mn-Ga magnetic shape memory alloys.

**Keywords:** magnetic shape memory alloy; martensitic transformation; Fe-Mn-Ga; incomplete phase transformation

## 1. Introduction

Ferromagnetic shape memory alloys (FSMAs) have attracted considerable attention due to their multifunctional properties, such as magnetic-field-induced strain [1–4], magnetoresistance [5–7], magnetocaloric effect [8–10], magnetothermal conductivity [11], and exchange bias behavior [12,13]. These properties make them promising materials for applications such as rapid actuators [1], efficient magnetic refrigerators [14], and recording materials [15]. In general, most FSMAs have interesting structural and magnetic properties resulting from their first-order phase transformation induced by external fields [1–13]. Typically, for the Ni-(Co)-Mn-X (X = In, Sn, Sb) alloys, magnetic fields may induce a structural transformation from weak magnetic martensite to ferromagnetic austenite at temperatures close to the austenitic transformation start temperature, leading to some pronounced magnetic-field-induced effects [2–7,16]. Another representative FSMA, i.e., Fe-Mn-Ga alloys displaying distinct and complex magnetic-field-induced martensitic transformation behavior in some out-stoichiometric compositions, are not yet completely understood.

Since Omori et al. reported that the $Fe_{43}Mn_{28}Ga_{29}$ alloy can go through martensitic transition from the paramagnetic (PM) cubic austenite phase to the ferromagnetic (FM) tetragonal martensite phase [17], the search for Fe-Mn-Ga alloys undergoing phase transformations and possessing related multifunctional properties has never stopped. Zhu et al. observed that a magnetic-field-induced transformation from the PM parent phase to the FM martensite phase takes place at 163 K for the slightly off-stoichiometric $Fe_{50}Mn_{22.5}Ga_{27.5}$ alloy, leading to large differences in magnetization between both phases and a huge shape memory strain of up to 3.6% [18]. Researchers found that the martensitic transition in Fe-Mn-Ga alloys can lead to significant changes in their magnetic and optical properties [19–21]. Recently, local symmetry-breaking behavior has been observed in $Fe_{50}Mn_{23}Ga_{27}$, which suppresses the martensitic transition while retaining the magnetic transition in the alloy [22]. In addition, giant exchange bias behavior resulting from the exchange coupling between the coexisting antiferromagnetic (AFM) and FM phase in the Fe-Mn-Ga alloys was achieved, resulting in an enhanced coercivity [23–25]. In the above works, martensitic transformation in Fe-Mn-Ga alloys always displayed significant transformation hysteresis across an incomplete process, which is considered as an obstacle to realize a recoverable and complete field-induced martensitic transformation [26,27]. To some extent, this could limit their applications as, for example, magnetocaloric and magnetostrain materials, and clarifying the mechanism of incomplete phase transformation is essential.

As a result, based on the previous studies, it is essential to carry out an in situ study on the structure and magnetic phase evolution of the Fe-Mn-Ga alloy across martensitic transformation under external fields, such as temperature and magnetic fields. In the present study, we prepared a $Fe_{41.5}Mn_{28}Ga_{30.5}$ alloy presenting martensitic transformation, and the temperature- and magnetic-field-induced phase transformation behaviors of this alloy were systematically studied. The potential reasons for this incomplete phase transformation behavior were discussed, which may be instructive for understanding the underlying mechanisms responsible for the multifunctional properties of Fe-Mn-Ga alloys.

## 2. Experimental Section

A $Fe_{41.5}Mn_{28}Ga_{30.5}$ magnetic shape memory alloy was prepared by repeated melting in an arc furnace under an argon atmosphere. The rod ingots were sealed in a quartz tube filled with high-purity argon gas, homogenized at 1273 K for 24 h, and finally quenched in cold water. The composition of $Fe_{41.5}Mn_{28}Ga_{28.5}$ was measured using an electron probe microanalyzer (EPMA-1720H, SHIMADZU, Tokyo, Japan). The composition was determined by averaging the compositions of five randomly measured points, and the result was $Fe_{41.5\pm0.1}Mn_{28\pm0.6}Ga_{28.5\pm0.3}$. Due to the fact that Fe-Mn-Ga polycrystalline samples are very brittle, a single crystal with a roughly ellipsoid shape was obtained using the grain growth method [17] by annealing at 1273 K for 168 h, followed by quenching in ice water.

The microstructure was characterized by scanning electron microscopy (SEM, Zeiss Supra 55, Oberkochen, Germany) and transmission electron microscopy (TEM, JEM-2100 F, JEOL, Tokyo, Japan). The phase transition temperatures of the polycrystalline sample and the single crystal sample were analyzed by differential scanning calorimetry (DSC, Netzsch DSC 214 Polyma, Selb, Germany) experiments performed with heating and cooling rates of 10 K·min$^{-1}$. The magnetic measurements and electrical resistivity measurements of the polycrystalline samples with sizes of $\Phi$ 3 × 1 mm$^3$ and 10 × 2 × 1 mm$^3$, respectively, were conducted using a physical property measurement system (PPMS, Quantum Design) with a cooling and heating rate of 5 K·min$^{-1}$. The external magnetic field is perpendicular to the plane of the polycrystalline sample. The resistivity test was based on a 4-point probe method. The strain, during martensitic transformation for the polycrystalline sample, was performed using a strain gauge glued to the sample and a data logger (TDS102, Tokyo Sokki Kenkyujo Co., Ltd, Tokyo, Japan) under zero field with a cooling and heating rate of 5 K·min$^{-1}$. Temperature-dependent in-situ synchrotron high-energy X-ray diffraction

(HEXRD) experiments were carried out at the 11-ID-C beamline at the Advanced Photon Source of Argonne National Laboratory (ANL). A monochromatic X-ray beam with a wavelength of 0.1173 Å was used. The diffraction Debye rings were collected using a two-dimensional (2D) large area detector. The polycrystalline samples were rotated at high speed to obtain full rings and eliminate the effects of preferred orientation during heating and cooling with a rough rate of 3 K·min$^{-1}$. The samples were cooled and heated in a step-wise manner between 380 K and 119 K, with an average temperature interval of 3 K. At each temperature step, the sample was soaked for five minutes to reach a thermally stabilized state. A small cryogenic detector and an infrared detector were applied during cooling and heating, respectively. Based on the X-ray diffraction intensity theory, the integral intensities of diffraction peaks are proportional to the volume fractions of the corresponding phases. As a result, the volume fractions of the martensite phase ($V_M$) and austenite phase ($V_A$) were obtained according to the following expression [28]:

$$V_A = (1 + \frac{I_{hkl}^M R_{hkl}^A}{I_{hkl}^A R_{hkl}^M})^{-1} \quad (1)$$

$$V_A + V_B = 1 \quad (2)$$

where $I_{hkl}^M$ and $I_{hkl}^A$ represent the integrated intensities of the considered {hkl} lattice plane, and $R_{hkl}^M$ and $R_{hkl}^A$ are the corresponding theoretical calculated intensities [28]. The crystal structural evolution during the increasing and decreasing of the magnetic fields was studied by in situ neutron diffraction experiments on the high intensity diffractometer WOMBAT [29] at the Australian Nuclear Science and Technology Organisation (ANSTO). The WOMBAT instrument was equipped with a 2D position-sensitive area detector that covered 120° in 2θ on the sample in the diffraction plane and about 15° in the vertical, out-of-plane direction. For the in situ measurements, WOMBAT was equipped with a vertical field magnet (field range of 0 T–11 T with 200 Oe·s$^{-1}$) with a temperature range of 1.5–300 K (2 K·min$^{-1}$ during cooling and heating). A polycrystalline sample and a single crystal sample were glued to pure aluminum bolts (where the bolts were used as is) separately for the test under the temperature field and that under the magnetic field. A wavelength of 2.41 Å was used for the measurements.

## 3. Results and Discussion

The phase transformation temperatures of the Fe$_{41.5}$Mn$_{28}$Ga$_{30.5}$ alloy are revealed by the DSC curves in Figure 1a,b. For the polycrystalline sample (Figure 1a), the martensitic transformation and its reverse transformation were revealed by multiple exothermal/endothermic peaks, which are considered to be jerky characteristics of martensitic transformations that have been reported in many alloy systems [30,31]. In contrast, the single-crystal sample showed exothermal/endothermic peaks which are easier to distinguish in in Figure 1b, indicating that grain boundaries may play important roles in presenting this "avalanche" behavior during phase transition [30]. $M_s$, $M_f$, $A_s$ and $A_f$ in Figure 1b are the martensitic and reverse transformation start and finish temperatures, and they were determined to be about 229 K, 201 K, 298 K, and 329 K, respectively. The temperature dependence of the resistivity $\rho(T)$ for the polycrystalline Fe$_{41.5}$Mn$_{28}$Ga$_{30.5}$ is shown in Figure 1c. Fe$_{41.5}$Mn$_{28}$Ga$_{30.5}$ showed a significant temperature hysteresis of resistivity on heating and cooling, which indicates a first-order phase transformation in the alloy. One can notice that the transformation temperature interval here was wider than that shown in the DSC results, which may result from the gradual phase transition beyond the peak temperature regions only releasing or absorbing a small fraction of heat that is insufficient to form visible DSC peaks. A significant decrease with a magnetoresistance $\Delta\rho/\rho_{224K}$ ($\Delta\rho = |\rho_{10K} - \rho_{224K}|$) of 8% during cooling can be seen in Fe$_{41.5}$Mn$_{28}$Ga$_{30.5}$, which is ascribed to the forward martensitic transformation. Actually, this is quite different from the temperature dependence of resistivity behavior across martensitic transformation in most Ni-Mn-based FSMAs, where the resistivity usually increases during cooling and decreases during heating [32–35]. Specifically, the electrical resistivity of the austenite phase decreases with the increase in temperature, which is opposite to the austenite phase with metallic behavior in most FSMAs, where the electrical resistivity increases with increasing temperatures [36–38]. Figure 1d shows that a remarkable s strain up to 0.47% during cooling and heating can be achieved in the polycrystal sample.

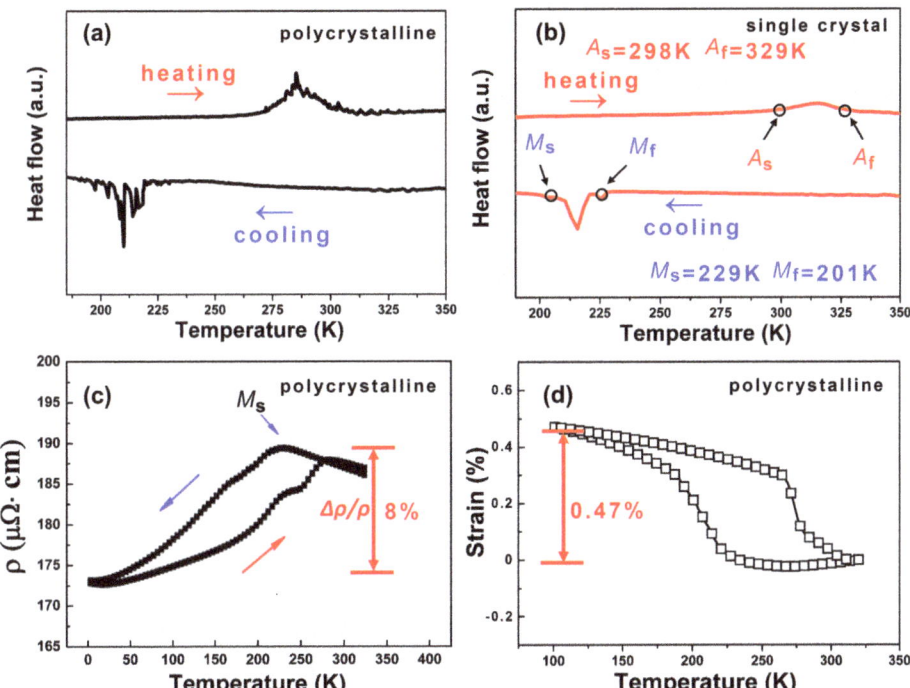

**Figure 1.** DSC curves for (**a**) the $Fe_{41.5}Mn_{28}Ga_{30.5}$ polycrystalline sample and (**b**) the $Fe_{41.5}Mn_{28}Ga_{30.5}$ single-crystal sample. (**c**) Temperature dependence of electrical resistivity during cooling and heating for polycrystalline $Fe_{41.5}Mn_{28}Ga_{30.5}$. (**d**) Shape memory effect of polycrystalline $Fe_{41.5}Mn_{28}Ga_{30.5}$, measured without preloading. $M_s$, $M_f$, $A_s$, and $A_f$ in Figure 1b denote the martensitic and reverse transformation starting and finishing temperatures, respectively.

To trace how the structure evolves when temperature changes, the variable-temperature HEXRD patterns for $Fe_{41.5}Mn_{28}Ga_{30.5}$ during the cooling and heating processes are displayed in Figure 2a,b. The results clearly indicate that the samples underwent a structural transformation from the $L2_1$ cubic structure (space group No. 225, $Fm\bar{3}m$) to the tetragonal structure (space group No. 139, $I4/mmm$) during the cooling process. It can be noticed that the competition and coexistence of the two phases are always presented in the whole temperature range during cooling and heating, as indicated by the gradual changes in peak intensities for the two phases. In Figure 2a, at 380 K, the cubic austenite is dominant in the sample, and only a tiny amount of martensite can be observed; a sudden increase in the intensities of tetragonal martensite phase peaks appears when cooling to 238 K. One should notice that there is a discrepancy in the martensitic transformation temperature obtained by DSC (Figure 1a) and HEXRD (here, we all used polycrystalline alloys for comparison), which may result from the different parts of the ingot being used for DSC and HEXRD tests. Different cooling rates may also affect the phase transformation temperatures as determined by two methods. When decreasing the temperature, the intensities of martensite peak continued to increase at the cost of austenite until the sample reached equilibrium, where both phases were found to be continuously co-existent down to 119 K. The variable-temperature HEXRD patterns for $Fe_{41.5}Mn_{28}Ga_{30.5}$ during heating, shown in Figure 2b, presented a trend nearly opposite to that in Figure 2a, where the intensities of austenite suddenly increased at 271 K. Generally, the lattice volumes of austenite and martensite decrease upon cooling. However, when the martensitic transformation occurs, the phase fraction of martensite increases at the cost of austenite. This can lead to a negative thermal expansion effect in the present alloy [27]. Figure 2c shows that the phase fractions of both phases almost remained unchanged until the temperature decreased to 235 K, where the martensitic transformation started. The austenite phase fraction then decreased from 95% to 25%, while the martensite phase fraction increased from 5% to 75%. Further cooling failed to bring additional obvious changes in phase fractions at temperatures below 140 K, indicating completion of the martensitic transformation. Typically, a large lattice distortion of $(c - a)/a = 32.3\%$ was obtained during the martensitic transformation, where $a = \sqrt{2}a_M = 5.208$ Å and $c = c_M = 6.891$ Å were calculated at 238 K from Figure 2a. Such a significant lattice distortion leads to a considerable shape memory effect, as displayed in Figure 1d. To detect the phase compositions at cryogenic temperatures, an in situ neutron diffraction experiment was performed at 4 K, and the results are shown in Figure 2d. One can see that, even at such a low temperature, the two phases still coexisted stably. At 4 K, the transformation was essentially stopped, and the crystal consisted of two phases. Overall, the picture of the martensitic transformation of the average structure in $Fe_{41.5}Mn_{28}Ga_{30.5}$ in this work is nucleation-driven [21], continuous, and incomplete.

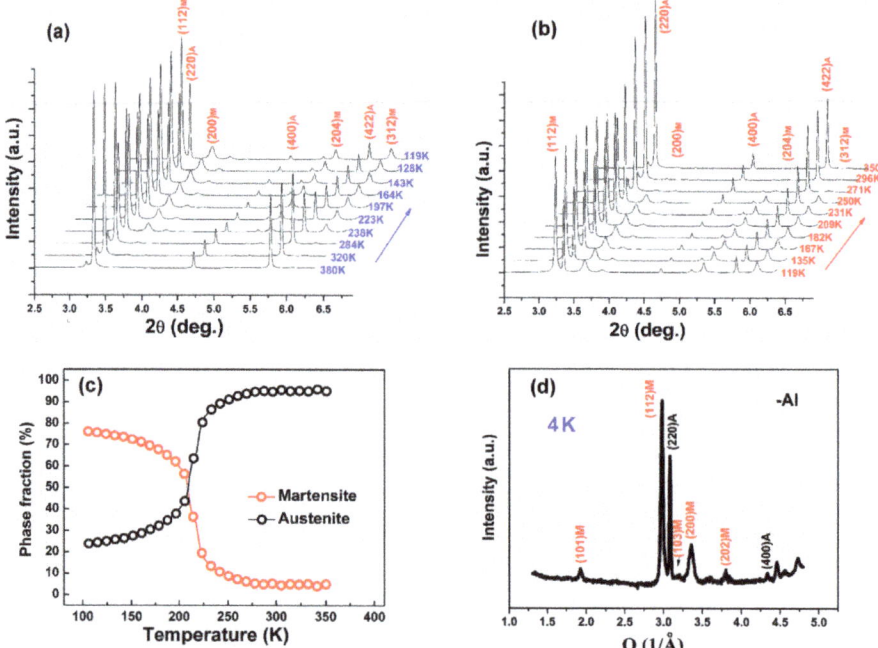

**Figure 2.** HEXRD patterns measured for the $Fe_{41.5}Mn_{28}Ga_{30.5}$ polycrystalline sample during (**a**) cooling and (**b**) heating. (**c**) Temperature dependence of the phase fractions, as obtained using Equations (1) and (2) during cooling for $Fe_{41.5}Mn_{28}Ga_{30.5}$. (**d**) One-dimensional (1D) diffraction patterns at 4 K without applying magnetic field for $Fe_{41.5}Mn_{28}Ga_{30.5}$. The letters "A" and "M" in the indices in (**a**) and (**d**) denote austenite and martensite, respectively.

The temperature dependence of the phase fraction growth rate data for $Fe_{41.5}Mn_{28}Ga_{30.5}$ (Figure 3) illustrates the hysteresis effect and the broad temperature range of transformation on both the heating and cooling processes. As shown in Figure 3, at 150 K and 330 K, the martensite and austenite growth rates were essentially below 0.5%, and the martensite and austenite phases accounted for about 75% and 25% of the structure at 150 K, as discussed in Figure 2c; the martensitic transformation was incomplete. The phase fraction growth rates showed two peaks at 213 ± 2 K during cooling and 278 ± 2 K during heating, respectively, further indicating a significant phase transformation hysteresis of 65 K in the $Fe_{41.5}Mn_{28}Ga_{30.5}$ alloy. This phase transformation hysteresis was larger than that shown in Figure 2, which was calculated to be about 33 K. This is because the phase fraction growth rate may be more sensitive in detecting the phase transformation start and stop timestamps.

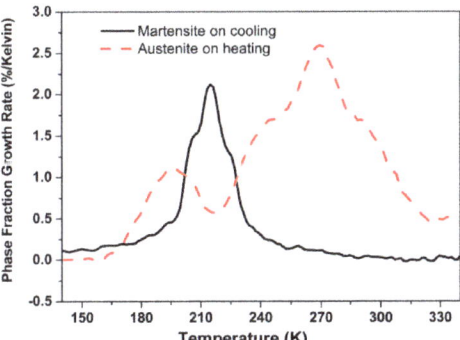

**Figure 3.** The temperature dependence of the phase fraction growth rate data for polycrystalline $Fe_{41.5}Mn_{28}Ga_{30.5}$.

The thermomagnetization curves ($M(T)$ curves) of $Fe_{41.5}Mn_{28}Ga_{30.5}$ during cooling and heating under magnetic fields of 0.03 T, 0.5 T, and 2 T are shown in Figure 4a. The magnetization increased significantly in certain temperature windows during the first cooling under 0.03 T, corresponding to the martensitic transformation, and significantly decreased in the magnetization during heating, with a result of under 0.03 T due to the reverse martensitic transformation. One can notice from the $M(T)$ curves that the martensite and austenite are considered to be ferromagnetic and paramagnetic, respectively [17]. When applying a magnetic field of 0.5 T, a large $\Delta M$ increased across martensitic transformation. Such a relatively large $\Delta M$ raised the transformation temperatures as compared to the phase transformation temperatures under 0.03 T. When it came to 2 T, $\Delta M$ became larger, and the phase transformation temperatures shifted towards higher temperatures. Usually, in the Ni-Mn-based FSMAs, the shift of the transformation temperatures under high magnetic fields is towards lower temperature areas compared with that under low magnetic fields [39–43]. However, the direction of the shift of phase transformation temperature under magnetic fields in Fe-Mn-Ga alloys was reversed. The difference in the magnetism of austenite and martensite could account for the abnormal shifting direction in Fe-Mn-Ga. Figure 4b shows the magnetization curves at different temperatures during cooling ($M(H)$ curves) for $Fe_{41.5}Mn_{28}Ga_{30.5}$. The austenite remained in the paramagnetism at 380 K, but still showed apparent magnetization behavior, which is caused by the residual martensite. The $M(H)$ curve showed quite significant hysteresis between the field-up and field-down at 250 K, 210 K, 130 K, and 60 K (taken one after the other), indicating a magnetic field-induced transformation within a wide temperature range in this alloy system. When the temperature was decreased to 4 K, the curve exhibited typical ferromagnetic properties [17]. The saturate magnetization of the martensite was estimated to be 72 emu·$g^{-1}$ at 4 K. Notably, the magnetic-field-induced transformation in the present alloy occurred from the paramagnetic austenite phase to the ferromagnetic martensitic phase, which is also the opposite of that of most other FSMAs [44–47]. This is caused when the martensitic transformation scuffles with the spontaneous magnetization transition in this sample [18]. Figure 4c shows the saturation magnetization as a function of the various temperatures at which the $M(H)$ loops were taken. One can see that the saturation magnetization values presented a sharp increase from 250 K and 130 K, and then remained unchanged during further cooling.

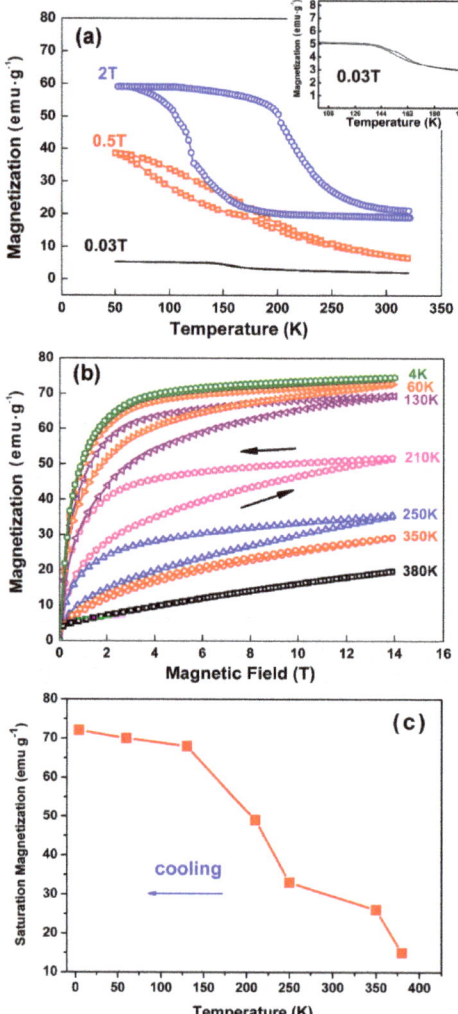

**Figure 4.** (**a**) Magnetization under 0.03 T, 0.5 T, and 2 T [$M(T)$] during cooling and heating for polycrystalline $Fe_{41.5}Mn_{28}Ga_{30.5}$, respectively. The inset shows the enlarged $M(T)$ curve of 0.03 T. (**b**) Isothermal magnetization [$M(H)$] curves were recorded at 380 K, 350 K, 250 K, 210 K, 130 K, 60 K and 4 K for polycrystalline $Fe_{41.5}Mn_{28}Ga_{30.5}$. (**c**) The saturation magnetization as a function of the various temperatures at which the M-H loops were taken.

To trace how the structure develops under magnetic fields, in situ neutron diffraction experiments were further used to monitor the structural evolution along with increasing and decreasing magnetic fields in a $Fe_{41.5}Mn_{28}Ga_{30.5}$ single-crystal alloy. For the in situ neutron diffraction experiments, the sample was first cooled from 400 K (furnace) to 250 K, and then at 250 K, reciprocal space mapping was presented along with increasing and decreasing magnetic fields in the sequence of 0 T–2 T–6 T–11 T–0 T (field ramped from 0 to 2 T, then measurement was performed, then it was ramped to 6 T and the 2nd measurement was performed, etc.). The reciprocal space maps measured at different magnetic fields are shown in Figure 5a–d. The coordinates of these figures are expressed with respect to the scattering vector **Q** with the length $Q = 2\pi/d = 4\pi \sin\theta/\lambda$, with $d$ being the interplanar spacing and $2\theta$ the diffraction angle. One-dimensional (1D) diffraction patterns from azimuthal integration of the reciprocal space maps in the **Q** range from 2.0 Å$^{-1}$ to 4.0 Å$^{-1}$ with the increasing and decreasing magnetic fields are shown in Figure 5e. The reciprocal space map shown in Figure 5a indicates that, at 250 K, the sample was in the austenite state, and all the diffraction spots were well indexed according to the L2$_1$ cubic structure with lattice parameter a = 5.675 Å. The two continuous diffraction rings in Figure 5a stem from the reflections of the polycrystalline aluminum bolt to which the sample was glued. One can see from Figure 5b that for Fe$_{41.5}$Mn$_{28}$Ga$_{30.5}$ under the magnetic field of 2 T, no obvious change in the diffraction patterns could be observed, and only the diffraction spot of (220)$_A$ (A denotes austenite) could be detected. In contrast, when the field was increased to 11 T (Figure 5c), diffraction patterns of martensite appeared [(112)$_M$ and (200)$_M$], while the intensity of the diffraction pattern for austenite significantly decreased. This process corresponds well to the *M(H)* curve at 250 K in Figure 4b. It should be noted that under 11 T, and even decreasing the field to 0 T (Figure 5d), both the transformation from austenite to martensite and its recovery transition were still incomplete. This could be attributed to the significant hysteresis in the Fe-Mn-Ga alloy system, and this will be discussed later.

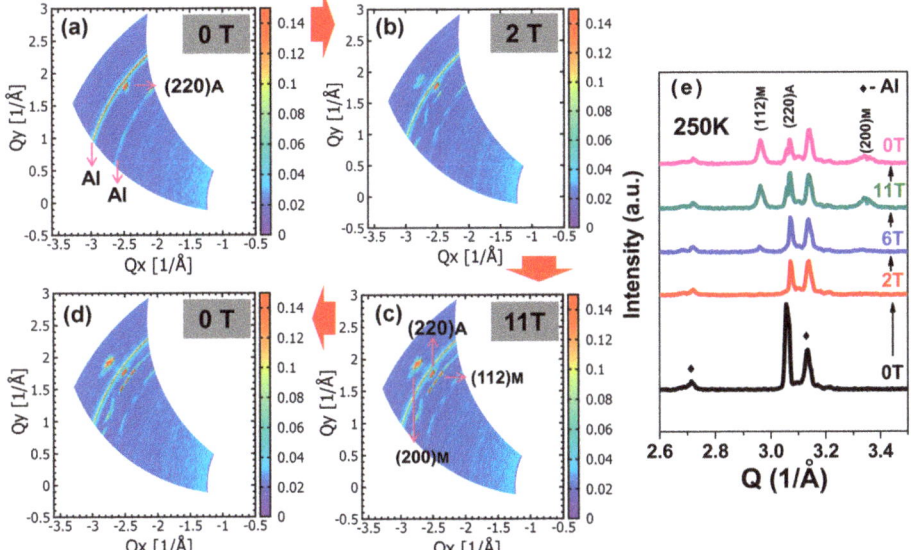

**Figure 5.** Reciprocal space maps measured at 250 K for the polycrystalline Fe$_{41.5}$Mn$_{28}$Ga$_{30.5}$ alloy with a magnetic field increasing from (**a**) 0 T to (**b**) 2 T and (**c**) 11 T, and then with a magnetic field decreasing to (**d**) 0 T; (**e**) 1D diffraction patterns, obtained by azimuthal integration of the reciprocal space maps (in the Q range from 2.0 Å$^{-1}$ to 4.0 Å$^{-1}$), for different magnetic field values at 250 K.

Figure 6a shows the evolution of the intensity ratios of the specific diffraction spots of martensite and austenite, namely, I(112)$_M$/I(220)$_A$ and I(200)$_M$/I(220)$_A$, with increasing and decreasing magnetic fields in the Fe$_{41.5}$Mn$_{28}$Ga$_{30.5}$ alloy at 250 K. One can see that there was no obvious change in the intensity ratios I(112)$_M$/I(220)$_A$ and I(200)$_M$/I(220)$_A$ when the magnetic field changed from 0 T to 2 T and 6 T. Strikingly, there was a sharp increase in the intensity ratios I(112)$_M$/I(220)$_A$ and I(200)$_M$/I(220)$_A$ when the magnetic field increased to 11 T, as seen from Figure 6a. These results indicate that the volume fraction of martensite became higher, as is consistent with the increase in magnetization from 0 T to 11 T on the *M(H)* curve at 250 K. When the magnetic field was removed, the intensity ratios I(112)$_M$/I(220)$_A$ and I(200)$_M$/I(220)$_A$ almost remained unchanged, indicating the incomplete recovery transition. Figure 6b shows the evolution of lattice parameters for martensite and austenite with the increasing and decreasing magnetic field. With the increasing and decreasing magnetic field, the lattice parameters for martensite and austenite only presented slight changes when martensitic transformation and its reverse transformation occurred, indicating that at 250 K, the magnetic field was unable to provide sufficient energy to easily overcome atomic connection energies to vary both phases' lattice parameters, although partial austenite still continued transforming into martensite. An abrupt unit cell volume increase, i.e., $\Delta V/V = +1.29\%$, could be achieved when increasing the magnetic field to 2 T, which is very close to that reported previously [17,18].

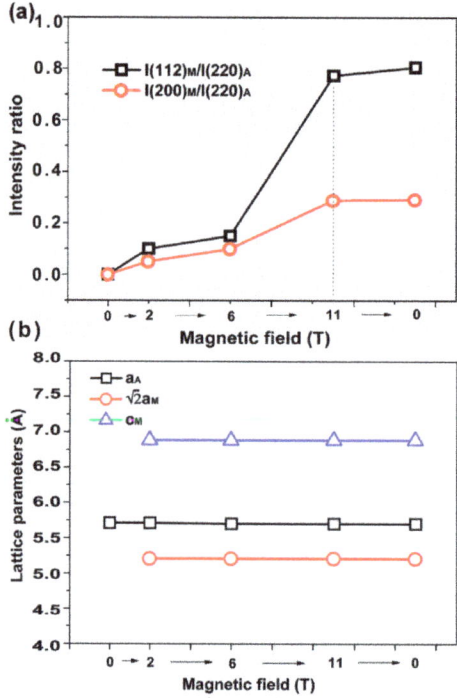

**Figure 6.** (a) Evolution of the intensity ratios, I(112)$_M$/I(220)$_A$ and I(200)$_M$/I(220)$_A$, as a function of magnetic field at 250 K. (b) Evolution of lattice parameters for Fe$_{41.5}$Mn$_{28}$Ga$_{30.5}$ as a function of magnetic field at 250 K.

Since the temperature- and magnetic-field-induced incomplete phase transformation behaviors in the Fe$_{41.5}$Mn$_{28}$Ga$_{30.5}$ alloy were systematically studied by means of in situ experiments, a further microstructural analysis by TEM is shown in Figure 7. Figure 7a presents a TEM image for Fe$_{41.5}$Mn$_{28}$Ga$_{30.5}$, taken at 300 K (cooling from 400 K), and Figure 7b shows an enlarged view of a local area in Figure 7a. No obvious characteristic morphology of martensite-like minor long strip-type grain can be seen in Figure 7a. Instead, some anti-phase boundaries [48] (APBs, marked by red arrows in Figure 7a) can be detected. The APBs' thickness was around 20–40 nm, and it was reported that in the Fe-Mn-Ga alloys, the APBs' crystal structure may appear while changing from L2$_1$ to lower-ordered B2 phases by quenching (B2 shares the same fundamental spots with L2$_1$) [28]. The corresponding selected area diffraction patterns of Figure 7a are shown in Figure 7a′, confirming the present area to be the L2$_1$ austenite phase state. In addition, some weak reflections, which may come from a lower-ordered B2 phase, appear at the positions between the austenite spots, as indicated with the red arrow in Figure 7a′, which represents the precipitates. Similar precipitates were found previously by Omori et al. [49] in Fe-Mn-Al alloys. It is reported that in the classical Fe-Ga alloy systems, the cubic-to-tetragonal phase transformation can develop within the B2 precipitates, leading to the multidomain structure being confined within a fixed-shape particle and to an incomplete transformation [50]. A similar episode could have occurred during the martensitic transformation process in Fe-Mn-Ga for the present work. When the sample was cooled to 150 K, as shown in Figure 7c, some long strip-type grains could be seen, some of which referred to the martensite. The width of the long strip-type grain was about 70 nm, as measured in Figure 7d. The corresponding selected area diffraction patterns of Figure 7c are shown in Figure 7c′, and one can see that, except for the $L2_1$ austenite phase (body-centered-cubic, bcc), some reflections can be identified as variants of [112] martensite (body-centered-tetragonal, bct), while the ordered B2 precipitates with {100} superlattice reflection spots can be detected within in the matrix along the [011] zone axis.

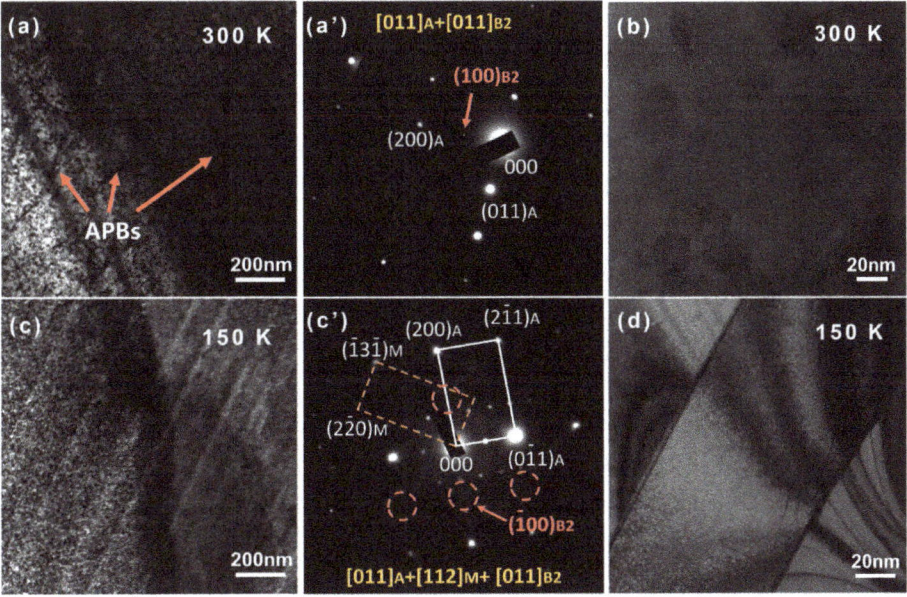

**Figure 7.** TEM observations of the $Fe_{41.5}Mn_{28}Ga_{30.5}$ alloy at (**a**) 300 K and (**c**) 150 K, with the selected area diffraction patterns at these temperatures shown in (**a′**,**c′**), respectively. (**b**,**d**) are two enlarged views of a local area in in (**b**,**c**), respectively.

In the present work, the existence of APBs was able to weaken the degree of local ordering, further leading to a decrease in the kinetics during martensitic transformation and an increase in internal stress around APBs to inhibit the subsequent development of martensitic transformation. Furthermore, the existence of B2 precipitates and their nucleation may only result from the vacancy absorption [51]. As precipitates, they may be formed by certain element enrichment of the Fe-Mn-Ga alloy, leading to a difference in their unit cell volume and an increase in local stress. In this case, precipitation is possible only if the absorbed excess vacancies annihilate this extra volume and, thus, eliminate the transformation-induced stress preventing decomposition. To some extent, the field-induced martensitic transformation of the $L2_1$ phase is restricted by the ordered B2 phase, either in the actual sample space or characterized by the reciprocal space. This incomplete martensitic transformation can also be discussed from a geometric compatibility (measured by the middle eigenvalue $\lambda_2$ of the transformation stretch matrix **U**) point of view [52,53], and the middle eigenvalue $\lambda_2$ for the Fe-Mn-Ga alloys usually seriously deviates from the perfect geometric compatibility of $\lambda_2 = 1$ [54]. This indicates that the present alloy with

$\lambda_2 = 0.92$ has poor geometric compatibility between austenite and martensite, making the phase transformation between the two phases difficult. The above combined factors collectively affect the temperature- and magnetic-field-induced incomplete phase transformation behaviors of the $Fe_{41.5}Mn_{28}Ga_{30.5}$ alloy.

## 4. Conclusions

In this work, the temperature- and magnetic-field-induced incomplete martensitic transformation of the $Fe_{41.5}Mn_{28}Ga_{30.5}$ magnetic shape memory alloy was systematically investigated. The temperature-field-induced incomplete phase transformation was directly evidenced by the crystal structure evolution during cooling and heating using the in situ synchrotron high-energy X-ray diffraction technique. The magnetic-field-induced phase transformation was revealed by the crystal structure evolution while increasing and decreasing the magnetic fields by means of in situ neutron diffraction experiments. The results show that even at 4 K, the alloy still presented a two-phase coexistence state. When changing the magnetic fields at 250 K, the phase transformation, which is always accompanied by the competing between martensite and austenite, cannot be accomplished. The variation in the magnitude of the applied magnetic field leads to an irreversible effect of changing the phase composition. TEM observation indicates that the existence of anti-phase boundaries and B2 precipitates may lead to difficulties during field-induced phase transformation. This work may help us to understanding the complex phase transition under external fields in Fe-Mn-Ga alloy systems, and may provide suggestions for the development of applicable Fe-Mn-Ga magnetic shape memory alloys with novel functionalities.

**Author Contributions:** Conceptualization, X.S. and Z.C.; methodology, K.-D.L.; software, R.L.; validation, Z.M.; formal analysis, R.L. and Z.M.; investigation, X.S. and J.C.; resources, X.S. and S.L.; data curation, X.S. and S.L.; writing—original draft preparation, X.S. and Z.C.; writing—review and editing, X.S. and Z.C., visualization, X.S.; supervision, X.S. and Z.C.; project administration, X.S.; funding acquisition, X.S. All authors have read and agreed to the published version of the manuscript.

**Funding:** This research was funded by the National Natural Science Foundation of China (Nos. 52101237 and 52201020).

**Data Availability Statement:** Not applicable.

**Acknowledgments:** Gene Davidson and Richard Spence are gratefully acknowledged for their technical support on the sample environments in ANSTO and ANL. We are grateful to Dennis E. Brown for his valuable discussions.

**Conflicts of Interest:** The authors declare no conflict of interest.

## References

1. Ullakko, K.; Huang, J.K.; Kantner, C.; O'handley, R.C.; Kokorin, V.V. Large magnetic-field-induced strains in $Ni_2MnGa$ single crystals. *Appl. Phys. Lett.* **1996**, *69*, 1966. [CrossRef]
2. Kainuma, R.; Imano, Y.; Ito, W.; Sutou, Y.; Morito, H.; Okamoto, S.; Kitakami, O.; Oikawa, K.; Fujita, A.; Kanomata, T.; et al. Magnetic-field-induced shape recovery by reverse phase transformation. *Nature* **2006**, *439*, 957–960. [CrossRef]
3. Saren, A.; Laitinen, V.; Vinogradova, M.; Ullakko, K. Twin boundary mobility in additive manufactured magnetic shape memory alloy 10M Ni-Mn-Ga. *Acta Mater.* **2023**, *246*, 118666. [CrossRef]
4. Krenke, T.; Duman, E.; Acet, M.; Wassermann, E.F.; Moya, X.; Mañosa, L.; Planes, A.; Suard, E.; Ouladdiaf, B. Magnetic superelasticity and inverse magnetocaloric effect in Ni-Mn-In. *Phys. Rev. B* **2007**, *75*, 104414. [CrossRef]
5. Bachaga, T.; Zhang, J.; Khitouni, M.; Sunol, J.J. NiMn-based Heusler magnetic shape memory alloys: A review. *Int. J. Adv. Manuf. Technol.* **2019**, *103*, 2761–2772. [CrossRef]
6. Koyama, K.; Okada, H.; Watanabe, K.; Kanomata, T.; Kainuma, R.; Ito, W.; Oikawa, K.; Ishida, K. Observation of large magnetoresistance of magnetic Heusler alloy $Ni_{50}Mn_{36}Sn_{14}$ in high magnetic fields. *Appl. Phys. Lett.* **2006**, *89*, 182510. [CrossRef]
7. Yu, S.Y.; Ma, L.; Liu, G.D.; Liu, Z.H.; Chen, J.L.; Cao, Z.X.; Wu, G.H.; Zhang, B.; Zhang, X.X. Magnetic field-induced martensitic transformation and large magnetoresistance in NiCoMnSb alloys. *Appl. Phys. Lett.* **2007**, *90*, 242501. [CrossRef]
8. Yang, J.; Li, Z.; Yang, B.; Yan, H.; Cong, D.; Zhao, X.; Zuo, L. Strain manipulation of magnetocaloric effect in a $Ni_{39.5}Co_{8.5}Mn_{42}Sn_{10}$ melt-spun ribbon. *Scr. Mater.* **2023**, *224*, 115141. [CrossRef]
9. Khan, M.; Ali, N.; Stadler, S. Inverse magnetocaloric effect in ferromagnetic $Ni_{50}Mn_{37+x}Sb_{13-x}$ Heusler alloys. *J. Appl. Phys.* **2007**, *101*, 053919. [CrossRef]

10. Liu, J.; Gottschall, T.; Skokov, K.P.; Moore, J.D.; Gutfleisch, O. Giant magnetocaloric effect driven by structural transitions. *Nat. Mater.* **2012**, *11*, 620–626. [CrossRef]
11. Zhang, B.; Zhang, X.X.; Yu, S.Y.; Chen, J.L.; Cao, Z.X.; Wu, G.H. Giant magnetothermal conductivity in the Ni-Mn-In ferromagnetic shape memory alloys. *Appl. Phys. Lett.* **2007**, *91*, 012510. [CrossRef]
12. Li, Z.; Jing, C.; Chen, J.; Yuan, S.; Cao, S.; Zhang, J. Observation of exchange bias in the martensitic state of $Ni_{50}Mn_{36}Sn_{14}$ Heusler alloy. *Appl. Phys. Lett.* **2007**, *91*, 112505. [CrossRef]
13. Wang, B.M.; Liu, Y.; Ren, P.; Xia, B.; Ruan, K.B.; Yi, J.B.; Ding, J.; Li, X.G.; Wang, L. Large exchange bias after zero-field cooling from an unmagnetized state. *Phys. Rev. Lett.* **2011**, *106*, 077203. [CrossRef]
14. Liu, J.; Woodcock, T.G.; Scheerbaum, N.; Gutfleisch, O. Influence of annealing on magnetic field-induced structural transformation and magnetocaloric effect in Ni-Mn-In-Co ribbons. *Acta Mater.* **2009**, *57*, 4911–4920. [CrossRef]
15. Takagishi, M.; Koi, K.; Yoshikawa, M.; Funayama, T.; Iwasaki, H.; Sahashi, M. The applicability of CPP-GMR heads for magnetic recording. *IEEE Trans. Magn.* **2002**, *38*, 2277–2282. [CrossRef]
16. Cong, D.Y.; Roth, S.; Schultz, L. Magnetic properties and structural transformations in Ni-Co-Mn-Sn multifunctional alloys. *Acta Mater.* **2012**, *60*, 5335–5351. [CrossRef]
17. Omori, T.; Watanabe, K.; Xu, X.; Umetsu, R.Y.; Kainuma, R.; Ishida, K. Martensitic transformation and magnetic field-induced strain in Fe-Mn-Ga shape memory alloy. *Scr. Mater.* **2011**, *64*, 669–672. [CrossRef]
18. Zhu, W.; Liu, E.K.; Feng, L.; Tang, X.D.; Chen, J.L.; Wu, G.H.; Liu, H.Y.; Meng, F.B.; Luo, H.Z. Magnetic-field-induced transformation in FeMnGa alloys. *Appl. Phys. Lett.* **2009**, *95*, 222512. [CrossRef]
19. Yang, H.; Chen, Y.; Bei, H.; dela Cruz, C.R.; Wang, Y.D.; An, K. Annealing effects on the structural and magnetic properties of off-stoichiometric Fe-Mn-Ga ferromagnetic shape memory alloys. *Mater. Des.* **2016**, *104*, 327–332. [CrossRef]
20. Zhang, Y.J.; Wu, Z.G.; Hou, Z.P.; Liu, Z.H.; Liu, E.K.; Xi, X.K.; Wang, W.H.; Wu, G.H. Magnetic-field-induced transformation and strain in polycrystalline FeMnGa ferromagnetic shape memory alloys with high cold-workability. *Appl. Phys. Lett.* **2021**, *119*, 142402. [CrossRef]
21. Jenkins, C.A.; Scholl, A.; Kainuma, R.; Elmers, H.J.; Omori, T. Temperature-induced martensite in magnetic shape memory $Fe_2MnGa$ observed by photoemission electron microscopy. *Appl. Phys. Lett.* **2012**, *100*, 032401. [CrossRef]
22. Ma, T.Y.; Liu, X.L.; Yan, M.; Wu, C.; Ren, S.; Li, H.Y.; Feng, M.X.; Qiu, Z.Y.; Ren, X.B. Suppression of martensitic transformation in $Fe_{50}Mn_{23}Ga_{27}$ by local symmetry breaking. *Appl. Phys. Lett.* **2015**, *106*, 211903. [CrossRef]
23. Tang, X.D.; Wang, W.H.; Zhu, W.; Liu, E.K.; Wu, G.H.; Meng, F.B.; Liu, H.Y.; Luo, H.Z. Giant exchange bias based on magnetic transition in γ-$Fe_2MnGa$ melt-spun ribbons. *Appl. Phys. Lett.* **2010**, *97*, 242513. [CrossRef]
24. Tang, X.D.; Wang, W.H.; Wu, G.H.; Meng, F.B.; Liu, H.Y.; Luo, H.Z. Tuning exchange bias by thermal fluctuation in $Fe_{52}Mn_{23}Ga_{25}$ melt-spun ribbons. *Appl. Phys. Lett.* **2011**, *99*, 222506. [CrossRef]
25. Shih, C.W.; Zhao, X.G.; Chang, H.W.; Chang, W.C.; Zhang, Z.D. The phase evolution, magnetic and exchange bias properties in $Fe_{50}Mn_{24+x}Ga_{26-x}$ (x = 0–3) melt-spun ribbons. *J. Alloys Compd.* **2013**, *570*, 14–18. [CrossRef]
26. Sun, X.M.; Cong, D.Y.; Ren, Y.; Liss, K.-D.; Brown, D.E.; Ma, Z.Y.; Hao, S.J.; Xia, W.X.; Chen, Z.; Ma, L.; et al. Magnetic-field-induced strain-glass-to-martensite transition in a Fe-Mn-Ga alloy. *Acta Mater.* **2020**, *183*, 11. [CrossRef]
27. Sun, X.M.; Cong, D.Y.; Ren, Y.; Brown, D.E.; Li, R.G.; Li, S.H.; Yang, Z.; Xiong, W.X.; Nie, Z.H.; Wang, L.; et al. Giant negative thermal expansion in Fe-Mn-Ga magnetic shape memory alloys. *Appl. Phys. Lett.* **2018**, *113*, 041903. [CrossRef]
28. E04 Committee. *Practice for X-ray Determination of Retained Austenite in Steel with Near Random Crystallographic Orientation*; ASTM International: West Conshohocken, PA, USA, 2013.
29. Studer, A.J.; Hagen, M.E.; Noakes, T.J. Wombat: The high-intensity powder diffractometer at the OPAL reactor. *Phys. B Condens. Matter* **2006**, *385*, 1013–1015. [CrossRef]
30. Gallardo, M.C.; Manchado, J.; Romero, F.J.; del Cerro, J.; Salje, E.K.H.; Planes, A.; Vives, E.; Romero, R.; Stipcich, M. Avalanche criticality in the martensitic transition of $Cu_{67.64}Zn_{16.71}Al_{15.65}$ shape-memory alloy: A calorimetric and acoustic emission study. *Phys. Rev. B* **2010**, *81*, 174102. [CrossRef]
31. Soto-Parra, D.E.; Moya, X.; Mañosa, L.; Flores-Zúñiga, A.H.; Alvarado-Hernández, F.; Ochoa-Gamboa, R.A.; Matutes-Aquino, J.A.; Ríos-Jara, D. Fe and Co selective substitution in $Ni_2MnGa$: Effect of magnetism on relative phase stability. *Philos. Mag.* **2010**, *90*, 2771–2792. [CrossRef]
32. Yu, S.Y.; Liu, Z.H.; Liu, G.D.; Chen, J.L.; Cao, Z.X.; Wu, G.H.; Zhang, B.; Zhang, X.X. Large magnetoresistance in single-crystalline $Ni_{50}Mn_{50-x}In_x$ alloys (x = 14–16) upon martensitic transformation. *Appl. Phys. Lett.* **2006**, *89*, 162503. [CrossRef]
33. Han, Z.D.; Wang, D.H.; Zhang, C.L.; Xuan, H.C.; Zhang, J.R.; Gu, B.X.; Du, Y.W. The phase transitions, magnetocaloric effect, and magnetoresistance in Co doped Ni-Mn-Sb ferromagnetic shape memory alloys. *J. Appl. Phys.* **2008**, *104*, 053906. [CrossRef]
34. Sun, X.M.; Cong, D.Y.; Li, Z.; Zhang, Y.L.; Chen, Z.; Ren, Y.; Liss, K.-D.; Ma, Z.Y.; Li, R.G.; Qu, Y.H.; et al. Manipulation of magnetostructural transition and realization of prominent multifunctional magnetoresponsive properties in NiCoMnIn alloys. *Phys. Rev. Mater.* **2019**, *3*, 034404. [CrossRef]
35. Sharma, V.K.; Chattopadhyay, M.K.; Shaeb, K.H.B.; Chouhan, A.; Roy, S.B. Large magnetoresistance in $Ni_{50}Mn_{34}In_{16}$ alloy. *Appl. Phys. Lett.* **2006**, *89*, 222509. [CrossRef]
36. Huang, X.M.; Zhao, Y.; Yan, H.L.; Jia, N.; Yang, B.; Li, Z.; Zhang, Y.; Esling, C.; Zhao, X.; Zuo, L. Giant magnetoresistance, magnetostrain and magnetocaloric effects in a Cu-doped<001>-textured $Ni_{45}Co_5Mn_{36}In_{13.2}Cu_{0.8}$ polycrystalline allo. *J. Alloys Compd.* **2021**, *889*, 161652.

37. Zheng, T.T.; Liu, K.; Chen, H.X.; Wang, C. Large magnetocaloric and magnetoresistance effects during martensitic transformation in Heusler-type $Ni_{44}Co_6Mn_{37}In_{13}$ alloy. *J. Magn. Magn. Mater.* **2022**, *563*, 170034. [CrossRef]
38. Huang, L.; Cong, D.Y.; Ma, L.; Nie, Z.H.; Wang, M.G.; Wang, Z.L.; Suo, H.L.; Ren, Y.; Wang, Y.D. Large magnetic entropy change and magnetoresistance in a $Ni_{41}Co_9Mn_{40}Sn_{10}$ magnetic shape memory alloy. *J. Alloys Compd.* **2015**, *647*, 1081–1085. [CrossRef]
39. Karaca, H.E.; Karaman, I.; Basaran, B.; Ren, Y.; Chumlyakov, Y.I.; Maier, H.J. Magnetic field-induced phase transformation in NiMnCoIn magnetic shape-memory alloys-a new actuation mechanism with large work output. *Adv. Funct. Mater.* **2009**, *19*, 983–998. [CrossRef]
40. Nayak, A.K.; Suresh, K.G.; Nigam, A.K. Giant inverse magnetocaloric effect near room temperature in Co substituted NiMnSb Heusler alloys. *J. Phys. D Appl. Phys.* **2009**, *42*, 035009. [CrossRef]
41. Monroe, J.A.; Karaman, I.; Basaran, B.; Ito, W.; Umetsu, R.Y.; Kainuma, R.; Koyama, K.; Chumlyakov, Y.I. Direct measurement of large reversible magnetic-field-induced strain in Ni-Co-Mn-In metamagnetic shape memory alloys. *Acta Mater.* **2012**, *60*, 6883–6891. [CrossRef]
42. Roca, P.L.; López-García, J.; Sánchez-Alarcosa, V.; Recartea, V.; Rodríguez-Velamazáne, J.A.; Pérez-Landazábal, J.I. Room temperature huge magnetocaloric properties in low hysteresis ordered Cu-doped Ni-Mn-In-Co alloys. *J. Alloys Compd.* **2022**, *922*, 166143. [CrossRef]
43. Recarte, V.; Pérez-Landazábal, J.I.; Sánchez-Alarcos, V.; Zablotskii, V.; Cesari, E.; Kustov, S. Entropy change linked to the martensitic transformation in metamagnetic shape memory alloys. *Acta Mater.* **2012**, *60*, 3168–3175. [CrossRef]
44. Lázpita, P.; Pérez-Checa, A.; Barandiarán, J.M.; Ammerlaan, A.; Zeitler, U.; Chernenko, V. Suppression of martensitic transformation in Ni-Mn-In metamagnetic shape memory alloy under very strong magnetic field. *J. Alloys Compd.* **2021**, *874*, 159814. [CrossRef]
45. Salas, D.; Wang, Y.; Duong, T.C.; Attari, V.; Ren, Y.; Chumlyakov, Y.; Arróyave, R.; Karaman, I. Competing interactions between mesoscale length-scales, order-disorder, and martensitic transformation in ferromagnetic shape memory alloys. *Acta Mater.* **2021**, *206*, 116616. [CrossRef]
46. Salazar-Mejía, C.; Devi, P.; Singh, S.; Felser, C.; Wosnitza, J. Influence of Cr substitution on the reversibility of the magnetocaloric effect in Ni-Cr-Mn-In Heusler alloys. *Phys. Rev. Mater.* **2021**, *5*, 104406. [CrossRef]
47. Kainuma, R.; Ito, W.; Umetsu, R.Y.; Oikawa, K.; Ishida, K. Magnetic field-induced reverse transformation in B2-type NiCoMnAl shape memory alloys. *Appl. Phys. Lett.* **2008**, *93*, 091906. [CrossRef]
48. Ahadi, A.; Sun, Q. Effects of grain size on the rate-dependent thermomechanical responses of nanostructured superelastic NiTi. *Acta Mater.* **2014**, *76*, 186–197. [CrossRef]
49. Omori, T.; Nagasako, M.; Okano, M.; Endo, K.; Kainuma, R. Microstructure and martensitic transformation in the Fe-Mn-Al-Ni shape memory alloy with B2-type coherent fine particles. *Appl. Phys. Lett.* **2012**, *101*, 231907. [CrossRef]
50. Bhattacharyya, S.; Jinschek, J.R.; Li, J.F.; Viehland, D. Nanoscale precipitates in magnetostrictive $Fe_{1-x}Ga_x$ alloys for $0.1 < x < 0.23$. *J. Alloys Compd.* **2010**, *501*, 148–153.
51. Khachaturyan, A.G.; Viehland, D. Structurally heterogeneous model of extrinsic magnetostriction for Fe-Ga and similar magnetic alloys: Part I. decomposition and confined displacive transformation. *Metall. Mater. Trans. A* **2007**, *38A*, 2308–2316.
52. Hane, K.F.; Shield, T.W. Microstructure in the cubic to monoclinic transition in titanium-nickel shape memory alloys. *Acta Mater.* **1999**, *47*, 2603–2617. [CrossRef]
53. Song, Y.T.; Chen, X.; Dabade, V.; Shield, T.W.; James, R.D. Enhanced reversibility and unusual microstructure of a phase-transforming material. *Nature* **2013**, *502*, 85–88. [CrossRef]
54. Sun, X.M.; Cong, D.Y.; Ren, Y.; Brown, D.E.; Gallington, L.C.; Li, R.G.; Cao, Y.X.; Chen, Z.; Li, S.H.; Nie, Z.H.; et al. Enhanced negative thermal expansion of boron-doped $Fe_{43}Mn_{28}Ga_{28.97}B_{0.03}$ alloy. *J. Alloys Compd.* **2021**, *857*, 157572. [CrossRef]

**Disclaimer/Publisher's Note:** The statements, opinions and data contained in all publications are solely those of the individual author(s) and contributor(s) and not of MDPI and/or the editor(s). MDPI and/or the editor(s) disclaim responsibility for any injury to people or property resulting from any ideas, methods, instructions or products referred to in the content.

Article

# The Tensile Properties and Fracture Toughness of a Cast Mg-9Gd-4Y-0.5Zr Alloy

Zhikang Ji [1], Xiaoguang Qiao [1,*], Shoufu Guan [1], Junbin Hou [1], Changyu Hu [1], Fuguan Cong [2], Guojun Wang [2] and Mingyi Zheng [1,*]

[1] School of Materials Science and Engineering, Harbin Institute of Technology, Harbin 150001, China; jzktom@163.com (Z.J.)
[2] Northeast Light Alloy Company Limited, Harbin 150060, China
* Correspondence: xgqiao@hit.edu.cn (X.Q.); zhenghe@hit.edu.cn (M.Z.)

**Abstract:** Low fracture toughness has been a major barrier for the structural applications of cast Mg-Gd-Y-Zr alloys. In this work, the tensile properties and fracture toughness of a direct-chill-cast Mg-9Gd-4Y-0.5Zr (VW94K) alloy were investigated in different conditions, including its as-cast and as-homogenized states. The results show that the tensile properties of the as-cast VW94K alloy are greatly improved after the homogenization treatment due to the strengthening of the solid solution. The plane strain fracture toughness values $K_{Ic}$ of the as-cast and as-homogenized VW94K alloys are 10.6 ± 0.5 and 13.8 ± 0.6 MPa·m$^{1/2}$, respectively, i.e., an improvement of 30.2% in $K_{Ic}$ is achieved via the dissolution of the $Mg_{24}(Gd, Y)_5$ eutectic phases. The initiation and propagation of microcracks in an interrupted fracture test are observed via an optical microscope (OM) and scanning electron microscope (SEM). The fracture surfaces of the failed samples after the fracture toughness tests are examined via an SEM. The electron backscatter diffraction (EBSD) technique is adopted to determine the failure mechanism. The results show that the microcracks are initiated and propagated across the $Mg_{24}(Gd, Y)_5$ eutectic compounds in the as-cast VW94K alloy. The propagation of the main cracks exhibits an intergranular fracture pattern and the whole crack propagation path displays a zigzag style. The microcracks in the as-homogenized alloy are initiated and propagated along the basal plane of the grains. The main crack in the as-homogenized alloy shows a more tortuous fracture characteristic and a trans-granular crack propagation behavior, leading to the improvement of the fracture toughness.

**Keywords:** Mg-Gd-Y-Zr alloy; homogenization treatment; tensile properties; fracture toughness; crack initiation and propagation

## 1. Introduction

To reduce energy consumption and carbon emissions, low-density magnesium alloys have attracted great attention for their use in lightweight structures in the aerospace, automobile and electronic industries [1–3]. Unfortunately, the low levels of strength, ductility and fracture toughness of Mg alloys limit their commercial applications [4–6]. Recently, Mg-Gd-Y-Zr alloys exhibiting higher levels of strength and ductility than conventional Mg alloys have been developed [7,8]. Rare earth (RE) elements have unique physical and chemical properties due to their special extranuclear electronic structures and have become the most effective and promising alloying elements in magnesium alloys [9,10]. Among all the RE elements, Gd and Y have larger solid solubilities in Mg matrix at high temperatures, and the solid solubilities of Gd and Y decrease rapidly with a decrease in temperature [11,12]. Therefore, the addition of Gd and Y can provide significant solid-solution-strengthening and precipitation-strengthening effects. In addition, adding Gd and Y elements at the same time can reduce the solid solubility of both in the Mg matrix; thus, more second phases can be precipitated [13,14].

Some investigations have been reported concerning the microstructures and mechanical properties of Mg-Gd-Y-Zr alloys. Wang et al. [15] investigated the effect of Y for enhancing the age hardening response and mechanical properties of Mg-10Gd-$x$Y-0.4Zr ($x$ = 1, 3, 5 wt.%) alloys. The results showed that both the age hardening response and the tensile properties of the alloys were enhanced with an increasing Y content. Liu et al. [16] investigated the high-temperature mechanical behavior of a low-pressure sand-cast Mg-10Gd-3Y-0.5Zr alloy, which indicated that both the ultimate tensile strength and yield strength of the tested alloy firstly increased and then decreased as the temperature increased. Its elongation increased monotonously with temperature. Jiang et al. [17] investigated the effects of different Gd contents on the mechanical properties of sand-cast Mg-$x$Gd-3Y-0.5Zr alloys. With increase in the Gd content from 9 to 11 wt.%, the amount of eutectic phase was increased, while the tensile properties were slightly decreased. The research studies on the tensile properties of as-cast Mg-Gd-Y-Zr alloys with different compositions indicated that the Mg-9Gd-4Y-0.5Zr alloy exhibited greater strength and ductility compared with the other Mg-Gd-Y-Zr alloys [17–19].

High-performance Mg-Gd-Y-Zr alloys have great potential for use in the complicated structural components of the aerospace and aircraft industries [20,21]. In order to satisfy both the reliability and safety requirements of these structural components, Mg alloys should have high fracture toughness, as well as high strength and ductility. Somekawa et al. found that the grain refinement, texture and precipitate shapes exhibited significant effects on the plane strain fracture toughness $K_{Ic}$ of wrought magnesium alloys [22–24]. The fracture toughness of extruded pure magnesium was increased from 12.7 MPa·m$^{1/2}$ to 17.8 MPa·m$^{1/2}$ by refining the grain size from 55 μm to 1 μm due to the effect of the plastic zone [22]. An extruded AZ31 alloy with a pre-crack normal to its basal plane distribution was found to have a higher fracture toughness compared to an alloy with a pre-crack parallel to its basal plane distribution due to the difference in surface energy on the basal and non-basal planes [23]. Spherically shaped precipitates were more effective than rod-shaped precipitates for improving fracture toughness since they were more effective at pinning dislocations [24]. Lu et al. [25] reported a Mg-5Gd-2Y-0.4Zr alloy prepared via multidirectional impact forging with a tensile yield strength of 337 MPa and a static toughness of 50.4 MJ/m$^3$. The enhanced mechanical properties were attributed to the grain refinement in the Mg-5Gd-2Y-0.4Zr alloy. However, only a few studies on the fracture toughness $K_{Ic}$ of cast Mg alloys have been reported in the literature. Liu et al. [26] investigated the fracture toughness and crack initiation mechanisms of a sand-cast Mg-10Gd-3Y-0.5Zr alloy. The plane strain fracture toughness of the sand-cast Mg-10Gd-3Y-0.5Zr alloy increased from 12.1 MPa·m$^{1/2}$ to 16.3 MPa·m$^{1/2}$ after T6 heat treatment (a solution treatment followed by artificial aging). For the sand-cast sample, the microcracks mainly were initiated in eutectic compounds, and the microcracks grew to become a main crack. Comparatively, the microcracks in the san-cast-T6 sample were probably initiated at the twins/α-Mg matrix and grain boundaries and propagated along the twin boundaries. The fracture morphologies indicated that the fracture mechanisms changed from the trans-granular fracture pattern of the sand-cast alloy to a mixture of intergranular and trans-granular modes in the sand-cast-T6 alloy. Wang et al. [27] found that the fracture toughness of a sand-cast Mg-6Gd-3Y-0.5Zr alloy was increased by 9.3% after it underwent a solution treatment. Compared with the as-cast sample, the as-homogenized alloy exhibited a mixture of trans-granular and intergranular fracture patterns. The as-cast alloy displayed many cleavage steps but more secondary cracks than in the as-homogenized alloy. However, the mechanisms of the initiation and propagation of microcracks in cast Mg-Gd-Y alloys remain unclear.

In this work, the microstructure, tensile properties and fracture toughness of as-cast and as-homogenized Mg-9Gd-4Y-0.5Zr alloys were investigated. And the crack initiation and propagation behaviors in interrupted fracture toughness tests were discussed in order to clarify the effect of the eutectic phase on the fracture mechanism of the alloy. The innovation of this work is that it presents the first systematic study of the fracture toughness, crack initiation and propagation mechanisms of as-cast and as-homogenized Mg-9Gd-4Y-0.5Zr alloys, and the influence of the eutectic phase on the fracture behavior was also analyzed.

## 2. Materials and Experimental Procedures

### 2.1. Material Preparation

The magnesium alloy used in the present work was a Mg-9Gd-4Y-0.5Zr (wt.%) alloy, designated as VW94K alloy henceforth, which prepared via direct chill casting [28,29] from pure Mg (99.9 wt.%), Mg-30Gd (wt.%), Mg-30Y (wt.%) and Mg-25Zr (wt.%) in an electric resistance furnace under a mixed atmosphere of $CO_2$ and $SF_6$ at a ratio of 100:1 [30,31]. An inductively coupled plasma (ICP) analyzer was employed to analyze the chemical composition, and the actual composition of the as-cast alloy was Mg-9.12Gd-3.93Y-0.52Zr (wt.%). The homogenization treatment was performed at 510 °C for 12 h, followed by immediate warm water quenching at a temperature of ~80 °C [32,33].

### 2.2. Mechanical Properties

Tensile tests were conducted at a crosshead speed of 1 mm/min on a universal testing machine (Instron 5569, Norwood, MA, USA) at room temperature. The gauge length of the tensile specimens was 15 mm, with cross-sectional area of 3 × 2 $mm^2$. A 0.2% offset strength was used as the yield strength $\sigma_{ys}$. To guarantee repeatability, three tensile samples were conducted under the same conditions.

The plane strain fracture toughness tests were conducted, and the load–displacement curves were obtained using an MTS 810 servo-hydraulic fatigue tester (Eden Prairie, MN, USA). Compact tension C(T) samples (see Figure 1) with a width $W$ = 30 mm, thickness $B$ = 15 mm and notch depth $a_0$ = 12 mm were prepared according to the ASTM E399 standard [34]. Before the fracture toughness tests, a sharp fatigue pre-crack was produced under cyclic tension–tension loading and stopped until a total crack length of $a$ = 15 mm was obtained. All samples were machined with side grooves on both surfaces, and the grooving depth was 0.1 times the sample thickness. The fracture tests were implemented at a speed of 2.5 $MPa \cdot m^{1/2} \cdot s^{-1}$. The stress intensity factor $K_Q$ was calculated as in the following equation [34]:

$$K_Q = \frac{P_Q}{\sqrt{B \times B_N} \times \sqrt{W}} \cdot f\left(\frac{a}{W}\right) \quad (1)$$

where $P_Q$ is the conditional load [34], $B_N$ is the thickness measured at the side grooves, and $f\left(\frac{a}{W}\right)$ is the geometrical factor [34]. According to the ASTM E399 standard, $K_Q$ can be considered the size-independent fracture toughness $K_{Ic}$ if the test meets the following two requirements:

$$\frac{P_{max}}{P_Q} \leq 1.10 \quad (2)$$

$$2.5\left(\frac{K_Q}{\sigma_{ys}}\right)^2 \leq (W - a) \quad (3)$$

where $P_{max}$ is the maximum load on the load–displacement curves. Three fracture toughness samples were tested for each state.

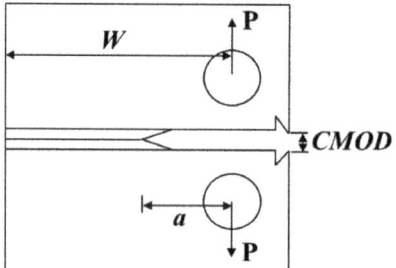

**Figure 1.** Schematic of the compact tension sample with CMOD.

### 2.3. Microstructure Characterization

The microstructural characteristics and crack propagation paths of the alloys were observed via an optical microscope (OM, Olympus PMG3, Olympus Corporation, Shinjuku, Japan) and a scanning electron microscope (SEM, ZEISS Supra 55, Carl Zeiss AG, Oberkochen, Germany) equipped with an electron backscatter diffraction instrument (EBSD, Oxford Instrument HKL, Oxford Instrument, Oxfordshire, UK). The OM and SEM samples were mechanically polished and then etched in a 4 vol% nitric acid alcohol solution. The EBSD samples were electropolished in a solution of ethanol and phosphoric acid, and the step size was 2 μm. The ESBD data were analyzed using Channel 5 software. The fracture surfaces were examined via the SEM and their three-dimensional (3D) topography was examined using a confocal laser scanning microscope (CLSM, Olympus LEXT OLS 3000, Olympus Corporation, Shinjuku, Japan). A slip trace analysis was performed via MATLAB, using an MTEX code [35].

## 3. Results

### 3.1. Microstructures

Figure 2 shows the OM and SEM microstructures of the as-cast VW94K alloy. It can be seen that the as-cast VW94K alloy is composed of an α–Mg matrix and network-shaped eutectic compounds at the grain boundaries. The average grain size, as determined via the linear intercept method [36], is about 75 μm. Based on the previous experimental results [7], the eutectic compounds in the as-cast VW94K are identified as $Mg_{24}(Gd, Y)_5$ phases.

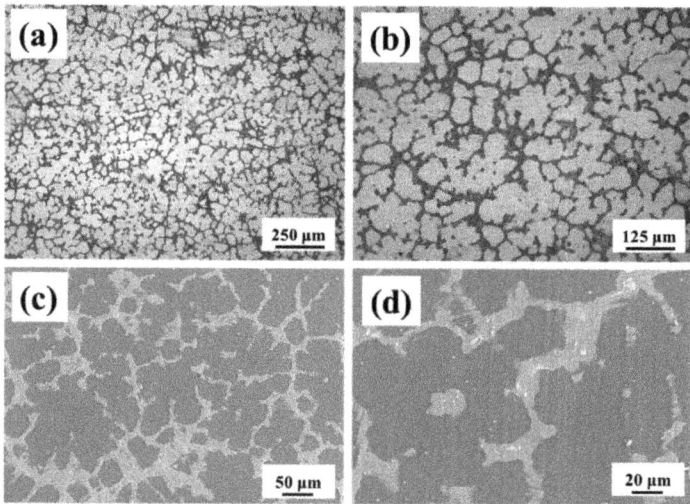

**Figure 2.** Microstructures of the as-cast VW94K alloy: (**a**,**b**) OM images and (**c**,**d**) SEM images.

Figure 3 shows the microstructures of the as-homogenized VW94K alloy. After the homogenization treatment at 510 °C for 12 h, the majority of the eutectic compounds of the $Mg_{24}(Gd, Y)_5$ phases were dissolved into the matrix. As shown in Figure 3d, some unevenly distributed small granular phases with white contrast are observed at the grain boundaries and within grains. These phases are determined to be RE-rich phases [32]. And the average grain size was slightly increased to 81 μm after the solution treatment.

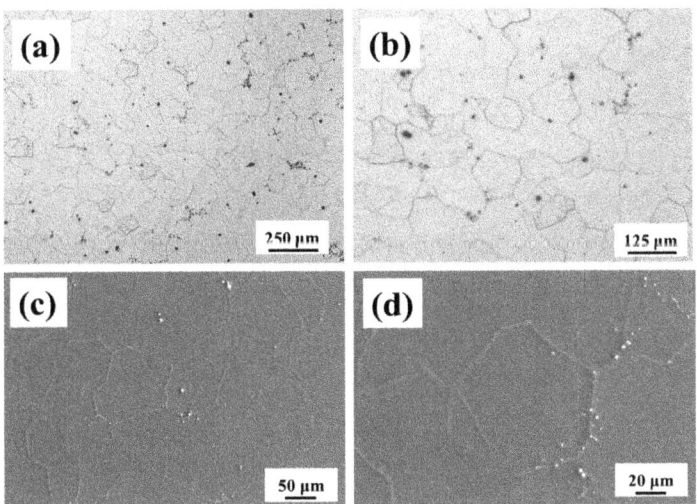

**Figure 3.** Microstructures of the as-homogenized VW94K alloy: (**a,b**) OM images and (**c,d**) SEM images.

*3.2. Tensile Properties*

Figure 4 shows the tensile engineering stress–strain curves of the as-cast and as homogenized VW94K alloys, and the values of yield strength (*YS*), ultimate tensile strength (*UTS*) and elongation to failure are summarized in Table 1. It can be seen that the as-cast VW94K alloy exhibits a *YS* of 141 MPa, a *UTS* of 205 MPa and an elongation-to-failure value of 2.5%, while the as-homogenized VW94K alloy has a YS of 165 MPa, a UTS of 233 MPa and an elongation-to-failure value of 4.6%. It can be noted that compared with the as-cast VW94K alloy, the *YS, UTS* and elongation-to-failure values of the as-homogenized alloy are improved remarkably. After the homogenization treatment, the $Mg_{24}(Gd, Y)_5$ eutectic compounds at the grain boundaries were dissolved into the matrix and formed a supersaturated solid solution, which led to a highly strengthening effect on the solid solution, reduced the microcrack initiation sites and improved the mechanical properties of the as-homogenized alloy. In a comparison of the mechanical properties of the solutioned WE43 alloy [29], the as-homogenized VW94K alloy exhibits higher yield strength and ultimate tensile strength values but shows a lower elongation-to-failure value.

**Table 1.** Tensile properties and fracture toughness values of the as-cast and as-homogenized VW94K alloys.

| Samples | YS/MPa | UTS/MPa | Elongation to Failure/% | $K_{Ic}$ (MPa·m$^{1/2}$) |
|---|---|---|---|---|
| As-cast | 141 ± 3 | 205 ± 2 | 2.5 ± 0.5 | 10.6 ± 0.5 |
| As-homogenized | 165 ± 2 | 233 ± 1 | 4.6 ± 0.3 | 13.8 ± 0.6 |

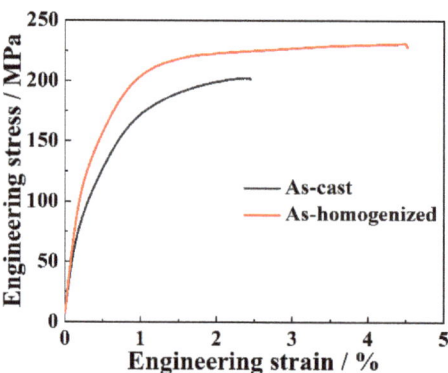

**Figure 4.** Tensile engineering stress–strain curves of the as-cast and as-homogenized VW94K alloys.

*3.3. Fracture Toughness*

The fracture toughness experiments were carried out with a displacement control as a function of the $CMOD$, which was obtained via the clip gauge installed on the pre-designed knife edge. The load–crack mouth opening displacement ($CMOD$) curves of the as-cast and as-homogenized VW94K alloys are presented in Figure 5a. The slopes of the linear regions in the *load–CMOD* curves are quite similar. After a linear increase in the load, both *load–CMOD* curves exhibit non-linear relationships, indicating that the crack tip underwent obvious passivation and plastic deformation. $P_{max}$ is the resistance to the initiation of crack growth. After initiating the crack growth at $P_{max}$, the load gradually decreased. The corresponding conditional load $P_Q$ and maximum load $P_{max}$ are listed in Table 2. The maximum load for the as-homogenized VW94K alloy is greater than the maximum load for the as-cast alloy, which are 4.39 kN and 3.62 kN, respectively. In addition, the $P_Q$ values of the as-cast and as-homogenized VW94K alloys are 2.92 kN and 3.57 kN, respectively. According to Equation (1), the stress intensity factors, $K_Q$, of the as-cast and as-homogenized VW94K alloys are 14.4 MPa·m$^{1/2}$ and 15.3 MPa·m$^{1/2}$, respectively. Obviously, both samples cannot meet the conditions that are expressed in Equation (2). According to the ASTM E399 [34], the stress intensity factors $K_Q$ of the as-cast and as-homogenized VW94K alloys are invalid. Alternatively, the plane strain fracture toughness $K_{Ic}$ can be estimated via a stretched zone ($SZ$) analysis and can be calculated according to the following equation [30,37]:

$$K_{Ic} = \sqrt{\frac{2 \times \lambda \times SZH \times E \times \sigma_{ys}}{1 - v^2}} \quad (4)$$

where $\lambda$ is a constant (=2 [38]), $v$ is Poisson's ratio (=0.35 [39]), and $SZH$, $E$ and $\sigma_{ys}$ are the stretched zone height, elastic modulus and yield strength, respectively. Typical cross-section profiles and three-dimensional (3D) CLSM observations of the fracture surface of the as-cast and as-homogenized VW94K alloys are shown in Figure 5b,c. It can be found that the $SZH$ values of the as-cast and as-homogenized VW94K alloys are 3.9 μm and 5.6 μm, respectively. According to Equation (4), the plane strain fracture toughness (named as $K_{cal}$) values can be calculated to be 10.6 ± 0.5 MPa·m$^{1/2}$ and 13.8 ± 0.6 MPa·m$^{1/2}$, respectively. The values of $SZH$ and fracture toughness $K_{cal}$ are summarized in Table 2. The $K_{cal}$ is smaller than the $K_Q$; hence, the $K_{cal}$ is regarded as the plane strain fracture toughness $K_{Ic}$. Therefore, after the homogenization treatment, the plane strain fracture toughness $K_{Ic}$ value increased from 10.6 ± 0.5 MPa·m$^{1/2}$ up to 13.8 ± 0.6 MPa·m$^{1/2}$.

Table 2. Results of the fracture toughness tests for the as-cast and as-homogenized VW94K alloys.

| Sample | $P_Q$ (kN) | $P_{max}$ (kN) | $K_Q$ (MPa·m$^{1/2}$) | $P_Q/P_{max}$ | SZH (μm) | $K_{cal}$ (MPa·m$^{1/2}$) | $K_{Ic}$ (MPa·m$^{1/2}$) |
|---|---|---|---|---|---|---|---|
| As-cast | 2.92 | 3.62 | 14.4 ± 0.2 | 1.24 | 3.9 ± 0.01 | 10.6 ± 0.5 | 10.6 ± 0.5 |
| As-homogenized | 3.57 | 4.39 | 15.3 ± 0.1 | 1.23 | 5.6 ± 0.01 | 13.8 ± 0.6 | 13.8 ± 0.6 |

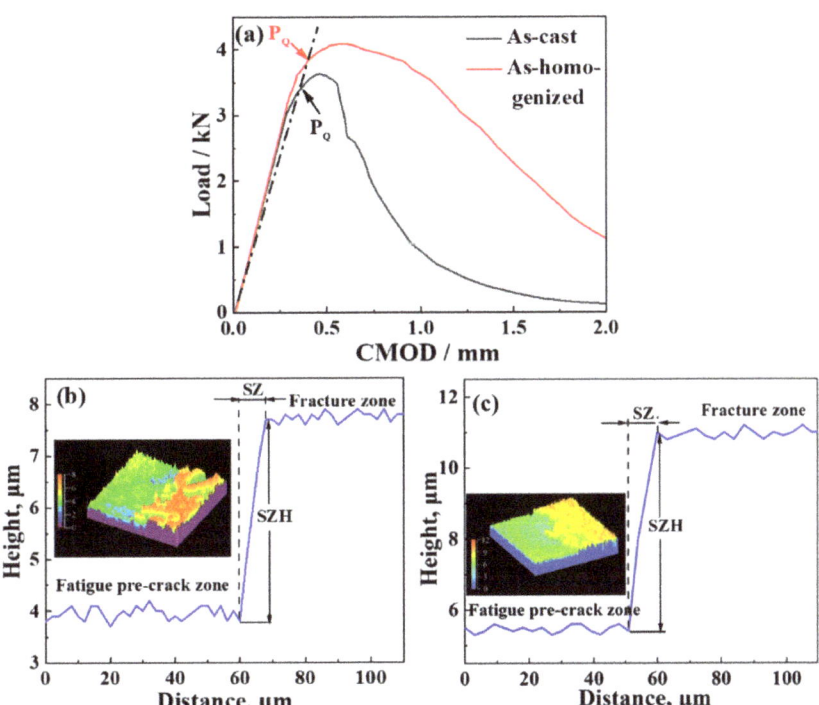

**Figure 5.** (a) Load–CMOD curves of the as-cast and as-homogenized VW94K alloys and the cross-section profiles and 3D observations of the fracture surfaces after fracture toughness tests of (b) the as-cast and (c) the as-homogenized VW94K alloys.

## 4. Discussion

### 4.1. Fractography

The fracture surfaces of the as-cast and as-homogenized VW94K alloys after the plane strain fracture toughness tests are shown in Figure 6. It can be seen that the overall fracture surfaces of the two samples comprise pre-crack regions and fracture regions (Figure 6a). For the pre-crack region, as shown in Figure 6b, there are many cleavage planes and cleavage steps on the fracture surface. It can be seen in Figure 6c,d that the fracture region of the as-cast alloy is typical of massive cleavage steps and secondary cracks. Comparatively, there are many cleavage planes, secondary cracks and tear ridges on the fracture surface of the as-homogenized alloy (Figure 6e,f). Therefore, both the as-cast and as-homogenized samples mainly exhibit brittle fracture characteristics.

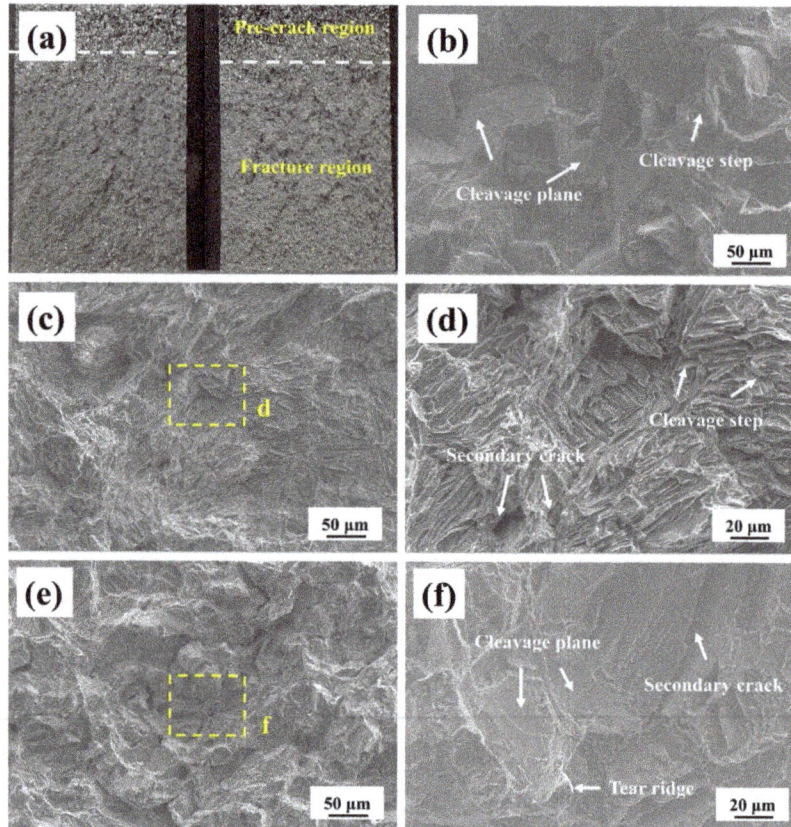

**Figure 6.** Fracture surfaces of the as-cast and as-homogenized VW94K alloys after fracture toughness tests: (**a**) macro-surfaces of two samples, (**b**) the fatigue pre-crack region, (**c**–**f**) the fracture regions, (**c**,**d**) the as-cast sample and (**e**,**f**) the as-homogenized sample.

*4.2. Crack Propagation Mechanism*

Figure 7 shows SEM images of the crack propagation path of the as-cast VW94K alloy in an interrupted fracture toughness test. From Figure 7a, it can be seen that the whole crack propagation path displays a zigzag style. The microcracks mainly initiated in the $Mg_{24}(Gd, Y)_5$ eutectic compounds around the main crack (Figure 7b). The main crack propagates along the brittle eutectic phases at the grain boundaries near the crack tip (Figure 7c). Meanwhile, at the ahead of the crack tip, several secondary cracks initiated and propagated across the $Mg_{24}(Gd, Y)_5$ eutectic phases (Figure 7d).

Figure 8 shows the OM images of the crack propagation path of the as-homogenized VW94K alloy in an interrupted fracture toughness test. It can be observed in Figure 8a that the as-homogenized alloy displays a more tortuous crack path than that of the as-cast alloy, which implies that more energy was consumed during crack propagation and a high level of crack propagation resistance existed. Compared with Reference [40], the fracture toughness of the as-homogenized alloy is greater than that of the LZ91 alloy, which is due to the straight path of the crack propagation in LZ91. From Figure 8b, it can be seen that the microcracks are initiated inside the grain and penetrate through the entire grain near the main crack. At the crack tip, as shown in Figure 8c, the microcracks are apt to propagate in a variety of directions. The microcracks in the same grain are parallel to each other, and these microcracks show a trend of interconnection, which can improve the fracture toughness to some extent due to the additional increase in the crack propagation paths. In the region not far from the main crack tip (Figure 8d), it can be clearly seen that microcracks initiate inside the grain and propagate through the entire grain.

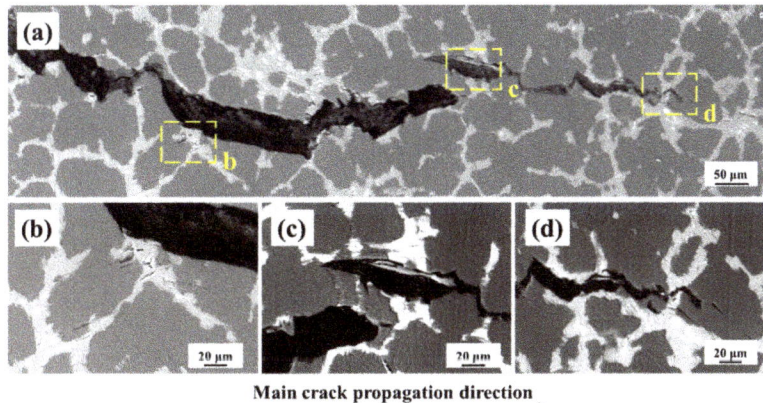

**Figure 7.** SEM images of the crack propagation in an interrupted fracture toughness test of as-cast VW94K alloy: (**a**) the whole crack propagation path, and in the magnification of the selected dashed rectangle: (**b**) region "b", (**c**) region "c" and (**d**) region "d" in (**a**).

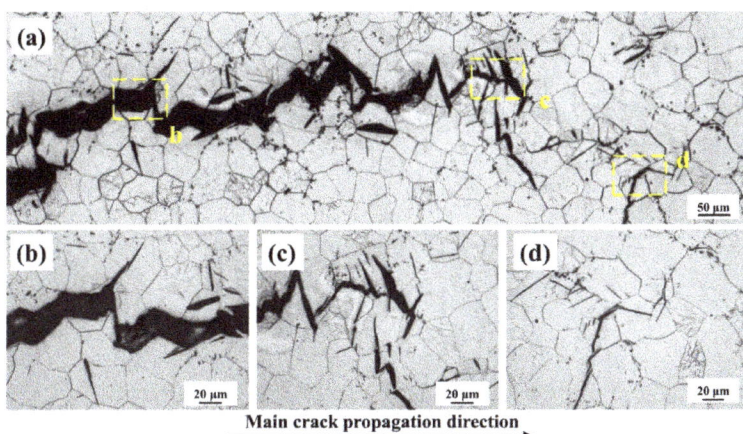

**Figure 8.** OM images of the crack propagation of the as-homogenized VW94K alloy in an interrupted fracture toughness test: (**a**) the whole crack propagation path, and in the magnification of the selected dashed rectangle: (**b**) region "b", (**c**) region "c" and (**d**) region "d" in (**a**).

Figure 9 shows the EBSD information and microcrack morphology of the as-homogenized VW94K alloy at the crack tip of an interrupted fracture toughness test. Figure 9a,b exhibit the band contrast map and inverse pole figure (IPF) map of the sample surface. It is remarkable that most of the microcracks tend to propagate in trans-granular ways. Figure 9c–e illustrate the slip trace analysis in Grains 1, 2 and 3, as marked in Figure 9b. The theoretical slip trace directions for the following slip systems (SSs) were computed using the grain orientation information of each grain: SS 1–3 for basal slip, SS 4–6 for prismatic <a> slip and SS 7–12 for pyramidal <c + a> slip, as shown in Table 3. For Grains 1, 2 and 3, all microcracks are parallel with their basal plane. This indicates that the plastic deformation is mainly dominated by the basal slip.

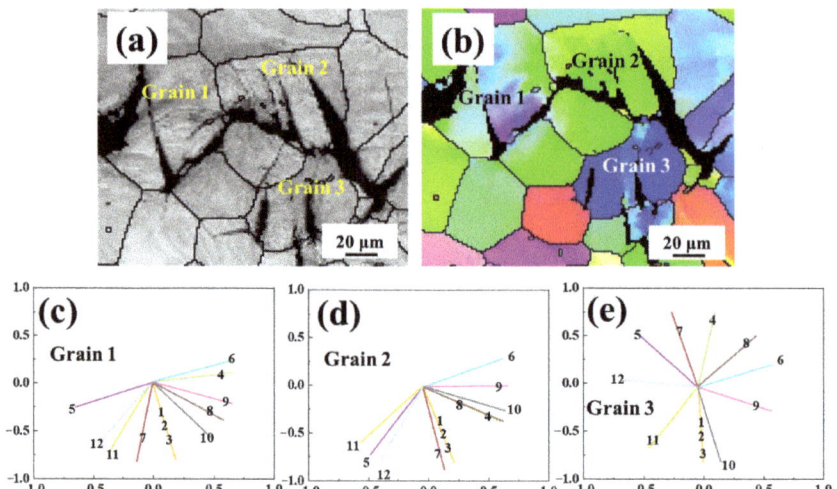

**Figure 9.** The typical microstructures of the as-homogenized VW94K alloy ahead of the crack tip. (**a**) Band contrast map, (**b**) inverse pole figure (IPF) map, (**c**) possible slip traces directions in Grain 1, (**d**) possible slip traces directions in Grain 2 and (**e**) possible slip traces directions in Grain 3.

**Table 3.** Calculated slip systems in Grains 1, 2 and 3.

| Slip System Number | Slip System | | |
|---|---|---|---|
| 1 | Basal <a> | (0001) | $[\bar{2}110]$ |
| 2 | | (0001) | $[\bar{1}2\bar{1}0]$ |
| 3 | | (0001) | $[\bar{1}\bar{1}20]$ |
| 4 | Prismatic <a> | $(01\bar{1}0)$ | $[2\bar{1}\bar{1}0]$ |
| 5 | | $(10\bar{1}0)$ | $[\bar{1}2\bar{1}0]$ |
| 6 | | $(\bar{1}100)$ | $[11\bar{2}0]$ |
| 7 | Pyramidal <c + a> | $(11\bar{2}2)$ | $[\bar{1}\bar{1}23]$ |
| 8 | | $(\bar{1}2\bar{1}2)$ | $[1\bar{2}13]$ |
| 9 | | $(\bar{2}112)$ | $[2\bar{1}\bar{1}3]$ |
| 10 | | $(\bar{1}\bar{1}22)$ | $[11\bar{2}3]$ |
| 11 | | $(1\bar{2}12)$ | $[\bar{1}2\bar{1}3]$ |
| 12 | | $(2\bar{1}\bar{1}2)$ | $[\bar{2}113]$ |

*4.3. Fracture Mechanism*

After the homogenization treatment, the majority of the $Mg_{24}(Gd, Y)_5$ eutectic compounds were dissolved into the matrix, and the tensile strength, ductility and fracture toughness $K_{Ic}$ are improved. Figure 10 shows schematic illustrations of the fracture mechanisms of the as-cast and as-homogenized VW94K alloys in interrupted fracture toughness tests. As illustrated in Figure 10a, the initiation of microcracks in the as-cast alloy was primarily caused by the $Mg_{24}(Gd, Y)_5$ phases. The microcracks then rapidly propagated along the eutectic compounds at the grain boundaries, and the intergranular microcracks were initiated. Figure 10b shows the fracture initiation and propagation of the as-homogenized alloy in an interrupted fracture toughness test. It can be seen from Figure 10b that the fracture initiation in the as-homogenized alloy during the fracture toughness test was caused by the generation of cleavage of microcracks along the basal slip bands, which is consistent with the cleavage facet features observed on the fracture surface. Then the microcracks coalesced to form longer trans-granular cracks.

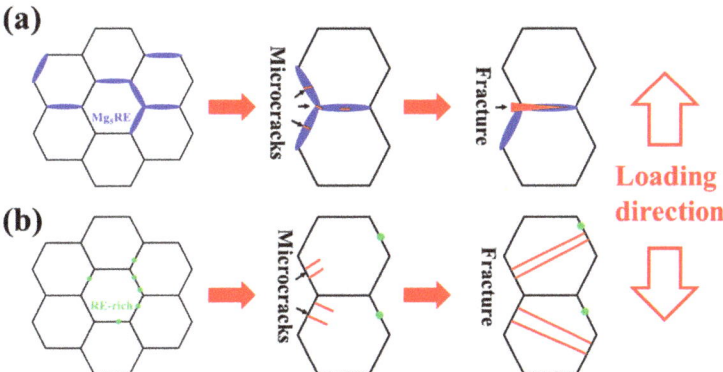

**Figure 10.** Schematic illustration of the fracture mechanism of the (**a**) as-cast and (**b**) as-homogenized VW94K alloys.

The coarse $Mg_{24}(Gd, Y)_5$ phases in the as-cast alloy led to dislocation accumulation and stress concentration at the grain boundaries [41]. The nucleation of microcracks first initiated at the interfaces between the eutectic phases and the α–Mg matrix, and the further coalescence of these microcracks promoted the formation of cracks [28]. After the homogenization treatment, the coarse $Mg_{24}(Gd, Y)_5$ eutectic phases were dissolved into the matrix, so the crack nucleation sites were reduced and the intrinsic resistance to crack propagation was increased, leading to the improvement of the fracture toughness.

## 5. Conclusions

Using OM, SEM and EBSD techniques, the present work clarified in detail the effect of $Mg_{24}(Gd, Y)_5$ eutectic compounds on the fracture toughness and fracture behaviors of the as-cast Mg-9Gd-4Y-0.5Zr (VW94K) alloy. The tensile properties and plane strain fracture toughness of the as-cast and as-homogenized VW94K alloys were investigated. The crack propagation and fracture mechanisms were discussed and the conclusions are summarized as follows:

1. Network-distributed $Mg_{24}(Gd, Y)_5$ eutectic compounds are dissolved into the matrix, and a supersaturated solid solution is obtained after homogenization treatment.
2. The as-homogenized VW94K alloy exhibits greater tensile properties than the as-cast alloy.
3. The plane strain fracture toughness $K_{Ic}$ of the as-cast VW94K alloy is $10.6 \pm 0.5$ MPa·m$^{1/2}$, while that of the as-homogenized alloy is $13.8 \pm 0.6$ MPa·m$^{1/2}$. The improvement of 30.2% in $K_{Ic}$ was achieved via the dissolution of the $Mg_{24}(Gd, Y)_5$ phases.
4. For the as-cast VW94K alloy, the microcracks were initiated and propagated across the $Mg_{24}(Gd, Y)_5$ eutectic compounds. The propagation of the cracks exhibits an intergranular fracture pattern, and the whole crack propagation path displays a zigzag style. Comparatively, the microcracks of the as-homogenized alloy were initiated and propagated along the basal plane. The main crack shows more tortuous fracture characteristics and a transgranular crack propagation pattern.

**Author Contributions:** Conceptualization, Z.J.; methodology, Z.J.; validation, Z.J. and X.Q.; investigation, Z.J., F.C. and G.W.; data curation, Z.J. and C.H.; writing—original draft, Z.J., S.G. and J.H.; writing—review & editing, X.Q. and M.Z.; supervision, X.Q. and M.Z. All authors have read and agreed to the published version of the manuscript.

**Funding:** This research was supported by the National Natural Science Foundation of China (grants no. U21A2047, no. 52071115 and no. 51971076).

**Data Availability Statement:** Not applicable.

**Conflicts of Interest:** The authors declare no conflict of interest.

## References

1. Zhang, Z.; Zhang, J.; Xie, J.; Liu, S.; Fu, W.; Wu, R. Developing a Mg alloy with ultrahigh room temperature ductility via grain boundary segregation and activation of non-basal slips. *Int. J. Plast.* **2023**, *162*, 103548. [CrossRef]
2. Chen, W.; Hou, H.; Zhang, Y.; Liu, W.; Zhao, Y. Thermal and solute diffusion in α-Mg dendrite growth of Mg-5wt.%Zn alloy: A phase-field study. *J. Mater. Res. Technol.* **2023**, *24*, 8401–8413. [CrossRef]
3. Chen, L.; Zhao, Y.; Li, M.; Li, L.; Hou, L.; Hou, H. Reinforced AZ91D magnesium alloy with thixomolding process facilitated dispersion of graphene nanoplatelets and enhanced interfacial interactions. *Mater. Sci. Eng. A* **2021**, *804*, 140793. [CrossRef]
4. Song, J.; She, J.; Chen, D.; Pan, F. Latest research advances on magnesium and magnesium alloys worldwide. *J. Magnes. Alloys* **2020**, *8*, 1–41. [CrossRef]
5. Ahmad, R.; Yin, B.; Wu, Z.; Curtin, W. Designing high ductility in magnesium alloys. *Acta Mater.* **2019**, *172*, 161–184. [CrossRef]
6. Wu, S.; Qiao, X.; Zheng, M. Ultrahigh strength Mg-Y-Ni alloys obtained by regulating second phases. *J. Mater. Sci. Technol.* **2020**, *45*, 117–124. [CrossRef]
7. Pan, J.; Fu, P.; Peng, L.; Hu, B.; Zhang, H.; Luo, A. Basal slip dominant fatigue damage behavior in a cast Mg-8Gd-3Y-Zr alloy. *Int. J. Fatigue* **2019**, *118*, 104–116. [CrossRef]
8. Pan, J.; Peng, L.; Fu, P.; Zhang, H.; Miao, J.; Yue, H.; Luo, A. The effects of grain size and heat treatment on the deformation heterogeneities and fatigue behaviors of GW83K magnesium alloys. *Mater. Sci. Eng. A* **2019**, *754*, 246–257. [CrossRef]
9. Rokhlin, L. *Magnesium Alloys Containing Rare Earth Metals: Structure and Properties*; CRC Press: London, UK, 2003.
10. Zeng, Z.; Stanford, N.; Davies, C.; Nie, J.; Birbilis, N. Magnesium extrusion alloys: A review of developments and prospects. *Int. Mater. Rev.* **2019**, *64*, 27–62. [CrossRef]
11. Wu, S.; Nakata, T.; Tang, G.; Xu, C.; Wang, X.; Li, X.; Qiao, X.; Zheng, M.; Geng, L.; Kamado, S.; et al. Effect of forced-air cooling on the microstructure and age-hardening response of extruded Mg-Gd-Y-Zn-Zr alloy full with LPSO lamella. *J. Mater. Sci. Technol.* **2021**, *73*, 66–75. [CrossRef]
12. Sun, W.; Qiao, X.; Zheng, M.; He, Y.; Hu, N.; Xu, C.; Gao, N.; Starink, M. Exceptional grain refinement in a Mg alloy during high pressure torsion due to rare earth containing nanosized precipitates. *Mater. Sci. Eng. A* **2018**, *728*, 115–123. [CrossRef]
13. Sun, W.; Qiao, X.; Zheng, M.; Xu, C.; Kamado, S.; Zhao, X.; Chen, H.; Gao, N.; Starink, M. Altered ageing behaviour of a nanostructured Mg-8.2Gd-3.8Y-1.0Zn-0.4Zr alloy processed by high pressure torsion. *Acta Mater.* **2018**, *151*, 260–270. [CrossRef]
14. Sun, W.; Qiao, X.; Zheng, M.; Zhao, X.; Chen, H.; Gao, N.; Starink, M. Achieving ultra-high hardness of nanostructured Mg-8.2Gd-3.2Y-1.0Zn-0.4Zr alloy produced by a combination of high pressure torsion and ageing treatment. *Scr. Mater.* **2018**, *15*, 21–25. [CrossRef]
15. Wang, J.; Meng, J.; Zhang, D.; Tang, D. Effect of Y for, enhanced age hardening response and mechanical properties of Mg-Gd-Y-Zr alloys. *Mater. Sci. Eng. A* **2007**, *456*, 78–84. [CrossRef]
16. Liu, W.; Zhou, B.; Wu, G.; Zhang, L.; Peng, X.; Cao, L. High temperature mechanical behavior of low-pressure sand-cast Mg-Gd-Y-Zr magnesium alloy. *J. Magnes. Alloys* **2019**, *7*, 597–604. [CrossRef]
17. Jiang, L.; Liu, W.; Wu, G.; Ding, W. Effect of chemical composition on the microstructure, tensile properties and fatigue behavior of sand-cast Mg-Gd-Y-Zr alloy. *Mater. Sci. Eng. A* **2014**, *612*, 293–301. [CrossRef]
18. Wang, C.; Li, H.; He, Q.; Wu, J.; Wu, G.; Ding, W. Improvements of elevated temperature tensile strengths of Mg-Gd-Y-Zr alloy through squeeze cast. *Mater. Charact.* **2022**, *184*, 111658. [CrossRef]
19. Li, S.; Li, D.; Zeng, X.; Ding, W. Microstructure and mechanical properties of Mg-6Gd-3Y-0.5Zr alloy processed by high-vacuum die-casting. *Trans. Nonferrous Met. Soc. China* **2014**, *24*, 3769–3776. [CrossRef]
20. Li, L.; Zhang, X. Hot compression deformation behavior and processing parameters of a cast Mg-Gd-Y-Zr alloy. *Mater. Sci. Eng. A* **2011**, *528*, 1396–1401. [CrossRef]
21. Ding, Z.; Zhao, Y.; Lu, R.; Yuan, M.; Wang, Z.; Li, H.; Hou, H. Effect of Zn addition on microstructure and mechanical properties of cast Mg-Gd-Y-Zr alloys. *Trans. Nonferrous Met. Soc. China* **2019**, *29*, 722–734. [CrossRef]
22. Somekawa, H.; Mukai, T. Effect of grain refinement on fracture toughness in extruded pure magnesium. *Scr. Mater.* **2005**, *53*, 1059–1064. [CrossRef]
23. Somekawa, H.; Mukai, T. Effect of texture on fracture toughness in extruded AZ31 magnesium alloy. *Scr. Mater.* **2005**, *53*, 541–545. [CrossRef]
24. Somekawa, H.; Singh, A.; Mukai, T. Effect of precipitate shapes on fracture toughness in extruded Mg-Zn-Zr magnesium alloys. *J. Mater. Res.* **2007**, *22*, 965–973. [CrossRef]
25. Lu, S.; Wu, D.; Yan, M.; Chen, R. Achieving high-strength and toughness in a Mg-Gd-Y alloy using multidirectional impact forging. *Materials* **2022**, *15*, 1508. [CrossRef] [PubMed]
26. Liu, W.; Jiang, L.; Cao, L.; Mei, J.; Wu, G.; Zhang, S.; Xiao, L.; Wang, S.; Ding, W. Fatigue behavior and plane-strain fracture toughness of sand-cast Mg-10Gd-3Y-0.5Zr magnesium alloy. *Mater. Des.* **2014**, *59*, 466–474. [CrossRef]
27. Wang, Q.; Xiao, L.; Liu, W.; Zhang, H.; Cui, W.; Li, Z.; Wu, G. Effect of heat treatment on tensile properties, impact toughness and plane-strain fracture toughness of sand-cast Mg-6Gd-3Y-0.5Zr magnesium alloy. *Mater. Sci. Eng. A* **2017**, *705*, 402–410. [CrossRef]
28. Xu, C.; Zheng, M.; Chi, Y.; Chen, X.; Wu, K.; Wang, E.; Fan, G.; Yang, P.; Wang, G.; Lv, X.; et al. Microstructure and mechanical properties of the Mg-Gd-Y-Zn-Zr alloy fabricated by semi-continuous casting. *Mater. Sci. Eng. A* **2012**, *549*, 128–135. [CrossRef]
29. Jiang, H.; Zheng, M.; Qiao, X.; Wu, K.; Peng, Q.; Yang, S.; Yuan, Y.; Luo, J. Microstructure and mechanical properties of WE43 magnesium alloy fabricated by direct-chill casting. *Mater. Sci. Eng. A* **2017**, *684*, 158–164. [CrossRef]

30. Ji, Z.; Qiao, X.; Hu, C.; Yuan, L.; Cong, F.; Wang, G.; Xie, W.; Zheng, M. Effect of aging treatment on the microstructure, fracture toughness and fracture behavior of the extruded Mg-7Gd-2Y–1Zn-0.5Zr alloy. *Mater. Sci. Eng. A* **2022**, *849*, 143514. [CrossRef]
31. Ji, Z.; Qiao, X.; Yuan, L.; Cong, F.; Wang, G.; Zheng, M. Exceptional fracture toughness in a high-strength Mg alloy with the synergetic effects of bimodal structure, LPSO, and nanoprecipitates. *Scr. Mater.* **2023**, *236*, 115675. [CrossRef]
32. Chi, Y.; Xu, C.; Qiao, X.; Zheng, M. Effect of trace zinc on the microstructure and mechanical properties of extruded Mg-Gd-Y-Zr alloy. *J. Alloys Compd.* **2019**, *789*, 416–427. [CrossRef]
33. Chi, Y.; Liu, J.; Zhou, Z.; Wu, S.; Liu, W.; Zheng, M. Investigation on the microstructure, texture and mechanical properties of Mg-Gd-Y(-Zn)-Zr alloys under indirect extrusion. *J. Alloys Compd.* **2023**, *943*, 169061. [CrossRef]
34. *ASTM E399*; Standard Test Method for Linear-Elastic Plane-Strain Fracture Toughness of Metallic Materials. ASTM: West Conshohocken, PA, USA, 2017; pp. 1–38.
35. Zhu, G.; Wang, L.; Zhou, H.; Wang, J.; Shen, Y.; Tu, P.; Zhu, H.; Liu, W.; Jin, P.; Zeng, X. Improving ductility of a Mg alloy via non-basal <a> slip induced by Ca addition. *Int. J. Plast.* **2019**, *120*, 164–179.
36. *ASTM E112*; Standard Test Methods for Determining Average Grain Size. ASTM: West Conshohocken, PA, USA, 2012.
37. Taisuke, S.; Somekawa, H.; Takara, A.; Nishikawa, Y.; Higashi, K. Plane-strain fracture toughness on thin AZ31 wrought magnesium alloy sheets. *Mater. Trans.* **2003**, *24*, 986–990.
38. Higashi, K.; Ohnishi, T.; Komatsu, K.; Nakatani, Y. Evaluation of fracture toughness of 5083 and 7075 alloys by stretched zone analysis. *J. Jpn. Inst. Light Met.* **1981**, *31*, 720. [CrossRef]
39. Handbook, A. *Magnesium and Magnesium Alloys, Materials Park*; ASM International: West Conshohocken, PA, USA, 1999.
40. Rahmatabadi, D.; Pahlavani, M.; Bayati, A.; Hashemi, R.; Marzbanrad, J. Evaluation of fracture toughness and rupture energy absorption capacity of as-rolled LZ71 and LZ91 Mg alloy sheet. *Mater. Res. Express.* **2019**, *6*, 36517. [CrossRef]
41. Lu, X.; Zhao, G.; Zhou, J.; Zhang, C.; Yu, J. Microstructure and mechanical properties of the as-cast and as-homogenized Mg-Zn-Sn-Mn-Ca alloy fabricated by semicontinuous casting. *Materials* **2018**, *11*, 703. [CrossRef]

**Disclaimer/Publisher's Note:** The statements, opinions and data contained in all publications are solely those of the individual author(s) and contributor(s) and not of MDPI and/or the editor(s). MDPI and/or the editor(s) disclaim responsibility for any injury to people or property resulting from any ideas, methods, instructions or products referred to in the content.

Article

# Numerical Simulation of Temperature Field during Electron Beam Cladding for NiCrBSi on the Surface of Inconel 718

Guanghui Zhao [1,2,3], Yu Zhang [1,2], Juan Li [1,2,*], Huaying Li [1,2], Lifeng Ma [1,2] and Yugui Li [1,2]

1. Engineering Research Center Heavy Machinery Ministry of Education, Taiyuan University of Science and Technology, Taiyuan 030024, China; zgh030024@163.com (G.Z.); zhangyu20230410@163.com (Y.Z.); huayne@163.com (H.L.); mlf060913@163.com (L.M.); lyg060913@163.com (Y.L.)
2. Shanxi Provincial Key Laboratory of Metallurgical Device Design Theory and Technology, Taiyuan University of Science and Technology, Taiyuan 030024, China
3. Upgrading Office of Modern College of Humanities and Sciences of Shanxi Normal University, Linfen 041000, China
* Correspondence: lijuanhello@163.com

**Abstract:** This study investigates the Inconel 718 alloy coated with NiCrBSi powder using the ABAQUS software. An accurate conical heat source model is constructed based on the three-dimensional Fourier heat conduction law. The heat source subroutine Dflux.for is successfully integrated to achieve a highly realistic simulation of the welding heat source. Using this model, the analysis focuses on the temperature distribution in electron beam melting. Furthermore, the accuracy and reliability of the simulation are validated through actual coating experiments. By examining the impact of various procedural factors on the temperature distribution, it is found that optimal coating results and a tightly formed elliptical molten zone are attained at an electron beam current of 18 mA, and the scanning speed is 300 mm/min. The peak temperature in the melt pool in the coating area is 5087 K, while the lowest temperature on the isothermal in the heat-affected zone is 1409 K. Over time, there is a swift rise in temperature for the data points taken along both the X and Z trajectories, followed by rapid cooling after rapid heating. Coating experiments conducted under the optimal parameters demonstrate a dense coating layer and good bonding with the substrate, thereby validating the accuracy of the simulation.

**Keywords:** Inconel 718 alloy; laser cladding; temperature field; numerical simulation

**Citation:** Zhao, G.; Zhang, Y.; Li, J.; Li, H.; Ma, L.; Li, Y. Numerical Simulation of Temperature Field during Electron Beam Cladding for NiCrBSi on the Surface of Inconel 718. *Crystals* **2023**, *13*, 1372. https://doi.org/10.3390/cryst13091372

Academic Editor: José L. García

Received: 7 August 2023
Revised: 23 August 2023
Accepted: 29 August 2023
Published: 14 September 2023

**Copyright:** © 2023 by the authors. Licensee MDPI, Basel, Switzerland. This article is an open access article distributed under the terms and conditions of the Creative Commons Attribution (CC BY) license (https://creativecommons.org/licenses/by/4.0/).

## 1. Introduction

The mechanical characteristics of Inconel 718 alloy are exceptional in both short-term and extended durations when exposed to cyclic atmospheres of elevated temperature with oxidation and carburization, rendering it especially fitting for applications demanding high-temperature endurance. Therefore, it is widely used in high-temperature structural components such as turbine blades and vanes [1,2]. However, because of the harsh conditions involving both high-temperature corrosion and wear, the part's performance might deteriorate. Predominantly, surface wear of materials happens, and the utilization of laser cladding techniques can notably enhance the surface characteristics of the Inconel 718 alloy [3].

Laser cladding is a direct metal deposition technology in which a layer or multiple layers of cladding material are melted and deposited onto the substrate using a laser electron beam to repair and rebuild worn parts [4]. By melting and swiftly cooling the surface coating of the base material, a metallurgically bonded cladding layer is produced, which adheres to the substrate [5,6]. This method greatly enhances the surface characteristics of metals and finds extensive application in sophisticated sectors like the aerospace, chemical, and pharmaceutical industries [7–10]. However, in the process of laser cladding, factors such as the size of the laser spot, scanning method, and powder layer thickness need to be considered, as these factors can significantly impact the quality of the final cladding layer [11,12]. Moreover, variables like the shape and dimensions of the powder particles, nozzle configuration, and the rate of powder feed throughout the cladding procedure can impact the dispersion of particle concentration between the nozzle and the substrate [13,14]. It is worth noting that variations in process parameters such as laser power and scanning speed [15,16] and adjustments to the scanning path directly affect the distribution of melt pool temperature and workpiece surface temperature. The increased concentration of transient heat input caused by these factors may lead to thermal stresses and deformations during and after cladding. These issues directly affect samples' quality and performance in practical applications.

Observing real-time alterations in the molten pool during conventional laser cladding experiments presents a formidable task [17]. Therefore, it is essential to accurately understand the temperature field and gradient distribution during the cladding process [18,19]. Numerical simulations are employed to construct 3D models that replicate the laser cladding procedure under various processing conditions [20]. Following this, the simulation outcomes are examined to ascertain appropriate processing variables, culminating in the execution of experiments. This approach proves advantageous in minimizing time and cost expenditures.

Currently, research on the simulation of temperature fields in electron beam cladding still needs to be completed. Yang et al. [21] analyzed the surface temperature distribution in the Inconel 718 single-clad model on 45 steel using the finite element method, investigating the influence of different laser powers and scanning speeds on the center temperature of the melt pool. On the other hand, Gan et al. [22] introduced a numerical model for laser cladding that simulates heat exchange, fluid movement, solidification, and multi-component mass transfer during the direct laser deposition of cobalt-based alloy onto steel. The analysis of transient heat dispersion facilitated the extraction of the solidified microstructure's morphology. He et al. [23] combined simulation and experimental methods to investigate the laser cladding of F102 nickel-based alloy powder on the surface of 40CrNi$_2$Si$_2$MoVA steel. Hoffman et al. [24] simulated the configuration and blending aspects of laser cladding, offering an in-process technique for managing dilution. This method has been adopted to enhance the overall workpiece quality. However, there still needs to be more research on controlling the temperature of the NiCrBSi cladding layer on the surface of the Inconel 718 alloy and selecting reasonable process parameters.

Electron beam cladding (EBC) and laser cladding (LC) are both directed energy deposition (DED) techniques that can be used to deposit NiCrBSi on the surface of Inconel 718. Both processes involve melting a powder or wire feedstock with a high-energy beam and creating a metallurgical bond with the substrate. However, there are some differences between EBC and LC that affect the quality and performance of the cladding layer. One of the main differences is the energy source and the working environment. EBC uses a focused electron beam that operates in a vacuum chamber, while LC uses a focused laser beam that operates in an inert gas atmosphere. The vacuum environment of EBC reduces the oxidation and contamination of the molten pool, resulting in a higher purity and lower porosity of the cladding layer [25]. The vacuum also enables a higher energy density and deeper penetration of the electron beam, which can improve the bonding strength and reduce the dilution of the substrate [26]. However, the vacuum also increases the cooling rate and thermal gradient of the cladding layer, which can induce higher residual stresses and distortions [27]. Another difference is the beam shape and distribution. EBC typically uses a circular or elliptical beam with a Gaussian intensity profile, while LC can use various beam shapes and profiles depending on the type of laser source. For example, solid-state lasers can produce rectangular or square beams with uniform or top-hat intensity profiles. The beam shape and profile can affect the geometry and uniformity of the cladding layer, as well as the heat input and melt pool dynamics. A rectangular or square beam can produce a wider and flatter cladding layer than a circular or elliptical beam, which can improve the surface quality and reduce the number of passes required. A uniform or top-hat intensity profile can also produce a more stable and symmetrical melt pool than a Gaussian profile, which can reduce the spatter and porosity of the cladding layer. Therefore, EBC and LC have different advantages and disadvantages for depositing NiCrBSi on Inconel 718. EBC can produce a higher quality cladding layer in terms of purity, porosity, bonding strength, and dilution, but it also requires a vacuum chamber and may cause higher residual stresses and distortions. LC can produce a better surface quality and geometry of the cladding layer, but it may also cause more oxidation, contamination, spatter, and porosity.

The objective of this research is to develop a 3D model for exploring how temperature is distributed within cladding layers using diverse process parameters. A sophisticated double-ellipsoid heat source for welding was devised for this purpose. A temperature-field numerical simulation of electron beam cladding of NiCrBSi powder on the surface of Inconel 718 was performed using ABAQUS. The simulation obtained reasonable processing parameters for laser cladding of NiCrBSi coating. The established 3D model was used to analyze the temperature distribution of the melt pool during single-track cladding under different process parameters. The temperature field distribution patterns of the model were analyzed using a temperature selection judging mechanism under foreign laser powers and scanning speeds.

## 2. Methods

Currently, commonly used electron beam heat source models include the Gaussian heat source model, the dual ellipsoid heat source model, and the conical heat source model. The outstanding feature of the conical heat source model is that its diameter decreases linearly in the longitudinal direction of the model. Any cross-section of the heat source has the distribution characteristics of a Gaussian heat source, and it can be approximately considered a rotating body heat source model with a diameter that attenuates in a specific regular pattern with the depth of heat flow [28]. This heat source model has a sizeable depth-to-width ratio and intense penetration and is generally used for the welding simulation of deep holes or medium-thick plates.

Figure 1 shows a schematic diagram of the conical heat source model. Its shape is conical, with a larger radius on the upper surface and a smaller radius on the lower surface. The cross-sections of the model at various positions on the Z-axis are all circles, and the heat flux density values at different places on the Z-axis are the same. The expression for the conical heat source model is as follows:

$$Q_v = \frac{9Q_0}{\pi(1-e^{-3})} \times \frac{1}{(Z_e - Z_i) \times (r_e^2 + r_e r_i + r_i^2)} \times \exp\left(-\frac{3r^2}{r_c^2}\right) \quad (1)$$

In the formula: $Q_0 = \eta p$

$$r_c = f(z) = r_i + (r_e - r_i)\frac{z - z_i}{z_e - z_i} \quad (2)$$

Here, $Q_v$ is the volumetric heat flux, $Q_0$ is the net heat flow, $P$ is the energy of the electron beam, $\eta$ is the efficiency value, $r$ is the radius function with respect to $x$ and $y$, $r_c$ is the thermal distribution coefficient with respect to depth $z$, $r_e$ and $r_i$ are the maximum and minimum radii, while $z_e$ and $z_i$ are the maximum and minimum values in the axial direction.

The choice of heat source model depends on various factors, such as the shape of the molten pool, the heat transfer zone, and the electron beam simulation process. Since this study focuses on vacuum electron beam welding, its penetrability is extremely strong. In this state, the finite element analysis of the molten pool often uses either a conical heat source model or a dual ellipsoid model. The conical heat source model can better simulate the energy distribution of the electron beam inside the powder layer, reflecting the scattering and absorption effects of the electron beam and thus more accurately calculating the depth distribution of the temperature field. This is significant for studying phase transformation, stress, cracks, and other phenomena in the electron beam cladding process.

The accuracy of the conical heat source model is also relatively high, and it can be validated and compared with experimental data or other theoretical models. Compared with the dual ellipsoid heat source model and the Gaussian heat source model, the conical heat source model exhibits higher consistency and stability in calculating the temperature field of electron beam cladding. Therefore, this study adopts the conical heat source model.

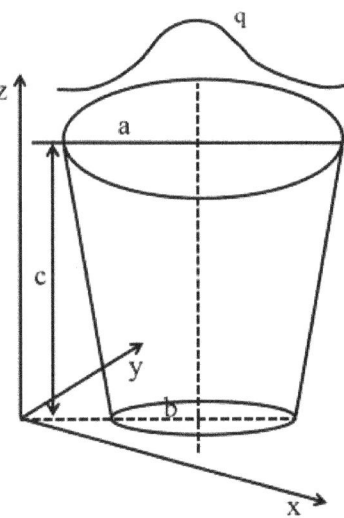

**Figure 1.** Model of a conical heat source.

## 3. Results

*3.1. Development of ABAQUS Finite Element Model*

The workpiece model was established using the ABAQUS finite element simulation software (2023 version, developed by Dassault Systèmes, a company headquartered in France), as shown in Figure 2. The experimental material used was a hot-rolled Inconel 718 sheet produced by Shanghai Baoshao Special Steel Co., Ltd. (Shanghai, China) A wire-cutting method was employed to obtain a plate specimen with dimensions of 50 mm × 30 mm × 10 mm (length × width × height). As shown in Figure 2 below, the lower part represents the Inconel 718 base material. In contrast, the upper part consists of a one-millimeter-thick sodium silicate binder and a mixture of NiCrBSi powder for the fused coating. The red circle indicates the starting point of electron beam scanning, and the red dashed line represents the scanning direction.

*3.2. Calculation of Thermal Property Parameters in Temperature Field*

This study simulates the temperature field during electron beam scanning using the finite element software ABAQUS. After constructing a three-dimensional model, it is necessary to set the thermal properties of the model and assign them to different objects. The thermal properties of materials vary with temperature. Common metallic materials can be obtained from the literature [29], while some uncommon materials can be calculated using Jmatpro software. The commonly used thermal properties in temperature field simulation include density, conductivity, specific heat, and latent heat [30,31].

The thermal properties of the Inconel 718 substrate are shown in Table 1. The thermal property curve of Inconel 718, a nickel-based alloy, is calculated based on the elemental mass content using Jmatpro and shown in Figures 3 and 4. The melting point of Inconel 718 is 1260–1320 °C, or 1533–1593 K. In the ABAQUS finite element software, the Kelvin scale is commonly used as the temperature unit.

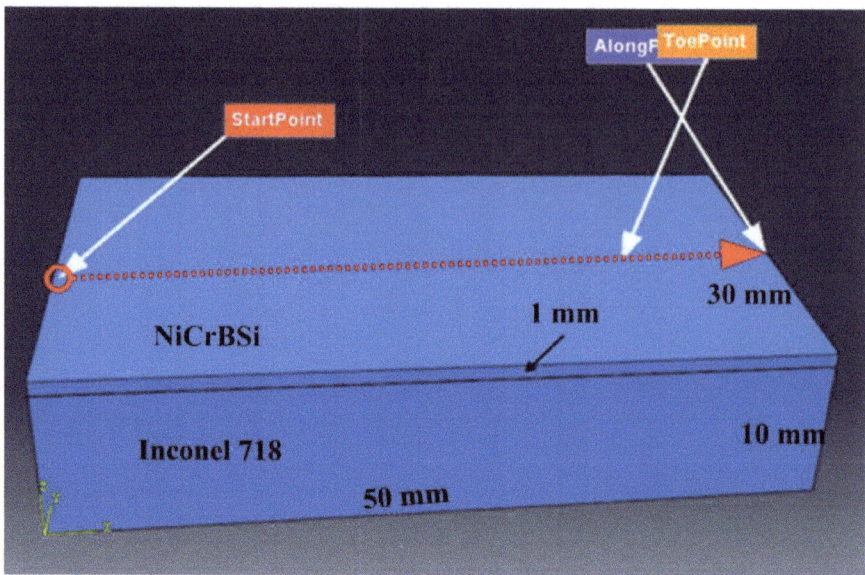

**Figure 2.** Electron beam cladding finite element model.

**Table 1.** Thermal physical parameters of the matrix Inconel 718.

| Temperature (k) | Thermal Conductivity (W/m·k) | Specific Heat (J·g$^{-1}$K$^{-1}$) | Density (kg/m$^3$) |
| --- | --- | --- | --- |
| 293 | 22 | 0.44 | 8240 |
| 473 | 25 | 0.5 | 8240 |
| 673 | 19 | 0.6 | 8240 |
| 873 | 20 | 0.65 | 8240 |
| 1073 | 23 | 0.72 | 8240 |
| 1273 | 26 | 0.8 | 8240 |
| 1473 | 29 | 0.62 | 8240 |
| 1673 | 32 | 0.75 | 8240 |

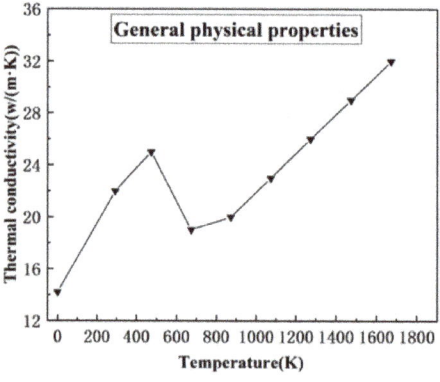

**Figure 3.** Thermal conductivity curve of Inconel 718.

### 3.3. Division of ABAQUS Mesh and Boundary Conditions

Grid division plays a vital role in finite element software's calculation accuracy and computational efficiency. Generally speaking, the finer the grid division, the higher the calculation accuracy. However, when the grid is refined to a certain level, further reducing the grid size increases the computer's computational time and has little contribution to the calculation results [32]. Therefore, grid division is also an essential step in finite element analysis. In this study, vacuum electron beam surface cladding demands substantial computational resources for the uppermost layer, with relatively reduced demands for regions farther from the substrate. Therefore, when dividing the grid, a finer 1 mm hexahedral grid is used for the cladding layer, while a 2 mm hexahedral grid is used for the substrate. This approach can improve computational efficiency. The grid division is shown in Figure 5.

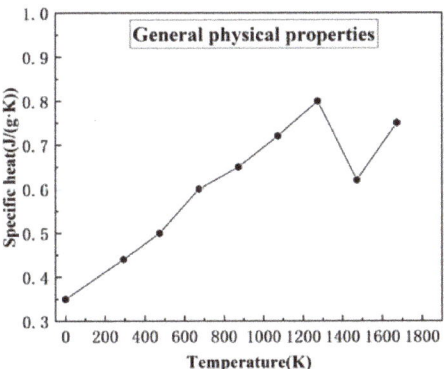

**Figure 4.** The specific heat curve of Inconel 718.

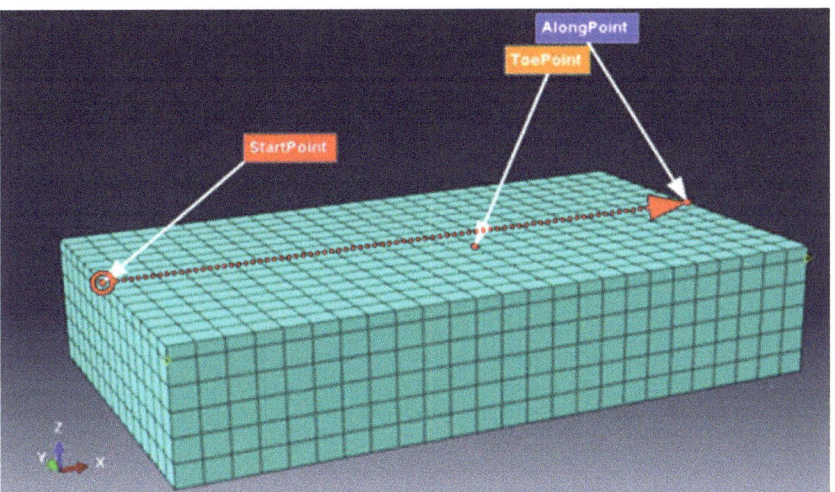

**Figure 5.** Model meshing.

Setting the boundary conditions for the electron beam cladding model:

(1) The starting temperature for the electron beam cladding is at room temperature, so the initial temperature for analysis is set at 20 °C, equivalent to 293 K. (2) During the movement of the electron beam welding heat source, conduction heat transfer is accounted for in both the sections exposed to radiation and those that are not. Temperature discrepancies occur across distinct regions of the workpiece, leading to heat transfer from higher to lower temperatures. This satisfies the expression:

$$K\frac{\partial T}{\partial X}n_x + K\frac{\partial T}{\partial Y}n_y + K\frac{\partial T}{\partial Z}n_z = \beta(T_\sigma - T_s) \qquad (3)$$

In the formula: $T_s$ represents the temperature in the irradiated region by the electron beam, $T_\sigma$ represents the temperature in the non-irradiated region, $\beta$ represents the overall heat transfer coefficient, $n_x$, $n_y$ and $n_z$ represents the cosine value at the outer normal.

## 4. Results and Discussion

The quality of electron beam surface cladding modification mainly depends on the degree of bonding between the cladding layer and the substrate. Given the selected coating powder and substrate, the parameter settings of the electron beam have the most significant impact during the cladding process. Through repeated practice, we have found that the parameters with the most considerable influence are the scan speed of the electron gun and the beam current. These two factors have a substantial impact on the temperature field's variation. At the same time, the degree of bonding between the cladding layer and the substrate largely depends on the interpretation of the molten pool temperature field [31]. Therefore, it is necessary to explore the effects of different scan speeds and beam currents on the molten pool temperature field to find the optimal parameters.

### 4.1. The Influence of Different Beam Distributions on the Temperature Field Distribution

When investigating the impact of various beam currents on the temperature distribution, the parameter scope was refined based on insights from the existing literature. Considering the equipment situation and experience, four different beam currents of 15 mA, 18 mA, 21 mA, and 24 mA were selected to investigate beam variation's effect on the cladding's quality. To ensure the accuracy of the experimental results, this simulation used the method of controlling variables to keep the other parameters outside the beam unchanged. The simulation parameters are shown in Table 2.

**Table 2.** Process parameters for electron beam cladding with different beam currents.

| Sample | Acceleration Voltage (kV) | Beam Current (mA) | Scanning Speed (mm/min) | Scanning Radius (mm) |
|---|---|---|---|---|
| 1 | 60 | 15 | 300 | 4 |
| 2 | 60 | 18 | 300 | 4 |
| 3 | 60 | 21 | 300 | 4 |
| 4 | 60 | 24 | 300 | 4 |

By using ABAQUS finite element simulation software and linking the Dflux.for subroutine, the conical heat source model is loaded onto the surface of the fusion layer finite element mesh segmentation. Taking the coordinate point (0, 12, 10) as the starting point of the electron beam, the electron beam was scanned along the positive x-axis for cladding. A distinct cladding trace, called the clad track, was formed where the electron beam was scanned. The analysis focused on temperature distribution maps of the molten pool under varying beam currents at t = 6.25 s, as depicted in Figure 6.

From Figure 6, it can be seen that as the beam current increases, the peak temperature of the workpiece surface also increases, indicating a positive correlation between the beam current and the temperature of the molten pool. At 15 mA, the temperature field diagram (depicted in Figure 6a) illustrates a maximum temperature of 4274 K within the molten pool, forming a molten pool on the surface of the overlay layer. The lowest temperature of the isotherm in the heat-affected zone is 1287 K. The temperature distribution map of the melted pool at 18 mA is displayed in Figure 6b, indicating a maximum temperature of 5087 K, forming a molten pool on the surface of the overlay layer. The lowest temperature of the isotherm in the heat-affected zone is 1409 K. In Figure 6c, the temperature mapping within the molten pool at 21 mA reveals a maximum temperature reaching 5667 K, forming a molten pool on the surface of the overlay layer. The lowest temperature of the isotherm in the heat-affected zone is 1635 K. The temperature distribution within the molten pool when the beam current is set at 24 mA is depicted in Figure 6d, highlighting a peak temperature of 6522 K, forming a molten pool on the surface of the overlay layer. The lowest temperature of the isotherm in the heat-affected zone is 1848 K. When the beam current is increased by three mA, the peak temperature increases by 813 K, and the lowest temperature in the heat-affected zone increases by 122 K. When the beam current is increased by six mA, the peak temperature increases by 1393 K, and the lowest temperature in the heat-affected zone rises by 348 K. When the beam current is increased by nine mA, the peak temperature increases by 2248 K, and the lowest temperature in the heat-affected zone increases by 561 K. The reason for this is that when the electron beam's beam current increases, the electron beam's power also increases, leading to an increase in heat flux density. This directly increases the temperature of the molten pool. With the rise of the beam current, the high temperature causes more intense reactions between the elements in the overlay layer, and the conversion rate from the solid phase to the liquid phase also increases, resulting in a denser microstructure. However, if the beam current of the electron beam is too large, the temperature in the molten pool region becomes too high. The dilution rate of NiCrBSi powder will be increased, causing the electron beam to load directly onto the substrate, leading to a decrease in the thickness of the bond layer and severe deformation of the entire overlayed sample. If the beam current of the electron beam is too small, the temperature in the molten pool is too low. The surface material of the substrate cannot fully dissolve in the NiCrBSi powder, failing to form a metallurgical bond between the powder and the substrate. The powder will melt into spherical material and fall directly off.

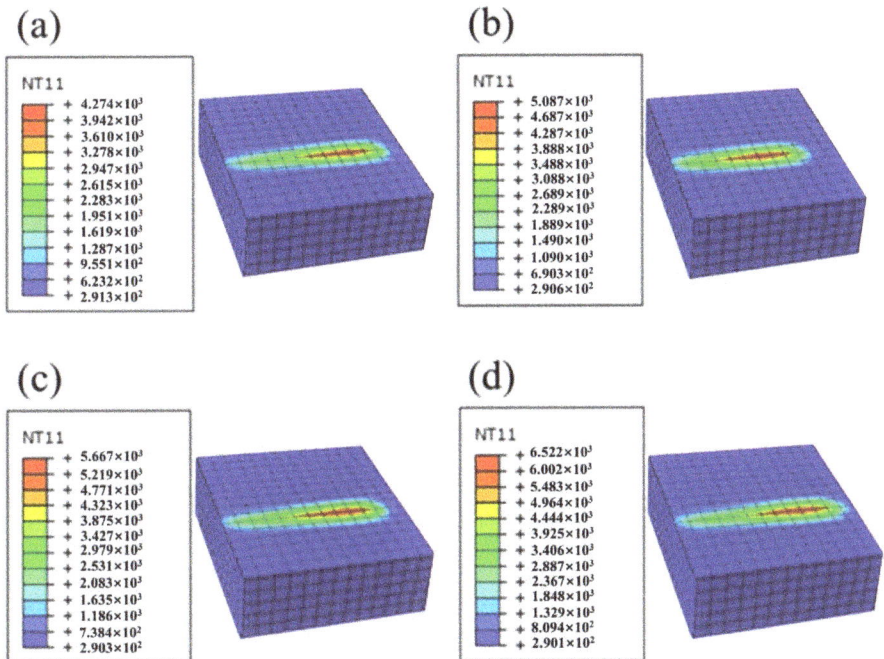

**Figure 6.** Temperature field clouds of the melt pool at different Iw at t = 6.25 s: (**a**) Iw = 15 mA; (**b**) Iw = 18 mA; (**c**) Iw = 21 mA; (**d**) Iw = 24 mA.

Since the conduction and variation of the junction temperature are directly related to the bonding quality between powder and substrate, a node set is established between the matrix and powder coating in ABAQUS, with the coordinates of the nodes as (8, 12, 9). The temperature of the junction area is then extracted for analysis, as shown in Figure 7. Figure 8 shows the temperature variation curve of the surface layer for different beam currents with the node coordinates of (8, 12, 10). When the beam current is 15 mA, the surface temperature reaches 3300 K, which is above the melting point of the powder NiCrBSi (1323 K). The temperature between the substrate and the overlay is approximately 1300 K, whereas the melting point of the substrate is around 1500 K. The optimal process parameters for electron beam cladding are the powder melting while the substrate is partially melted. When the beam current is 15 mA, excluding the influence of errors, the junction area is closer to the melting point of the substrate. This indicates that only a thin layer of overlay is formed on the surface of the substrate, with a good surface roughness but low bonding strength, making it not the best process parameter. When the beam current is 18 mA, the peak temperature of the surface layer is around 4200 K, and the NiCrBSi powder is completely melted into a liquid phase. The temperature of the junction area between the substrate and powder is around 1500 K, reaching the melting point temperature of the Inconel 718 substrate. The substrate surface undergoes a solid-to-liquid phase transition. This represents the best process condition, where the powder is melted while the substrate is partially melted. At this point, the metallurgical bonding layer formed by the powder and substrate is denser, and the temperature gradient distribution is reasonable. When the beam current is 21 mA, the surface temperature reaches 4700 K, but the temperature of the junction area is around 1690 K. Although the NiCrBSi powder is completely melted, the junction temperature exceeds the melting point of the Inconel 718 substrate. This will cause overheating of the substrate and excessive dilution of the powder, resulting in poor comprehensive mechanical properties of the cladding layer. It will also cause a deeper surface melt pool, which is considered to be overheating and not an ideal parameter. When the beam current is 24 mA, the temperature of the junction area reaches 1800 K, far higher than the melting point of the substrate. At this point, the workpiece is severely overheated. In summary, the temperature gradient distribution is more reasonable for the electron beam current of 18 mA.

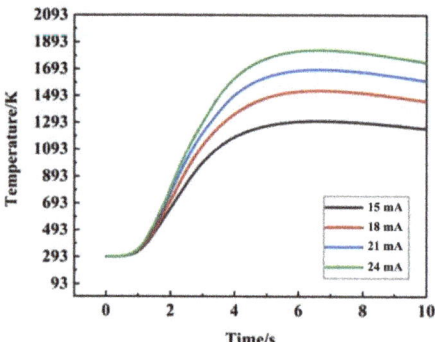

**Figure 7.** The temperature variation patterns of different beam merging region nodes.

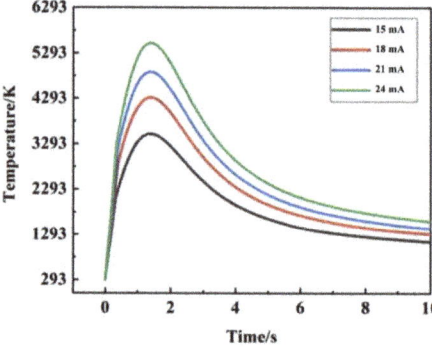

**Figure 8.** The variation pattern of the peak temperature in the molten pool under different beam currents.

When the beam current is 21 mA, the surface temperature reaches 4700 K, but the temperature in the binding zone only reaches around 1690 K. Although the NiCrBSi powder is completely melted, the high temperature in the critical area exceeds the melting point of the Inconel 718 substrate. This will lead to overheating of the substrate and excessive dilution of the powder, resulting in poor overall mechanical properties of the cladding layer. It will also cause deep surface melting, which is considered overheating and not an ideal parameter. When the beam current is 24 mA, the temperature in the binding zone reaches 1800 K, far exceeding the melting point of the substrate. At this point, the workpiece is severely overheated. In summary, the temperature gradient distribution is reasonable when the beam current of the electron beam is 18 mA.

### 4.2. The Influence of Different Scanning Speeds on the Temperature Field

When an electron beam is used for energy scanning, the scanning speed directly affects the aggregation of thermal energy, the absorption rate of thermal energy, and the rate of thermal energy loss. These factors directly impact the degree of the powder and substrate bonding. Therefore, the scanning speed is essential for the quality of fusion in the temperature field. In this simulation, other variables will be controlled while maintaining a beam current of 18 mA to explore and optimize the scanning speed. The simulation parameters are shown in Table 3.

**Table 3.** Process parameters for electron beam cladding at different speeds.

| Sample | Accelerating Voltage (kV) | Beam Current (mA) | Scanning Speed (mm/min) | Scanning Radius (mm) |
| --- | --- | --- | --- | --- |
| 1 | 60 | 18 | 180 | 4 |
| 2 | 60 | 18 | 240 | 4 |
| 3 | 60 | 18 | 300 | 4 |
| 4 | 60 | 18 | 360 | 4 |

Figure 9 shows the temperature field cloud map of the molten pool at the moment when the electron beam is scanned at different scanning speeds for t = 6.25 s. It can be observed from the figure that as the scanning speed increases, the peak temperature in the molten pool decreases continuously. This indicates that a faster scanning speed results in a shorter time for the energy to stay on the surface of the cladding, thus reducing the heat absorbed by the workpiece. When the scanning speed is 180 mm/min and the beam current is 18 mA, the peak temperature in the molten pool is 6677 K, and the lowest temperature in the heat-affected zone is 1887 K. As the scanning speed escalates to 240 mm/min, the peak temperature decreases by 1145 K, reaching 5532 K.

Further increasing the scanning speed to 300 mm/min, the peak temperature in the molten pool drops to 5144 K. When the scanning speed is 360 mm/min, the peak temperature in the molten pool decreases to 4014 K. The reason for this is that as the scanning speed of the electron beam increases, the heating time of each microelement in the finite element decreases, resulting in a decrease in the peak temperature in the molten pool. Excessive temperatures can cause the grains in the cladding area to grow too large, which is undesirable.

On the other hand, an appropriate scanning speed can accelerate the cooling rate of the area irradiated by the electron beam, resulting in a delicate and uniform microstructure in the cladding region. In terms of the quality of the cladding surface produced by the electron beam, further research is needed on the node temperature in the joint area to achieve powder melting and microfusion of the substrate. This can ensure parameter optimization and improve the comprehensive performance of the cladding while preserving the excellent properties of the original material.

Extracting the nodal coordinates (8, 12, 9) between the matrix and the powder from ABAQUS, the changing curve of the pool temperature at different time points is analyzed, as shown in Figure 10. Figure 11 shows the temperature variation curve of the material surface at coordinate points (8, 12, 10) at different time points. From the figure, it can be observed that when the scanning speed is 180 mm/min, the peak temperature of the pool reaches 5700 K, and the powder is completely melted into the liquid phase. The temperature in the joining zone reaches as high as 2300 K. This value significantly surpasses the matrix's melting point of 1533 K. This indicates that while the powder is melting, the matrix is also rapidly melting. This is considered overeating, leading to powder dilution and decreased bonding performance. At the same time, extensive melting of the matrix will result in a decrease in its own performance and severe surface deformation. When the scanning speed reaches 240 mm/min, the joining zone temperature reaches 1700 K, which is still above the melting point of the matrix, indicating over-melting. When the scanning speed is 300 mm/min, the joining zone temperature is around 1500 K, and the peak temperature on the surface is 4200 K. At this point, the matrix undergoes partial melting while the powder is fully melted. The liquid phase of the fully melted powder and the surface of the matrix form a closely bonded metallurgical layer, achieving a good overlay process. When the scanning speed is 360 mm/min, the joining zone temperature is only 1300 K. At this time, the overlay layer is shallow, the joining zone is not compact, and there is only a thin layer, indicating under-melting.

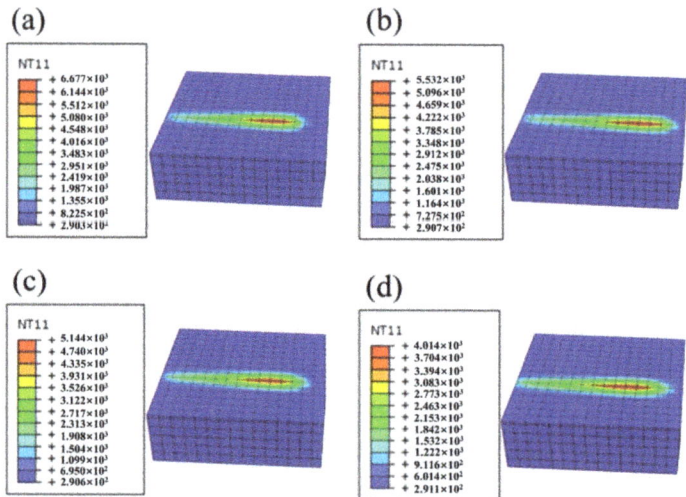

**Figure 9.** Temperature field clouds of the melt pool at different scanning speeds at t = 6.25 s: (**a**) 80 mm/min (**b**) 240 mm/min (**c**) 300 mm/min (**d**) 360 mm/min.

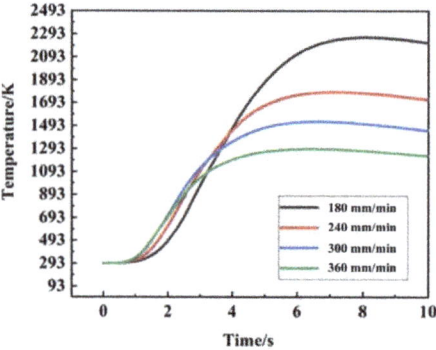

**Figure 10.** Combining different scanning speeds with the temperature variation patterns of zone nodes.

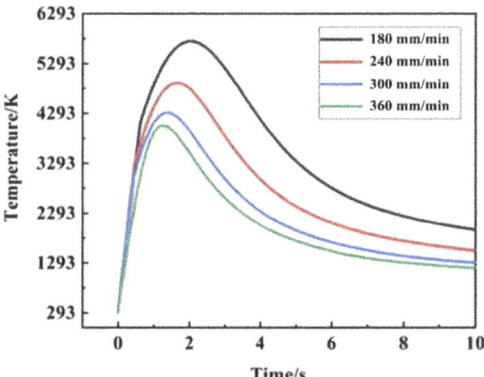

**Figure 11.** The variation pattern of peak temperature in the molten pool under different scanning speeds.

*4.3. Temperature Field Cloud Maps at Different Moments in the Process of Fusion Coating*

Figure 12 shows the temperature field overlay map of the molten pool scanned at different time points under the optimal process parameters of 18 mA and 300 mm/min with different beam currents and velocities. From Figure 12, it can be observed that at t = 0.384 s, the electron beam heat source just started to bombard the surface of the cladding layer, and the peak temperature at the center reached 2836 K. As the heat source moves forward, at t = 1.460 s, the highest temperature of the beam spot in the molten pool reaches 4845 K. This is because the preceding heat source had a preheating effect on the surface of the cladding layer, and the center of the workpiece has slower heat dissipation, resulting in temperature concentration. At t = 3.190 s and t = 4.921 s, the peak temperature of the beam spot in the molten pool hovers around 5100 K, and the temperature change in the molten pool tends to be relatively stable, indicating the best welding quality at this time. While the electron beam heat source moves, the workpiece undergoes rapid heating and rapid cooling. The surface temperature drops rapidly as the heat source moves away, and the trajectory of the cladding resembles a comet with a long tail. Heat is conducted in different directions within the substrate, and the powder and substrate undergo fusion and solidification. Ultimately, the surface of the substrate undergoes the formation of a compact cladding layer, successfully attaining the intended enhancement outcome.

*4.4. Experimental Verification*

Figure 13 depicts the macroscopic surface morphology of Inconel 718, which has been electron beam surface-clad with NiCrBSi powder. The process parameters used are outlined in Table 2. The illustration indicates that at lower electron beam currents, the energy is insufficient to achieve immediate coating melting. Consequently, due to the brittle nature of the layer and uneven heating, the coating easily detaches from the substrate's surface. Increasing the scanning beam current leads to a gradual widening of the melting zone and a decrease in the roughness of the cladding sample surface. This is because higher electron beam currents result in increased energy and temperature, leading to a broader range of heat conduction. Figure 13 reveals the presence of numerous fine particles at the edge of the sample's melting zone. This is due to the high temperature, which leads to the decomposition of $CO_2$ in the molten pool: $CO_2 \rightarrow CO + O$. The expansion of CO under high temperatures results in a rapid increase in pressure. The molten pool inhibits its escape, leading to a local explosion and the splattering of metal particles.

Figure 14 presents the macroscopic surface morphology of Inconel 718, which has been surface-clad with NiCrBSi powder using various scanning speeds, as documented in Table 3. The figure reveals that, with the beam current held constant, the scanning speed increase results in a narrower melting zone. This occurs when the scanning speed is low, as the prolonged exposure of the electron beam energy on the plate's surface leads to an increased accumulation of power. Consequently, this causes excessive remelting of material and ultimately widens the melting zone.

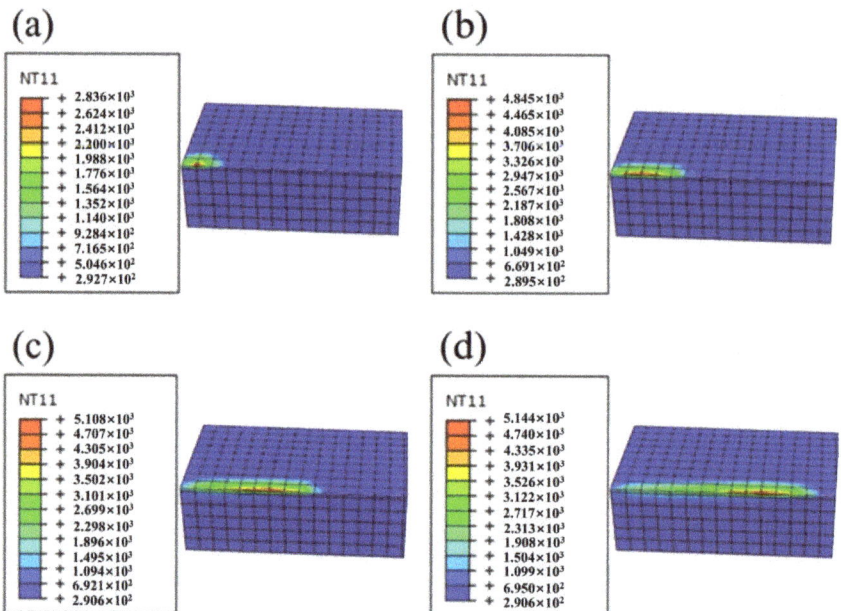

**Figure 12.** Subsurface temperature field cloud maps at different moments in the molten pool with a beam current of 18 mA and a scanning speed of 300 mm/min: (**a**) 0.384 s (**b**) 1.460 s (**c**) 3.190 s (**d**) 4.921 s.

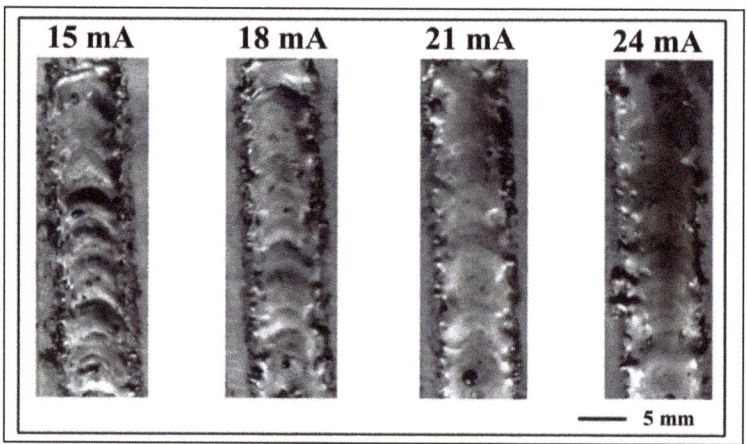

**Figure 13.** Macroscopic visualization of the surface of electron beam cladding samples under different beam conditions.

Energy spectrum analysis was conducted on the sub-surface of the overlay region to enhance the understanding of the bonding strength between the substrate and overlay powder. Figure 15 presents the cross-sectional morphology of the electron beam overlay layer and the line scan image captured at a beam current of 18 mA and a scan speed of 300 mm/min. The left side depicts the substrate region, whereas the right side illustrates the overlay region. In Figure 15a, with a beam current of 18 mA, there is a notable increase in the concentration of hard particles in the overlay region compared to the substrate. The corresponding EDS line data reveals that silicon (Si) has the highest content in the overlay region, indicating the excellent performance of the NiCrBSi powder process for electron beam surface overlay of Inconel 718 at a beam current of 18 mA. Figure 15b demonstrates an SEM image at a scan speed of 300 mm/min, revealing a narrow and well-bonded interface region that suggests the formation of an exceptional metallurgical bonding layer. The overlay region still exhibits the highest silicon element content based on the corresponding EDS line data, indicating positive overlay effects in the 300 mm/min process. Because Inconel 718 has a relatively soft nature, the enhanced silicon content in the metallurgical bonding layer can significantly improve the overall performance of the material.

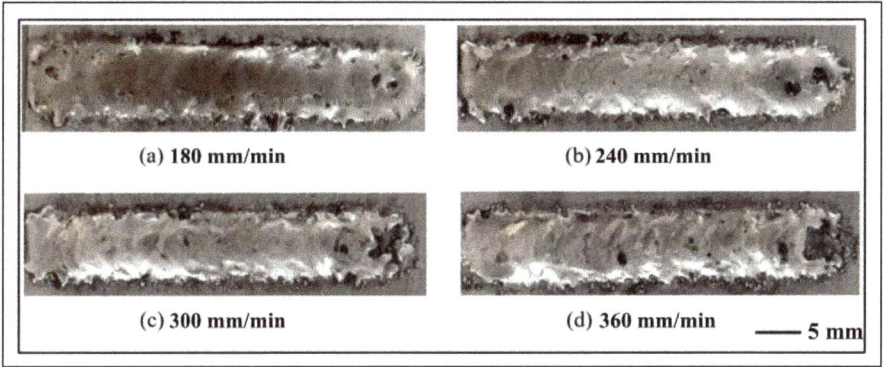

**Figure 14.** Macroscopic images of electron beam cladding test samples under different scanning speed conditions. (**a**)180 mm/min (**b**) 240 mm/min, (**c**) 300 mm/min (**d**) 360 mm/min.

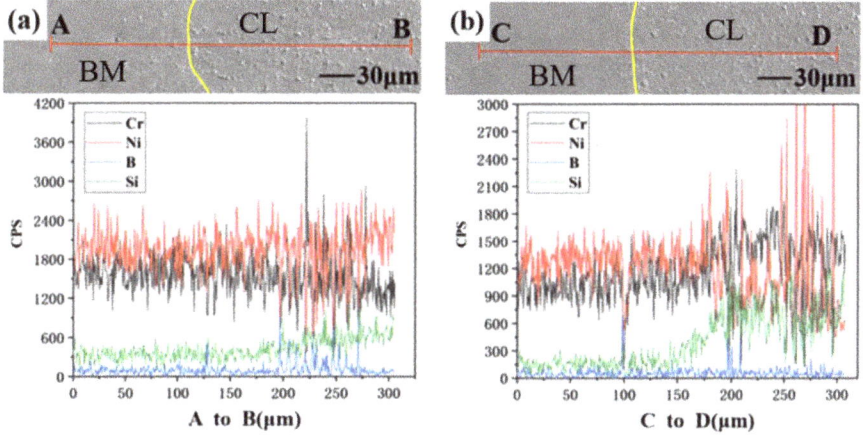

**Figure 15.** SEM micrographs and line scan EDS images of the cross-section morphology of the electron beam cladding layer at a beam current of 18 mA (**a**) and a scanning speed of 300 mm/min (**b**).

The experimental comparisons above demonstrate that the most effective overlay, characterized by a highly dense layer with strong bonding to the substrate, is achieved when using an electron beam current of 18 mA and a scan speed of 300 mm/min. The interface exhibits no visible pores or cracks. A wavy boundary between the overlay layer and the substrate suggests a favorable metallurgical bonding between the two. The simulated temperature field analysis corroborates these findings, thus validating the reliability and accuracy of the simulation.

## 5. Conclusions

This study established an electron beam cladding model using ABAQUS finite element simulation software. The heat source subroutine Dflux.for was used to simulate the conical heat source. Different size grids were used to divide the substrate and cladding powder, and the double ellipsoidal heat source was loaded onto the grid model according to the electron beam scanning path for simulation. The key issues affecting cladding quality, namely the density of the bonding layer and the dilution rate of the powder, were analyzed. Temperature field contour maps and temperature variation curves of the surface molten pool and bonding zone nodes were established. The main conclusions are as follows:

(1) According to the simulation results, during the electron beam scanning heating process, the surface temperature of the sample rapidly increased above the material's melting point. The center temperature of the electron beam focus was much higher than the average temperature of the molten pool, reaching a maximum of over 5000 K. When controlling other parameters, the electron beam current and molten pool temperature were positively correlated, while the scanning speed and molten pool temperature were negatively correlated.

(2) During the electron beam cladding process, the molten pool temperature changed dramatically. The temperature in the area irradiated by the electron beam heat source increased rapidly, while the temperature rapidly decreased in the area where the electron beam heat source left. The cladding process can be divided into three stages: the heating and melting stage, the temperature stabilization stage, and the cooling and solidification stage. In the direction of cladding thickness, the temperature decrease rate becomes faster as it goes down.

(3) When the electron beam scanning current was 18 mA and the scanning speed was 300 mm/min, the peak temperature of the molten pool in the cladding zone was around 5087 K, and the lowest temperature of the isotherm line in the heat-affected zone was 1409 K. The temperature of the bonding zone nodes between the substrate and powder was around 1500 K, reaching the melting point of Inconel 718 material. At this point, the NiCrBSi powder was completely melted, and the substrate was slightly dissolved, indicating relatively ideal electron beam cladding process parameters.

(4) According to the morphology and energy spectrum analysis of the electron microscopy, the cladding layer was dense, crack-free, and tightly connected to the substrate. The wave shape of the analysis was consistent with the temperature field, verifying the reliability and accuracy of the simulation.

**Author Contributions:** Conceptualization, G.Z.; analysis, Y.Z.; investigation, Y.Z.; data curation, Y.Z.; writing—original draft preparation, G.Z.; writing—review and editing, G.Z. and Y.Z.; super-vision, J.L. and Y.Z.; funding acquisition, H.L. and Y.L.; resource, L.M. All authors have read and agreed to the published version of the manuscript.

**Funding:** This work was supported by the National Natural Science Foundation of China (U1910213), the Fundamental Research Program of Shanxi Province (20210302123207 and 20210302124009), Scientific and Technological Innovation Programs of Higher Education Institutions in Shanxi (2021L292), Taiyuan University of Science and Technology Scientific Research Initial Funding (20212026).

**Data Availability Statement:** Not applicable.

**Acknowledgments:** The project was supported by the Fundamental Research Program of Shanxi Province (20210302123207 and 20210302124009), Taiyuan University of Science and Technology Scientific Research Initial Funding (20212026), the Shanxi Outstanding Doctorate Award Funding Fund (20222042), Taiyuan University of Science and Technology Graduate Innovation Project (BY2022004 and SY2022088) and the Coordinative Innovation Center of Taiyuan Heavy Machinery Equipment.

**Conflicts of Interest:** The authors declare no conflict of interest.

## References

1. Patel, V.; Sali, A.; Hyder, J.; Corliss, M.; Hyder, D.; Hung, W. Electron beam welding of inconel 718. *Procedia Manuf.* **2020**, *48*, 428–435. [CrossRef]
2. Hong, J.K.; Park, J.H.; Park, N.K.; Eom, I.S.; Kim, M.B.; Kang, C.Y. Microstructures and mechanical properties of Inconel 718 welds by $CO_2$ laser welding. *J. Mater. Process. Technol.* **2008**, *201*, 515–520. [CrossRef]
3. Liu, L.; Liu, H.; Zhang, X.; Wang, Y.; Hao, X. Corrosion Behavior of TiMoNbX (X = Ta, Cr, Zr) Refractory High Entropy Alloy Coating Prepared by Laser Cladding Based on TC4 Titanium Alloy. *Materials* **2023**, *16*, 3860. [CrossRef] [PubMed]
4. Thawari, N.; Gullipalli, C.; Vanmore, H.; Gupta, T. In-situ monitoring and modelling of distortion in multi-layer laser cladding of Stellite 6: Parametric and numerical approach. *Mater. Today Commun.* **2022**, *33*, 104751. [CrossRef]
5. Liverani, E.; Toschi, S.; Ceschini, L.; Fortunato, A. Effect of selective laser melting (SLM) process parameters on microstructure and mechanical properties of 316L austenitic stainless steel. *J. Mater. Process. Technol.* **2017**, *249*, 255–263. [CrossRef]
6. Aguilar-Hurtado, J.Y.; Vargas-Uscategui, A.; Paredes-Gil, K.; Palma-Hillerns, R.; Tobar, M.J.; Amado, J.M. Boron addition in a non-equiatomic $Fe_{50}Mn_{30}Co_{10}Cr_{10}$ alloy manufactured by laser cladding: Microstructure and wear abrasive resistance. *Appl. Surf. Sci.* **2020**, *515*, 146084. [CrossRef]
7. Huang, S.W.; Nolan, D.; Brandt, M. Pre-placed WC/Ni clad layers produced with a pulsed Nd: YAG laser via optical fibres. *Surf. Coat. Technol.* **2003**, *165*, 26–34. [CrossRef]
8. Dilip, D.G.; Ananthan, S.P.; Panda, S.; Mathew, J. Numerical simulation of the influence of fluid motion in mushy zone during micro-EDM on the crater surface profile of Inconel 718 alloy. *J. Braz. Soc. Mech. Sci. Eng.* **2019**, *41*, 107. [CrossRef]
9. Maremonti, P.; Crispo, F.; Grisanti, C.R. Navier–Stokes equations: A new estimate of a possible gap related to the energy equality of a suitable weak solution. *Meccanica* **2023**, *58*, 1141–1149. [CrossRef]
10. Liu, H.; Qi, Z.; Wang, B.; Zhang, G.; Wang, X.; Wang, D. Numerical simulation of temperature field during electron beam cladding for NbSi2 on the surface of Inconel617. *Mater. Res. Express* **2018**, *5*, 036528. [CrossRef]
11. Pandey, K.; Datta, S. Performance of Si-doped TiAlxN supernitride coated carbide tool during dry machining of Inconel 718 superalloy. *J. Manuf. Process.* **2022**, *84*, 1258–1273. [CrossRef]
12. Li, Y.; Su, K.; Bai, P.; Wu, L. Microstructure and property characterization of Ti/TiBCN reinforced Ti based composite coatings fabricated by laser cladding with different scanning speed. *Mater. Charact.* **2020**, *159*, 110023. [CrossRef]
13. Ashurova, K.; Vorobyov, M.; Koval, T. Numerical Simulation of the Effect of an Electron Beam on the Surface of Materials. In Proceedings of the 2020 7th International Congress on Energy Fluxes and Radiation Effects (EFRE), Tomsk, Russia, 14–26 September 2020; IEEE: Piscataway, NJ, USA, 2020; pp. 150–153.
14. He, X.; Song, R.G.; Kong, D.J. Effects of TiC on the microstructure and properties of TiC/TiAl composite coating prepared by laser cladding. *Opt. Laser Technol.* **2019**, *112*, 339–348. [CrossRef]
15. Wang, J.; Li, C.; Zeng, M.; Guo, Y.; Feng, X.; Tang, L.; Wang, Y. Microstructural evolution and wear behaviors of NbC-reinforced Ti-based composite coating. *Int. J. Adv. Manuf. Technol.* **2020**, *107*, 2397–2407. [CrossRef]
16. Ning, J.; Lan, Q.; Zhu, L.; Xu, L.; Yang, Z.; Xu, P.; Xue, P.; Xin, B. Microstructure and mechanical properties of SiC-reinforced Inconel 718 composites fabricated by laser cladding. *Surf. Coat. Technol.* **2023**, *463*, 129514. [CrossRef]
17. Liu, C.; Li, C.; Zhang, Z.; Sun, S.; Zeng, M.; Wang, F.; Guo, Y.; Wang, J. Modeling of thermal behavior and microstructure evolution during laser cladding of AlSi10Mg alloys. *Opt. Laser Technol.* **2020**, *123*, 105926. [CrossRef]
18. Markov, A.B.; Solovyov, A.V.; Yakovlev, E.V.; Pesterev, E.A.; Petrov, V.I.; Slobodyan, M.S. Computer simulation of temperature fields in the Cr (film)-Zr (substrate) system during pulsed electron-beam irradiation. *J. Phys. Conf. Ser.* **2021**, *2064*, 012058. [CrossRef]
19. Shao, Y.P.; Xu, P.; Tian, J.Y. Numerical Simulation of the Temperature and Stress Fields in Fe-Based Alloy Coatings Produced by Wide-Band Laser Cladding. *Met. Sci. Heat Treat.* **2021**, *63*, 327–333. [CrossRef]
20. Zuo-Jiang, S.; Yu, H.; Jiang, X.; Gao, W.; Sun, D. A thermal field FEM of titanium alloy coating on low-carbon steel by laser cladding with experimental validation. *Surf. Coat. Technol.* **2023**, *452*, 129113. [CrossRef]
21. Yang, S.; Bai, H.; Li, C.; Shu, L.; Zhang, X.; Jia, Z. Numerical Simulation and Multi-Objective Parameter Optimization of Inconel 718 Coating Laser Cladding. *Coatings* **2022**, *12*, 708. [CrossRef]
22. Gan, Z.; Yu, G.; He, X.; Li, S. Numerical simulation of thermal behavior and multicomponent mass transfer in direct laser deposition of Co-base alloy on steel. *Int. J. Heat Mass Transf.* **2017**, *104*, 28–38. [CrossRef]
23. He, F.; Zhou, H.; Li, K.; Zhu, Y.; Wang, Z. Numerical Analysis and Experimental Verification of Melt Pool Evolution During Laser Cladding of 40CrNi2Si2MoVA Steel. *J. Therm. Spray Technol.* **2023**, *32*, 1416–1432. [CrossRef]
24. Hofman, J.T.; De Lange, D.F.; Pathiraj, B.; Meijer, J. FEM modeling and experimental verification for dilution control in laser cladding. *J. Mater. Process. Technol.* **2011**, *211*, 187–196. [CrossRef]
25. Oane, M.; Mihailescu, I.N.; Ristoscu, C.G. Thermal Fields in Laser Cladding Processing: A "Fire Ball" Model. A Theoretical Computational Comparison, Laser Cladding Versus Electron Beam Cladding. In *Nonlinear Optics: From Solitons to Similaritons*; IntechOpen: Houston, TX, USA, 2021; pp. 137–147.
26. Naeem, M.; Lock, S.; Collins, P.; Hooley, G. Electron beam welding vs. laser beam welding for air bearing shaft. In Proceedings of the International Congress on Applications of Lasers & Electro-Optics, Miami, FL, USA, 31 October–3 November 2005; Laser Institute of America: Orlando, FL, USA, 2005; Volume 2005, p. M305.

27. Azinpour, E.; Darabi, R.; de Sa, J.C.; Santos, A.; Hodek, J.; Dzugan, J. Fracture analysis in directed energy deposition (DED) manufactured 316L stainless steel using a phase-field approach. *Finite Elem. Anal. Des.* **2020**, *177*, 103417. [CrossRef]
28. Nguyen, D.S.; Park, H.S.; Lee, C.M. Optimization of selective laser melting process parameters for Ti-6Al-4V alloy manufacturing using deep learning. *J. Manuf. Process.* **2020**, *55*, 230–235. [CrossRef]
29. Tadano, S.; Hino, T.; Nakatani, Y. A modeling study of stress and strain formation induced during melting process in powder-bed electron beam melting for Ni superalloy. *J. Mater. Process. Technol.* **2018**, *257*, 163–169. [CrossRef]
30. Singh, S.; Sachdeva, A.; Sharma, V.S. Optimization of selective laser sintering process parameters to achieve the maximum density and hardness in polyamide parts. *Prog. Addit. Manuf.* **2017**, *2*, 19–30. [CrossRef]
31. Mianji, Z.; Kholopov, A.A.; Binkov, I.I.; Kiani, A. Numerical simulation of thermal behavior and experimental investigation of thin walls during direct metal deposition of 316L stainless steel powder. *Lasers Manuf. Mater. Process.* **2021**, *8*, 426–442. [CrossRef]
32. Ahmad, A.S.; Wu, Y.; Gong, H.; Liu, L. Numerical simulation of thermal and residual stress field induced by three-pass TIG welding of Al 2219 considering the effect of interpass cooling. *Int. J. Precis. Eng. Manuf.* **2020**, *21*, 1501–1518. [CrossRef]

**Disclaimer/Publisher's Note:** The statements, opinions and data contained in all publications are solely those of the individual author(s) and contributor(s) and not of MDPI and/or the editor(s). MDPI and/or the editor(s) disclaim responsibility for any injury to people or property resulting from any ideas, methods, instructions or products referred to in the content.

Article

# The Relationship between Polishing Method and ISE Effect

Jozef Petrík [1], Peter Blaško [1,*], Dagmar Draganovská [2], Sylvia Kusmierczak [3], Marek Šolc [1], Miroslava Ťavodová [4] and Mária Mihaliková [1]

[1] Institute of Materials and Quality Engineering, Faculty of Materials, Metallurgy and Recycling, Technical University of Kosice, Letna 1/9, 04200 Košice, Slovakia; jozef.petrik@tuke.sk (J.P.); marek.solc@tuke.sk (M.Š.); maria.mihalikova@tuke.sk (M.M.)
[2] Faculty of Mechanical Engineering, Technical University of Kosice, Letna 1/9, 04200 Košice, Slovakia; dagmar.draganovska@tuke.sk
[3] Faculty of Mechanical Engineering, Jan Evangelista Purkyně University, Pasteurova 3334/7, 400 01 Usti nad Labem, Czech Republic; sylvia.kusmierczak@ujep.cz
[4] Faculty of Technology, Technical University in Zvolen, Tomáša Garrigue Masaryka 24, 96001 Zvolen, Slovakia; tavodova@tuzvo.sk
* Correspondence: peter.blasko@tuke.sk; Tel.: +421-55-602-2872

**Abstract:** The aim of the submitted work is to study the relationship between the method of polishing the metallurgical surface and the indentation size effect (ISE). The material of the sample was annealed 99.5% aluminum. The polishing time ranged between 300 and 3600 s. An aqueous emulsion of aluminum oxide (spineline) and diamond paste were used as the polishing agents. The surface quality of the samples was measured with roughness meters. Applied loads in the micro-hardness test were 0.0981, 0.2452, 0.4904, and 0.9807 N. The effect of polishing on micro-hardness, Meyer's index n, and ISE characteristics was evaluated using the PSR method and the Hays–Kendall approach. As the polishing time increases, the micro-hardness values decrease, and the value of Meyer's index n increases from "normal" to neutral, i.e., Kick's law applies. The finding was confirmed for both of the used polishing agents.

**Keywords:** aluminum; polishing; micro-hardness; ISE

## 1. Introduction

An indentation size effect (ISE), in general, is observed in shallow indentation tests of (micro-)hardness, which is manifested as an increase ("normal") or decrease (RISE) in hardness with penetration depth decreases.

Among other factors, the size and nature of the ISE are influenced by the sample preparation method (polishing time, polishing agent) to create the metallographic surface for micro-hardness measurement. The quality of the obtained surface is usually determined subjectively, so the goal is to achieve a mirror finish without visible scratches. The polishing process must be controlled so that the Beilby layer does not form. Although the quality of the metallographic surface did not change significantly during the polishing time in the observed range, this time had a statistically significant effect on the size, nature, and other parameters of the ISE. Thus, the polishing time, when inappropriately chosen, can distort the measurement results to a certain extent.

The measurement of micro-hardness makes it possible to determine the basic mechanical properties of a small volume of material using an almost non-destructive method. If, as in the case of (macro-)hardness, the Vickers method is used, the only difference is a lower load (lower than 1.691 N). Measurements of micro-hardness can be used for miniature components, thin surface layers, or a metallography. The shape of the indentation (pyramid) is geometrically similar for all test loads. It is therefore expected that the measured hardness will be over a broad load range if the tested sample is homogeneous.

Unfortunately, this statement only applies to the range of loads intended for measuring (macro-)hardness. If "a very low" test load is used, the measured value is influenced by other factors. However, the term "very low load" is not exactly defined. Standard ISO 6507-1:2018 lists loads (test forces) in the range between 0.009807 N (1 g) and 0.9807 N (100 g) [1]. However, according to standard ISO 14577-1:2015, the values of the loads for the micro-hardness tests are less than 2 N (~200 g), while the indentation depth h > 0.2 µm [2]. As stated Voyiadjis and Peters, "a very low" load results in indentation with a depth less than 10 µm (... but not less than 0.2 µm in ISO 14577-1:2015) [2–4].

The ISE may be caused by the following:

1. The testing equipment—it includes the characteristics of devices used to measure the dimensions of indentations and loads [4–6].
2. Intrinsic properties of the samples—work hardening during indentation, the load to initiate plastic deformation, the indentation elastic recovery and elastic resistance of the materials [5–7], and the influence of crystallographic orientation [8,9].
3. The method of preparing the samples—the cutting, the grinding, the polishing, and stresses in the samples resulting from their manufacture as well as many other factors such as the indenter/sample friction, the lubrication, the corrosion, and speed of the indenter's penetration [4,6,7,10,11].

In contrast to a "normal" ISE, a reverse (RISE) type of ISE, where the apparent micro-hardness increases with increasing test load, is also known. It mostly takes place in materials with predominant plastic deformation. As a rule, it is explained by the existence of a distorted zone near the crystal–medium interface, the effects of vibration and the bluntness of the indenter, the applied energy loss as a result of specimen chipping around the indentation, and the generation of the cracks [7].

As mentioned above, the quality and method of preparation of the metallographic surface, on which the measurement will be carried out, has an influence on the measured micro-hardness values and at the same time on the ISE parameters.

The mutual relationship between the quality of the measured surface, expressed by its roughness (for example, Ra), and the obtained values of micro-hardness (or nano-hardness) has already been addressed by several researchers in the past. As an example, ref. [12] studied the influence of the surface quality of nickel samples on hardness and, using the Nix–Gao model, derived the relationship between surface quality and critical contact depth. In [13], SCM21 steel samples were also analyzed in a similar way.

Xia et al. published the results of studying the impact of roughness on hardness for Ti alloy $AlTi_6V_4$, and Xia published the results of studying the impact of roughness on hardness for Ti alloy, without evaluating the impact on the ISE [14].

The relationship between the quality of the surface and the hardness of non-metallic materials (aluminum oxide and polystyrene) was studied [15]. The variation of Knoop micro-hardness (loads between 200 g/1.96 N and 1000 g/9.81 N) follows the reverse ISE trend, i.e., an increase in hardness on load in the low-load region beyond where it becomes relatively constant.

The influence of the surface roughness on the ISE in micro-indentation was examined using the proportional specimen resistance model [16]. Stainless steel, aluminum (6061-T6: 95.9–98.6% Al with 0.8–1.2 Mg and 0.4–0.8 Si), and copper surfaces were polished to different levels of roughness with spineline (alumina powder 5–0.05 µm) and subjected to HV micro-indentation. The load ranged between 0.147 and 1.962 N. To evaluate the factors of material elasticity and friction effect that make up the elastic proportional resistance, coefficient $a_1$, related to the elastic properties (it characterizes the load dependence of micro-hardness and describes the ISE in the PSR model), was plotted against the sample surface roughness Ra. As the roughness increases, the value of this coefficient for all three tested materials increases. As mentioned below, a similar relationship between roughness and parameter $a_1$ was observed by the authors of the paper. To evaluate the roughness effect on the ISE, an equation to predict the ISE was proposed, corresponding to the surface roughness factor for micro-indentation.

Another important area in the relationship between roughness and micro-hardness is the resin composite used in dental medicine, polished by one-step polishing systems or by a conventional multi-step system, studied, for example, by Edemir et al. or Korkmaz et al. [17,18]. As in the case of work focused on metal materials, for these materials, the authors focused only on the relationship of roughness–(micro-)hardness, without further study of the ISE.

As a follow-up to the mentioned works, the authors of the present contribution tried to influence the method of polishing the metallurgical surface and its surface roughness. They evaluate the impact of these factors not only on micro-hardness but also on the size and nature of the indentation size effect (ISE).

## 2. Materials and Methods

Tempered (400 °C/1 h) 99.5% Al (EN AW 1350) in form of the wire (diameter 9 mm) with yield strength (YS) = 25 MPa, ultimate tensile strength (UTS) = 73 MPa, total elongation (TE) = 59.7%, and the reduction of the area (or contraction Z) = 90.7% was an experimental material [19].

The wire was cut using a cooled diamond saw perpendicular to the axis. The pieces were in random order embedded in the resin (dentacryl) and ground with silicon-carbide papers in the sequences 80, 220, 240, 280, 500, 800, 1000, and 3000 ANSI/CAMI. The metallographic surface was subsequently polished 300, 600, 1200, 1800, 2400, 3000, and 3600 s with polishing agents by the same operator:

1. Aqueous emulsion of $Al_2O_3$ (alumina powder or spineline, 400 mL $H_2O$, and 25 g $Al_2O_3$ with grain size 10–40 μm); felt was used as a textile for the polishing wheel.
2. Diamond paste (product by Pramet/Urdiamant Šumperk, Czech Republic) in the 2–3 μm size ranges (corresponding with the D2 FEPA Fédération Européenne des Fabricants de Produits Abrasifs) moistened with kerosene; velvet was used as a textile for the polishing wheel.

In both cases (spineline and diamond), the circumferential speed of the polisher disc (ø 270 mm) was 11 revolutions per second.

When viewed with the naked eye for both agents used for polishing, after 300 s of polishing, the surface was mirror-like, without visible scratches or grooves. On the contrary, at a magnification of 200×, the scratches were visible regardless of the polishing time. Even polishing for 3600 s did not completely remove them. On the contrary, even with longer polishing times, the formation of the Beilby layer (over-polishing) was not observed [20]. This was probably the result of thorough wetting of the polishing wheel textile, the rotational movement of the samples, and a reasonably chosen pressing force (polishing was carried out manually, with all samples by one operator).

To objectify the quality of the ground (3000 ANSI/CAMI grit) and polished surface, its roughness Ra (arithmetical mean height; arithmetical mean height indicates the average of the absolute value along the sampling length) was measured using standard ISO 4287:1997 [21].

In the first stage, the roughness was measured with a contact tester Surftest SJ301 (Mitutuyo). N = 5 (5 sections of 0.5 mm each), i.e., the measured length l = 1.25 mm, and the radius of the sensor was 0.4 μm.

As found using microscopic examination of the surface of the samples after measuring the roughness, the arm of the tester device left traces on the surface of the sample. It means that due to the low hardness of the samples, the compressive force caused plastic deformation. Therefore, we can consider the measured values of Ra as indicative at best, even though the value of W (the smallest load that causes an indentation) is in the range of 0.0019 to 0.0384 N, and the pressing force of the tester (measuring force) is 0.00075 N [22].

The test material was too soft to use a contact roughness meter. Therefore, in the following, we will consider the roughness values measured in this way only as indicative.

The mentioned deficiency was eliminated by measuring roughness with a non-contact confocal laser scanning microscope Olympus LEXT 3100. The hardness values of the ground sample (3000 ANSI/CAMI grit) and the polished samples were measured in two mutually perpendicular directions (x–y). From the measured roughness values (e.g., Rp, Rv), the value of Ra was used in the next arithmetical mean height (Ra), which indicates the average of the absolute value along the sampling length.

A Hanemann tester, type Mod D32 fitted to microscope Neophot-32Micro-hardness, measured the micro-hardness with a magnification 480×.

The tester meets the requirements of the standard ISO 6507-2, 2005, due to its repeatability $r_{rel}$ = 4.22%, an error of tester $E_{rel}$ = −0.92%, and the relative expanded uncertainty of calibration $U_{rel}$ = 6.76% [23].

The values of the $U_{rel}$ uncertainty of the measured micro-hardness of the sample listed in Table 1 are overvalued, considering the relationship between $u_{CRM}$ (4.0 HV0.05) and the mean micro-hardness of the samples (29.3 for spineline and 26.1 for diamond); therefore, it should only be taken as an informative value.

Table 1. The values of polishing time, "the path of the sample on the polishing wheel" (km), the mean micro-hardness HV, micro-hardness HV0.05, the relative expanded uncertainty $U_{rel}$, and the speed of the indenter's penetration $v$.

| Polishing Time (s) | The "Path" of the Sample (km) | Spineline | | | | Diamond | | | |
|---|---|---|---|---|---|---|---|---|---|
| | | HV | HV0.05 | $U_{rel}$ (%) | $v$ (µm s$^{-1}$) | HV | HV0.05 | $U_{rel}$ (%) | $v$ (µm s$^{-1}$) |
| 0 | 0 | 36.20 | 33.21 | 50.37 | 1.97 | 36.20 | 33.21 | 50.37 | 1.97 |
| 300 | 2.4 | 31.71 | 31.84 | 51.20 | 2.24 | 28.53 | 26.99 | 65.62 | 3.63 |
| 600 | 4.8 | 29.87 | 30.17 | 53.92 | 2.56 | 27.03 | 26.91 | 65.87 | 3.14 |
| 1200 | 9.6 | 29.43 | 29.32 | 55.45 | 2.37 | 26.10 | 26.60 | 66.40 | 3.32 |
| 1800 | 14.4 | 28.37 | 28.65 | 56.72 | 2.25 | 28.36 | 29.63 | 59.87 | 3.21 |
| 2400 | 19.2 | 28.94 | 29.29 | 55.52 | 2.31 | 26.24 | 26.14 | 67.73 | 2.97 |
| 3000 | 24.0 | 28.12 | 27.87 | 58.46 | 2.36 | 26.16 | 26.45 | 66.98 | 2.98 |
| 3600 | 28.8 | 28.12 | 28.12 | 57.85 | 2.59 | 26.76 | 26.44 | 66.95 | 3.12 |

$U_{rel}$ is the expanded uncertainty (k = 2) of the result of the measurement expressed as a percentage. It is calculated according to the standard ISO 6507-2:2018 [24].

The same operator measured the micro-hardness of selected areas on the metallographic surface of the sample according to the standard ISO 6507-1:2018 [1]. The applied loads P were 0.09807 N, 0.24518 N, 0.49035 N, and 0.9807 N with a load duration of 15 s. The speed of the indenter's penetration was calculated using the method described in [25]. The values of the speed (v, in µm s$^{-1}$) are shown in Table 1.

The result of the measurement was a "cluster" of 20 indentations in one area. The mean of the micro-hardness of individual clusters HV, the micro-hardness HV0.05, and its relative expanded uncertainty $U_{rel}$ are shown in Table 1; the values of micro-hardness at individual loads are shown in Figure 1.

Grubbs' test (significance level $\alpha$ = 0.05) was used for the detection of statistical outliers. Their presence would indicate a measurement process suffering from special disturbances and out of statistical control. The normality was determined using Freeware Process Capability Calculator software (Anderson–Darling test). The normality and the outliers were determined for files involving values of one "cluster". The values of micro-hardness of all "clusters" have normal distribution without outliers. Their absence suggests the process is unimpeded by gross errors.

Analysis of variance (ANOVA) is a standard statistical technique and can be used to analyze the measurement error and other sources of variability of data in a measurement systems study. The authors used the procedure recommended by Chajdiak and reference manual MSA for calculating the significance of individual factors [26,27]. Two-way ANOVA compares the means of a single variable at different levels of two conditions (factors). A $p$ value is used to determine whether a certain pattern they have measured is statistically significant. Statistical significance is a way of saying that the $p$ value of a statistical test is small enough to reject the null hypothesis of the test. The most common threshold is $p < 0.05$, that is, when you would expect to find a test statistic as extreme as the one calculated by your test only 5% of the time. The threshold value for determining statistical significance is also known as the alpha value. According to two-way ANOVA without replication, for spineline, the polishing time ($p = 0.000443$) and the load ($p = 0.000111$) both have statistically significant effects on the measured value of the micro-hardness. For diamond paste, the polishing time ($p = 0.1896$) and the influence of the load ($p = 0.0633$) are not statistically significant at the significance level $\alpha = 0.05$.

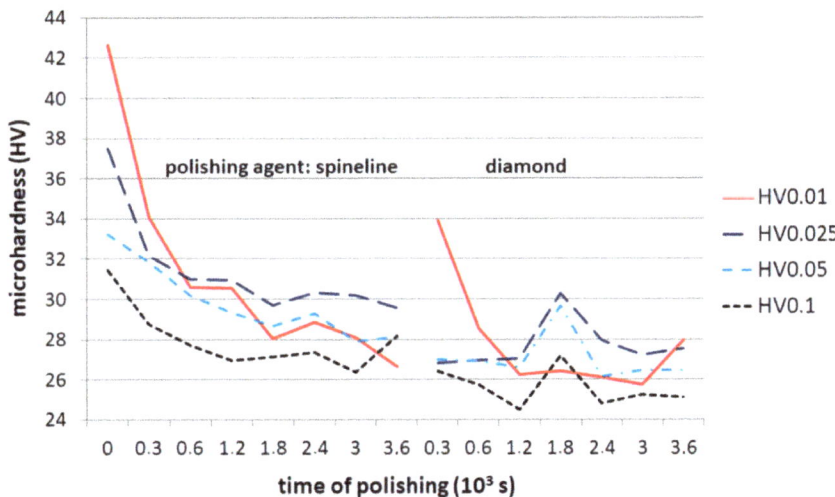

**Figure 1.** The values of micro-hardness.

The influence of polishing time (s) on the roughness Ra (μm) measured using a confocal microscope for both axes (x and y perpendicular to each other) is shown in Figure 2. Ra values are presented together for both axes since, by using an unpaired $t$-test, the difference between the Ra values in both axes is not statistically significant ($p = 0.9182$ for spineline and $p = 0.9727$ for diamond), and the scratches and grooves on the metallographic surface of the samples do not have a preferred direction. As the polishing time increases, the roughness decreases. But, as can be seen from the picture, the relationship is not completely ideal, especially with longer polishing times. This is evidenced by the values of the Pearson correlation coefficient r and the determination coefficient $R^2$ (Table 2). A correlation coefficient is a number between -1 and 1 that indicates the strength and direction of a relationship between variables. The coefficient of determination is a number between 0 and 1 that measures how well a statistical model predicts an outcome. Table 2 also shows indices (a—slope and b—intercept) describing the linear dependence. The parameter "b" is the value of y (x = 0) at the intersection of the line with the y-axis at x = 0. If the polishing time exceeds 1800 s, the roughness practically does not decrease. The particles of both polishing agents produce new scratches and grooves, which are partially removed by further polishing; at the same time, additional scratches and grooves are created. At the given granularity of the agent, the process stabilizes after a certain time, and its further continuation makes no sense. We do not state the dependence of roughness on the polishing time, measured with a contact tester. It is even less informative than when measured with a confocal microscope. The roughness values measured in this way are an order of magnitude higher (for example polishing time 1200 s, spineline: 50.0 μm contact tester, and 2.75 μm confocal microscope).

Table 2. Values of indices (a—slope and b—intercept) describing the linear dependence and coefficients of correlation r and determination $R^2$.

| Polishing Agent | Spineline | | | | Diamond | | | |
|---|---|---|---|---|---|---|---|---|
| Indices and Coefficients | a(x) | b | r | $R^2$ | a(x) | b | r | $R^2$ |
| Figure 2 | −0.7348 | 3.9259 | −0.659 | 0.4341 | −0.7182 | 4.3394 | −0.675 | 0.4559 |
| Figure 3 | −0.0502 | 2.0403 | −0.737 | 0.5439 | −0.0491 | 2.057 | −0.758 | 0.5739 |
| Figure 4 | 0.073 | 0.513 | - | 0.4038 | 0.0846 | 0.3681 | - | 0.8011 |
| Figure 5 | 0.2127 | 138.83 | - | 0.0421 | 0.1382 | 129.13 | - | 0.0159 |
| Figure 6 | 0.0039 | −0.0666 | - | 0.512 | 0.003 | −0.0566 | - | 0.1580 |
| Figure 7 | 0.0005 | 0.0038 | - | 0.377 | 0.0002 | 0.0041 | - | 0.2582 |
| Figure 8 | −0.0268 | 2.7249 | 0.833 | 0.8449 | −0.0181 | 2.4246 | 0.624 | 0.4923 |
| Figure 9 | −0.0012 | 0.0341 | - | 0.1816 | −0.0012 | 0.0315 | - | 0.1300 |

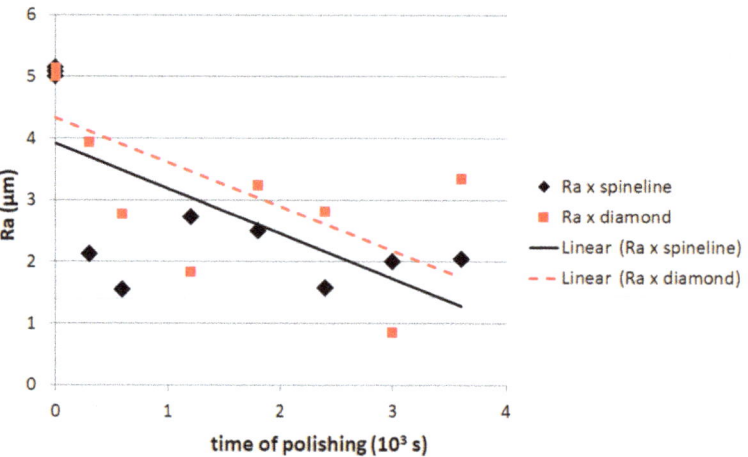

Figure 2. Relationship between polishing time and roughness Ra, measured with a confocal microscope.

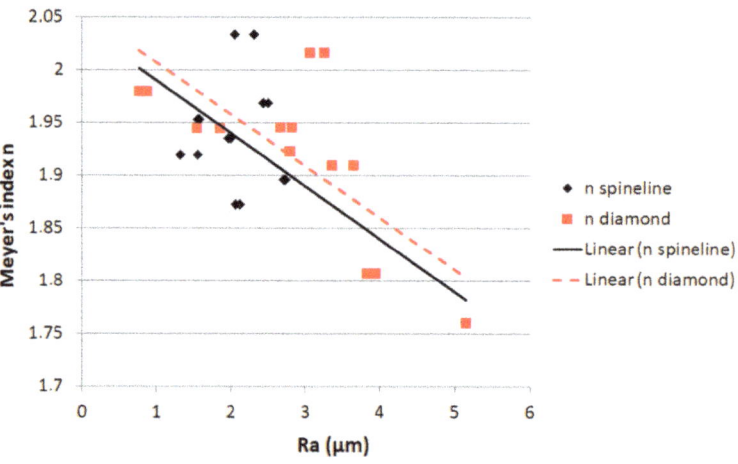

Figure 3. The relationship between Ra and Meyer's index n.

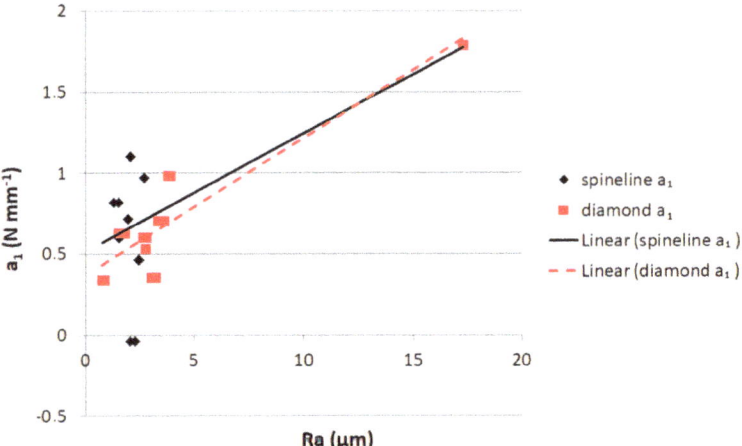

**Figure 4.** The relationship between Ra and parameter $a_1$.

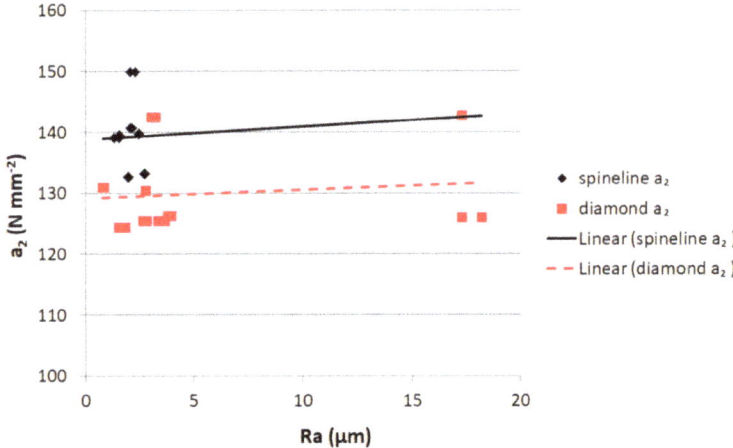

**Figure 5.** The relationship between Ra and parameter $a_2$.

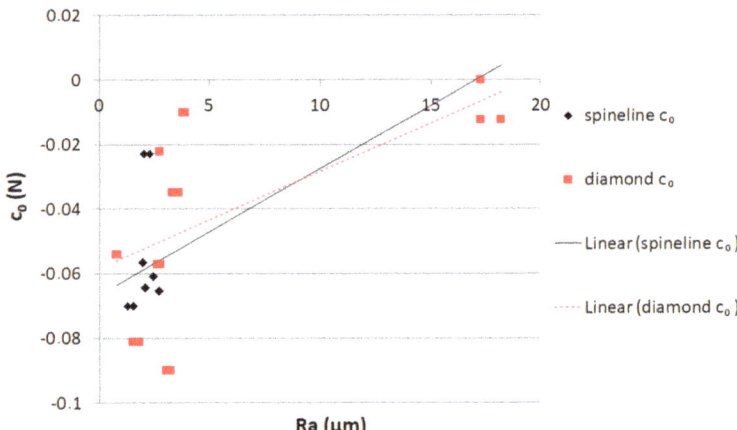

**Figure 6.** The relationship between Ra and $c_0$.

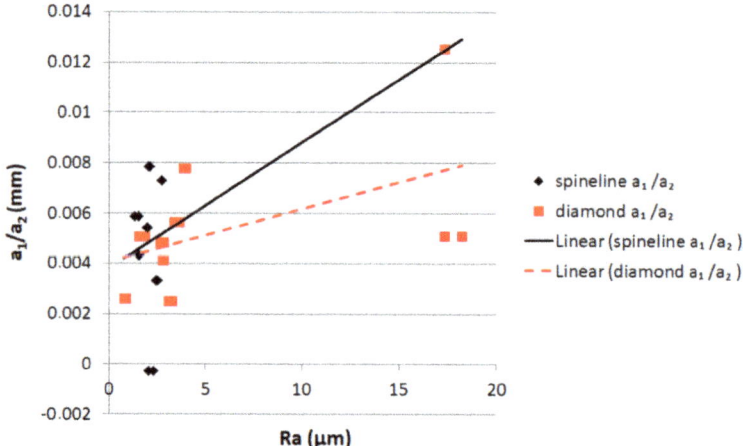

**Figure 7.** The relationship between Ra and $a_1/a_2$.

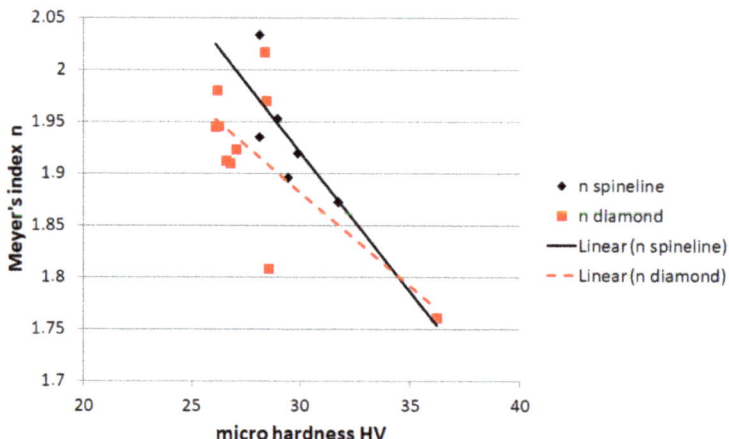

**Figure 8.** The relationship between micro-hardness HV and Meyer's index n.

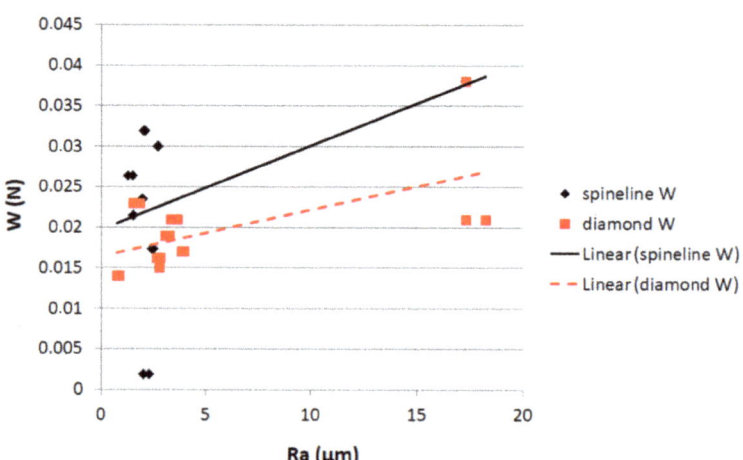

**Figure 9.** The relationship between Ra and W.

The points in the graphs represent measured values or parameter values calculated from measured values. The variance in the experimental points is really high, resulting in a low value (Figures 5 and 9 in Table 2). The chosen linear approximation used in the graphs serves to illustrate the trend.

## 3. Results

The calculation of Meyer's index and related parameters is detailed in, e.g., Petrík et al., 2023, or Šolc et al., 2023, using [4–6,12,28–30].

Meyer's power law and proportional specimen resistance (PSR) are two principal approaches to describe the ISE quantitatively. The value of Meyer's index n or the work hardening coefficient is n < 2 for "normal" ISE, and n > 2 for reverse ISE. If n = 2, the micro-hardness is independent of the load and is given by Kick's law. The relationship between Ra and Meyer's index n is shown in Figure 3. The value of Meyer's index n decreases with increases in the roughness. The values of the Pearson correlation coefficient r and the determination coefficient $R^2$ are shown in Table 2.

The nature of the ISE changes from "normal" ISE for ground (unpolished) samples with n = 1.7 to values close to the validity of Kick's law even after a short polishing time. As the polishing time increased, the value of n rose slightly above 2, i.e., into the reverse region (RISE) for samples polished with spineline.

Calculated using two-way ANOVA without replication, for spineline, the polishing time ($p$ = 0.0012738) has a statistically significant effect on the measured value of Meyer's index n but the polishing agent does not ($p$ = 0.786067) at the significance level $\alpha$ = 0.05. The fact that the agent does not have a statistically significant influence on the value of Meyer's index was also confirmed by a two-tailed $t$-test ($p$ = 0.8856).

Using a proportional specimen resistance model (PSR), a modified form of the Hays–Kendall approach, yields the parameters $a_1$ and $a_2$. Parameter $a_1$ (N mm$^{-1}$) is related to elastic properties and characterizes the load dependence of micro-hardness. It consists of two components: the elastic resistance of the sample and the friction resistance at the indenter facet/sample interface. Parameter $a_2$ (N mm$^{-2}$) is related to the elastic and plastic properties of the specimen.

The influence of the roughness on the values of parameters $a_1$ and $a_2$ can be seen in Figures 4 and 5, respectively. The value of both parameters increases with increasing roughness; the difference is more pronounced for $a_1$ with a significantly lower correlation for parameter $a_2$. The values of the determination coefficient $R^2$ are shown in Table 2.

As mentioned in the introduction, the growth of the $a_1$ parameter with the growth of roughness was also observed by Chuah and Ripin in the analysis of stainless steel, copper, and aluminum samples [16].

Due to the very low values of the coefficient of determination $R^2$ (Figure 5, Table 2) for the relationship between the roughness Ra and the parameter $a_2$, the determination of "true hardness" using $a_2$ has no meaning.

The parameter $c_0$ (N) can be calculated using an equation based on a modified form of the PSR model. It is associated with residual surface stress in the sample. The relationship between Ra and parameter $c_0$ can be seen in Figure 6. For both polishing agents, the residual surface stress increases with the polishing time and thus with the decrease in roughness, and its negative tensile component increases. As it follows from the values of the coefficients of determination $R^2$ in Table 2, the dependence is tighter for spineline than for diamond paste.

The ratio $a_1/a_2$ is a measure of the residual stress due to machining and polishing. As can be seen in Figure 7 and Table 2, residual stress decreases with a decrease in roughness for both polishing agents.

Meyer's index n decreases with increasing average micro-hardness HV, as can be seen in Figure 8. As it follows from Table 2, the values of the coefficients of determination $R^2$, the dependence is strong and tighter for spineline than for diamond paste. As the micro-hardness increases, the "normal" character of the ISE, typical for materials with lower plasticity, increases. Reverse ISE and an inverse relationship between the micro-hardness and n was observed for CRMs made of iron or heat-treated steel with micro-hardness between 195 HV0.05 and 519 HV0.05, heat-treated carbon steel and aluminum alloy EN 6082, or technically pure metals such as Al, Zn, Cu, Fe, Ni, and Co [31,32]. Except for grinding and polishing, the given examples were not deformed.

Hays and Kendall proposed the existence of minimum test load W (N) necessary to initiate plastic deformation; the relationship between the roughness Ra and load W can be seen in Figure 9. As the roughness increases, the load value increases, and the load value varies in the range between 0.0019 and 0.0384 N. As already mentioned above, despite the declared pressure force of the touch tester, traces of the tip of the tester's arm were visible (with a microscope) on the surface of the sample. Thus, plastic deformation occurred, although it was not expected. This anomaly regarding load W was already observed by Petrík et al. and deserves a more detailed analysis [33].

## 4. Discussion

As for the influence of the polishing time on the quality of the metallographic surface, with the use of both agents, it was possible to achieve a shiny, mirror-like surface when viewed with the naked eye. Even with long polishing times (extremely 3600 s), no affected (Beilby) layer was formed. As the polishing time increases to approx. 1800 s, the roughness of Ra decreases, after which time it stabilizes for both agents without the influence of further polishing. Microscopically, it looks like older scratches and grooves gradually smooth out and disappear. At the same time, due to the inhomogeneity of the agents (larger or harder particles), new scratches are created, which are gradually smoothed out. That is, the process will continue indefinitely, or until over-polishing. The agents used will probably never be able to completely remove the scratches, and extending the polishing time makes no sense. On the contrary, as mentioned below, the polishing time influences the parameters characterizing the ISE and must be taken into account when interpreting them.

As mentioned above by Chuah and Ripin [16], after grinding (no polishing) samples based on stainless steel, copper, and aluminum, an order of magnitude lower roughness Ra (0.0062–0.1328 μm) was achieved compared to the presented values (0.87–3.35 μm).

The samples whose ISE-related properties are shown below were polished with spine-line for 300 s. If the sample was deformed by tension (tensile test), then with the growth of the local degree of deformation (Z), the value of n decreases slightly from RISE (n = 2.1) at zero deformation to slightly "normal" (n = 1.95 at Z = 80%). The same course has a dependence between roughness and Meyer's index n. With a decrease in roughness, this goes from significantly "normal" (approx. n = 1.7 in the ground state) to neutral or reverse at lower roughness. Thus, a decrease in roughness has the same effect on Meyer's index as a decrease in the degree of tensile deformation. On the contrary, during deformation by compression, the value of Meyer's index increases from a slightly "normal" to the reverse region (approx. n = 2.2 at $\varepsilon$ = 80%) as the degree of deformation increases.

An increase in polishing time, a decrease in roughness, and an increase in the degree of deformation move the values of Meyer's index into the reversal region, characteristic of plastic materials.

For all three materials—stainless steel, copper, and aluminum alloy—the same course of the effect of roughness on Meyer's index was analyzed: with a decrease in roughness, it goes from significantly "normal" to neutral to reverse at lower roughness as stated by Chuah and Ripin [16], Table 1, which corresponds to the results of the authors of this paper (n = 1.72 to 1.98). The most significant changes were observed in the case of stainless steel.

Parameter $a_1$ characterizes the load dependence of micro-hardness. Its value decreases with increasing polishing time (and thus with decreasing roughness). Thus, for less rough (highly polished) samples, a smaller influence of load on micro-hardness values is expected. As the degree of deformation increases, its value decreases (positive for flat and negative for compressive deformation). The same decrease in the value of coefficient $a_1$ with a decrease in roughness was also noted by [16,28,33].

Parameter $c_0$ is associated with residual surface stress in the sample. With the polishing time, the values increase slightly, from zero to negative values at minimum roughness. Thus, with a decrease in roughness, surface tensions increase. As the deformation increases, the (negative) value of $c_0$ grows, both in compression and tension (more pronounced in the latter).

The ratio $a_1/a_2$ is the measure of the residual stress due to machining and polishing. Its value decreases with increasing polishing time (and thus with decreasing roughness). As the degree of tensile deformation increases, it rises from negative to positive values; with compression deformation, the trend is the opposite. If we take into account the values of $a_1$ and $a_2$ given by Chuah and Ripin for aluminum alloy and calculate the parameter $a_1/a_2$ from them, then this has the same dependence on roughness—with a decrease in roughness, the value of the parameter and therefore the residual stresses decrease [16].

The longer the sample is polished and the smaller the roughness, the smaller the values of $c_0$ (residual stresses) and $a_1/a_2$ (stresses stress due to machining and polishing). It is possible that these stresses induced by grinding and polishing are removed. So, the longer the sample is polished, the smaller they are.

## 5. Conclusions

1. The polishing time has a statistically significant effect on the size of the Meyer's index n, but the polishing agent does not.
2. If diamond paste is used as a polishing agent, the resulting micro-hardness is lower than when using spineline.
3. There is a correlation between the polishing time (and roughness Ra) and Meyer's index n; with a decrease in roughness, the value of n increases (from "normal" to the reverse character).
4. Extending the polishing time above 1800 s with the agents used is not important, as it cannot completely remove scratches.
5. When interpreting the parameters characterizing the ISE, it is necessary to take into account the polishing time and subsequent roughness.

**Author Contributions:** Quality, preparation of samples, J.P., D.D., M.Ť. and S.K.; ISE calculation, P.B. and J.P.; micro-hardness measurement, J.P. and P.B.; translation, statistics, J.P.; literature review, M.Š., M.M. and J.P. All authors have read and agreed to the published version of the manuscript.

**Funding:** This work was supported by the Cultural and Educational Grant Agency of the Ministry of Education of the Slovak Republic, KEGA project 009TUKE-4/2023.

**Data Availability Statement:** Data are contained within the article.

**Conflicts of Interest:** The authors declare no conflict of interest.

## References

1. ISO 6507-1; Metallic Materials—Vickers Hardness Test Part 1—Test Method. International Organization for Standardization (ISO): Geneva, Switzerland, 2018.
2. ISO 14577-1; Metallic Materials—Instrumented Indentation Test for Hardness and Materials Parameters Part 1—Test Method. International Organization for Standardization (ISO): Geneva, Switzerland, 2015.
3. Voyiadjis, G.Z.; Peters, R. Size Effects in Nanoindentation: An Experimental and Analytical Study. *Acta Mech.* **2010**, *211*, 131–153. [CrossRef]
4. Gong, J.; Wu, J.; Guan, Z. Examination of the Indentation Size Effect in Low-Load Vickers Hardness Testing of Ceramics. *J. Eur. Ceram. Soc.* **1999**, *19*, 2625–2631. [CrossRef]
5. Sangwal, K.; Surowska, B.; Błaziak, P. Analysis of the Indentation Size Effect in the Microhardness Measurement of Some Cobalt-Based Alloys. *Mater. Chem. Phys.* **2003**, *77*, 511–520. [CrossRef]
6. Ren, X.J.; Hooper, R.M.; Griffiths, C.; Henshall, J.L. Indentation Size Effect in Ceramics: Correlation with H/E. *J. Mater. Sci. Lett.* **2003**, *22*, 1105–1106. [CrossRef]
7. Sangwal, K. On the Reverse Indentation Size Effect and Microhardness Measurement of Solids. *Mater. Chem. Phys.* **2000**, *63*, 145–152. [CrossRef]

8. Hakamada, M.; Nakamoto, Y.; Matsumoto, H.; Iwasaki, H.; Chen, Y.; Kusuda, H.; Mabuchi, M. Relationship between Hardness and Grain Size in Electrodeposited Copper Films. *Mater. Sci. Eng. A* **2007**, *457*, 120–126. [CrossRef]
9. Şahin, O.; Uzun, O.; Kölemen, U.; Uçar, N. Mechanical Characterization for β-Sn Single Crystals Using Nanoindentation Tests. *Mater. Charact.* **2008**, *59*, 427–434. [CrossRef]
10. Navrátil, V.; Novotná, J. Some problems of microhardness of metals. *J. Appl. Math.* **2009**, *2*, 241–244.
11. Cepova, L.; Kovacikova, A.; Cep, R.; Klaput, P.; Mizera, O. Measurement System Analyses—Gauge Repeatability and Reproducibility Methods. *Meas. Sci. Rev.* **2018**, *18*, 20–27. [CrossRef]
12. Kim, J.-Y.; Kang, S.-K.; Lee, J.-J.; Jang, J.; Lee, Y.-H.; Kwon, D. Influence of Surface-Roughness on Indentation Size Effect. *Acta Mater.* **2007**, *55*, 3555–3562. [CrossRef]
13. Kim, J.-Y.; Kang, S.-K.; Greer, J.R.; Kwon, D. Evaluating Plastic Flow Properties by Characterizing Indentation Size Effect Using a Sharp Indenter. *Acta Mater.* **2008**, *56*, 3338–3343. [CrossRef]
14. Xia, Y.; Bigerelle, M.; Marteau, J.; Mazeran, P.; Bouvier, S.; Iost, A. Effect of Surface Roughness in the Determination of the Mechanical Properties of Material Using Nanoindentation Test. *Scanning* **2014**, *36*, 134–149. [CrossRef] [PubMed]
15. Alsoufi, M.S.; Alhazmi, M.W.; Ghulman, H.A.; Munshi, S.M.; Azam, S. Surface Roughness and Knoop Indentation Micro-Hardness Behavior of Aluminium Oxide ($Al_2O_3$) and Polystyrene (C8H8)n Materials. *Int. J. Mech. Mechatron. Eng.* **2016**, *16*, 43–49.
16. Chuah, H.G.; Ripin, Z.M. Quantifying the Surface Roughness Effect in Microindentation Using a Proportional Specimen Resistance Model. *J. Mater. Sci.* **2013**, *48*, 6293–6306. [CrossRef]
17. Erdemir, U.; Sancakli, H.S.; Yildiz, E. The Effect of One-Step and Multi-Step Polishing Systems on the Surface Roughness and Microhardness of Novel Resin Composites. *Eur. J. Dent.* **2012**, *06*, 198–205. [CrossRef]
18. Korkmaz, Y.; Ozel, E.; Attar, N.; Aksoy, G. The Influence of One-Step Polishing Systems on the Surface Roughness and Microhardness of Nanocomposites. *Oper. Dent.* **2008**, *33*, 44–50. [CrossRef] [PubMed]
19. *ISO 6892-1*; Metallic Materials—Tensile Testing. Part 1: Method of Test at Room Temperature. International Organization for Standardization (ISO): Geneva, Switzerland, 2016.
20. Cuff, T. Beilby Layer. 2016. Available online: https://www.researchgate.net/publication/297761245_Beilby_Layer?channel=doi&linkId=56e338f508ae98445c1b2d31&showFulltext=true (accessed on 30 September 2023).
21. *ISO 4287-1*; Surface Roughness—Terminology—Part 1: Surface and Its Parameters. International Organization for Standardization ISO: Geneva, Switzerland, 1997.
22. Mitutoyo Surftest-SJ-310. Available online: https://shop.mitutoyo.eu/web/mitutoyo/en/mitutoyo/1292249267209/Surftest%20SJ-310%20%5Bmm%5D/$catalogue/mitutoyoData/PR/178-570-01D/index.xhtml (accessed on 2 September 2023).
23. *ISO 6507-2*; Metallic Materials—Vickers Hardness Test Part 2—Verification and Calibration of Testing Machines. International Organization for Standardization (ISO): Geneva, Switzerland, 2005.
24. *ISO 6507-2*; Metallic Materials-Vickers Hardness Test—Verification and Calibration of Testing Machines. International Organization for Standardization (ISO): Geneva, Switzerland, 2018.
25. Petrík, J.; Palfy, P.; Blaško, P.; Girmanová, L.; Havlík, M. The Indentation Size Effect (ISE) and the Speed of the Indenter Penetration into Test Piece. *Manuf. Technol.* **2016**, *16*, 771–777. [CrossRef]
26. Chajdiak, J. *Štatistika v Exceli 2007*, 2nd ed.; STATIS: Bratislava, Slovakia, 2009.
27. Chrysler Group. *Measurement Systems Analysis: Reference Manual*, 4th ed.; Chrysler Group: Detroit, MI, USA, 2010; ISBN 978-1-60534-211-5.
28. Petrík, J.; Blaško, P.; Markulík, Š.; Šolc, M.; Palfy, P. The Indentation Size Effect (ISE) of Metals. *Crystals* **2022**, *12*, 795. [CrossRef]
29. Kim, H.; Kim, T. Measurement of Hardness on Traditional Ceramics. *J. Eur. Ceram. Soc.* **2002**, *22*, 1437–1445. [CrossRef]
30. Li, N.; Liu, L.; Zhang, M. The Role of Friction to the Indentation Size Effect in Amorphous and Crystallized Pd-Based Alloy. *J. Mater. Sci.* **2009**, *44*, 3072–3076. [CrossRef]
31. Karaca, I.; Büyükkakas, S. Microhardness Characterization of Fe- and Co-Based Superalloys. *Iran. J. Sci. Technol. Trans. Sci.* **2019**, *43*, 1311–1319. [CrossRef]
32. Cai, X.; Yang, X.; Zhou, P. Dependence of Vickers Microhardness on Applied Load in Indium. *J. Mater. Sci. Lett.* **1997**, *16*, 741–742. [CrossRef]
33. Petrík, J.; Blaško, P.; Petryshynets, I.; Mihaliková, M.; Pribulová, A.; Futáš, P. The Influence of the Degree of Tension and Compression of Aluminum on the Indentation Size Effect (ISE). *Metals* **2022**, *12*, 2063. [CrossRef]

**Disclaimer/Publisher's Note:** The statements, opinions and data contained in all publications are solely those of the individual author(s) and contributor(s) and not of MDPI and/or the editor(s). MDPI and/or the editor(s) disclaim responsibility for any injury to people or property resulting from any ideas, methods, instructions or products referred to in the content.

MDPI
St. Alban-Anlage 66
4052 Basel
Switzerland
www.mdpi.com

*Crystals* Editorial Office
E-mail: crystals@mdpi.com
www.mdpi.com/journal/crystals

Disclaimer/Publisher's Note: The statements, opinions and data contained in all publications are solely those of the individual author(s) and contributor(s) and not of MDPI and/or the editor(s). MDPI and/or the editor(s) disclaim responsibility for any injury to people or property resulting from any ideas, methods, instructions or products referred to in the content.

www.ingramcontent.com/pod-product-compliance
Lightning Source LLC
LaVergne TN
LVHW070426100526
838202LV00014B/1533